The Role of Phages in Early Molecular Genetics

In the shadow of global conflicts—from World War II through the tensions of the Cold War—an extraordinary scientific revolution was quietly unfolding. A brilliant collective of geneticists and microbiologists, known as the Phage Group, harnessed the power of bacterial viruses to unlock the very structure of genes, giving birth to Molecular Biology as we know it today.

As antimicrobial resistance threatens global health, phages have reemerged as critical subjects of scientific inquiry, yet their pivotal role in the genetic discoveries of the 1940s–1960s remains largely overlooked—until now. *The Role of Phages in Early Molecular Genetics: History of an Emerging Discipline* bridges history, physics, and virology, offering both scientists and general readers a compelling narrative of how these microscopic entities revolutionized our understanding of life itself.

Key Features:

- Explains how bacterial viruses became the unexpected heroes in humanity's quest to understand the genetic code
- Discusses the forgotten historical and social forces—Lysenkoism, Spencerism, and McCarthyism—that shaped scientific progress yet vanished from our collective memory
- Provides a fascinating parallel between the emergence of phage biology and quantum physics, two revolutionary fields developing in tandem

By illuminating the shared patterns of scientific discovery across disciplines, *The Role of Phages in Early Molecular Genetics: History of an Emerging Discipline* provides a timely reminder of how breakthrough knowledge emerges—and why the Phage Group's legacy deserves its rightful place in scientific history.

The Role of Phages in Early Molecular Genetics

History of an Emerging Discipline

Luigi Marongiu

CRC Press
Taylor & Francis Group
Boca Raton London New York

CRC Press is an imprint of the
Taylor & Francis Group, an **informa** business

Designed cover image: Shutterstock

First edition published 2026
by CRC Press
2385 NW Executive Center Drive, Suite 320, Boca Raton FL 33431

and by CRC Press
4 Park Square, Milton Park, Abingdon, Oxon, OX14 4RN

CRC Press is an imprint of Taylor & Francis Group, LLC

ISBN: 9781041109877 (hbk)
ISBN: 9781041109884 (pbk)
ISBN: 9781003657712 (ebk)

DOI: 10.1201/9781003657712

Typeset in Times
by codeMantra

To my family,
who sustained me; my mentor,
who coached me; and my colleagues, who bolstered me.

Contents

Preface

The ongoing spread of infections due to antimicrobial-resistant bacteria such as *Staphylococcus aureus* and *Pseudomonas aeruginosa* is a modern crisis demanding novel approaches to substitute antibiotics in preventing and treating these diseases. Bacterial viruses, better known as bacteriophages or phages, represent one of the most promising additions to the new arsenal of antibacterial products. Just like nuclear fusion's delivery is near but never quite there, the implementation of phages as an antibacterial treatment, a process known as phage therapy, is within reach but not quite accomplished yet. Contradictory results from clinical studies, the complexity of the interaction between phages and the immune system, genetic mutation in both phages and hosts, as well as the loss of viral titers due to the shelf life of phage preparation and their administration route all hinder the development of adequate phage preparations for clinical use. In veterinary settings, instead, phages have provided more clear-cut results.

The paternity of the discovery of phages has been hotly debated. Ernest Hankin is regarded as one of the first researchers to describe phages through his report on the bacteriolytic activity of river waters. About two decades later, Frederick Twort reported bacteriolysis in his culture, which he called "glassy transformation." None of them, though, could provide an explanation for these observations. It was Félix d'Hérelle who understood the viral origin of the factors leading to bacteriolysis, which he called "bacteriophages" (eaters of bacteria). Not only did d'Hérelle provide an in-depth characterization of these viruses but he also understood their value in treating bacterial infections, establishing the field of phage therapy.

The work of d'Hérelle was not smooth: his altercation with the Nobel Prize winner and direct superior at the Pasteur Institute, Jules Bordet, represents one of the most notable disputes in the history of science; one should just consider that Bordet began working with phages specifically to disprove his subordinate's findings. The conflict between d'Hérelle and Bordet exemplifies poor interpersonal relations in the laboratory.

Due to the current interest in phage therapy, this aspect of phage discovery has been covered in detail. There is, however, another lesser-known aspect of phage research that risks sliding into oblivion due to the current emphasis on phage therapy: the employment of phages in genetic research.

In the two decades spanning the 1950s, a group of geneticists started to use phages as a tool to understand the structure of the gene, an entity that, back in the day, there was no knowledge of whatsoever. After the seminal writing of Gregor Mendel in the late 1800s, genetics had advanced relatively slowly. In the first half of the 20th century, fruit flies were the primary genetic tool, pioneered by Thomas Morgan. These animals were pivotal in understanding the transmission of genetic information but did little to determine the actual structure of the gene. Back then, the gene was even considered just a theoretical construct. Microorganisms such as bacteria and yeasts provided more advancements in the field. It was phages, however, that provided the highest understanding in the field of genetics. Phages demonstrated the randomness of genetic mutation and that nucleic acids, not proteins, carried the genetic information. Phages also showed that bacterial genomes were more fluid than higher organisms.

An assemblage of geneticists that became known as the "Phage Group" pioneered the use of bacterial viruses in genetics, establishing the field of molecular biology. This ensemble of high-caliber scientists gathered at a training course held annually at the Cold Spring Harbor Laboratory in New York. The group's leading figures were Max Delbrück and Salvador Luria, both European émigrés with quantum physics backgrounds who relocated to the US. They were not the only ones. The "father of the atom bomb," Leo Szilard, was another example of a European émigré who moved from nuclear physicists to phage biology. The relocation of European scientists to the US in the middle 20th century constitutes probably one of the most remarkable aspects not only of the Phage Group but also of science as a whole, projecting this country to the forefront of research and technology.

These scientists did not conduct their research in the vacuum of their laboratories but lived during one of the most tumultuous epochs of human history, which included the Second World War and Cold War that almost determined the end of humanity. They experienced the revolution in physics brought forward by quantum mechanics but also endured the times of social oppression represented by the rise of fascism and McCarthyism. This aspect has been less represented by the literature.

The main goal of this book is to offer a timely and original update on phage research, while also providing historical context. It aims to present a clear understanding of the significant events that shaped the 20th century, alongside concepts from quantum physics and virology. Such knowledge is not only an end in itself but can provide modern readers, and young scientists in particular, with the understanding that the totalitarianism that characterized the "Age of Extremes," to use a term of the historian Eric Hobsbawm, was not a chance occurrence but a risk that civil society must continually be vigilant against. The involvement of the Phage Group members in civil rights emphasizes the vital role that researchers can have in science and society as a whole.

While phages are no longer used as genetic tools, their applications have expanded over the last century to include not only phage therapy but also food biocontrol, peptide screening, and even material science.

But that is another story.

LM, 2025.

Acknowledgments

I would like to acknowledge the researchers collaborating with *Stack Exchange Biology* to highlight features not readily evident in the literature and Roberto Faleri, University of Siena, Italy, to provide access to less-known manuscripts. Special thanks go to Mr. Alberto Curcio, Head of *TMT Portfolio*, for his valuable insights. I would also like to thank Prof. Ewa Brzozowska from the Hirszfeld Institute, Poland, and Dr. Maurizio Grilli from the Medical Faculty Mannheim, Germany, for their support to this manuscript. Lastly, I extend my deep appreciation to Dr. Chuck Crumly, Senior Acquisitions Editor at *Life Sciences*, for believing in this project.

About the Author

Dr. Luigi Marongiu studied Molecular Biology at the University of Roma Tre, Italy, and obtained a Ph.D. in Virology from the University College London, England. He studied viral infections at the Universities of Cambridge, Edinburgh, Heidelberg, and Tübingen. Currently, he is leading the Nutritional Virology Unit within the Department of Nutritional Biochemistry at the University of Hohenheim, Germany, where he is investigating the role of nutrition in altering the biology of bacteriophages and its effect on human health. He is the author of the manual *Machine Learning Analysis of qPCR Data Using R.*

1 Prologue

1.1 1980, WICHITA

The University of Kansas School of Medicine is not typically associated with cutting-edge research. It does not have 900 years of history, a list of alumni that includes heads of State, and it does not count 16 Nobel Prizes in medicine, as Oxford does. Unlike Harvard, its graduates do not include eight Presidents of the United States, and it did not teach the inventors of anesthesia or the smallpox vaccine. The University of Kansas School of Medicine was founded in 1880 simply to provide a "preparatory medical course." Over time, new campuses opened in Salina (1920) and Wichita (1971). Despite the humble origins, things were about to change on the 20th of December 1978: the University of Kansas would be making history by being the first institution to apply bacteria to fight diabetes.

Diabetes is a sneaky illness. It is not a disease because it is not a single condition but a collection of ailments with similar symptoms. It is not as deadly as pancreatic cancer, yet it can cut life expectancy by one-third. It is not infectious like the Ebola virus, yet one in 11 adults suffers from it, with the projection of reaching a prevalence of one in ten in the next few decades. It is not as disfiguring as being hit by a bombshell in a conflict zone, yet diabetic people are 25 times more likely to go blind.

There are three types of diabetes (more appropriately: diabetes mellitus, the "sweet siphon"). Type I represents about 9% of all cases of diabetes, type II is the most common (about 90% of cases), and the third type (gestational diabetes mellitus) occurs about 1% of the time but only during pregnancy. The so-called β-cells produce insulin in the pancreas. In normal conditions, this proteic hormone regulates the intake of dietary sugar glucose in the bodily cells. The attachment of insulin to its receptor on the surface of cells induces a molecular signal that stimulates the cells to express more surface pumps (glucose transporters) that transfer glucose from the bloodstream into the cells. Moreover, the insulin receptor boosts the activity of the cellular enzymes that store glucose into glycogen (a glucose polymer) and fat.

In type I diabetes, an auto-immune dysfunction (thus, akin to lupus erythematosus) causes the destruction of the β-cells by T lymphocytes wiping out insulin from the circulation. Without insulin, the glucose pumps are not activated, and glucose does not enter the cells. People with type I diabetes live a paradoxical condition where the more they eat, the more they are hungry. They also have polyuria; thus, they are very thirsty to compensate for dehydration. Like Tantalus, their hunger and their thirst are never quenched. However, their condition can be put at bay with insulin injections.

In type II diabetes, the dysfunction of any of the many enzymes involved in glucose metabolism causes cellular starvation: the cells cannot use glucose to produce the energy necessary to build all the biomolecules needed to sustain cellular biochemistry. The β-cells respond to such starvation by producing more insulin, but no matter how much insulin they pour into the bloodstream, the body cells keep starving. Eventually, the overworked β-cells establish a stress status and begin to die. Once the pancreas is

DOI: 10.1201/9781003657712-1

depleted of β-cells, type II diabetes converges to type I since there is no longer insulin production. In addition, in the absence of physical activity, muscle cells become resistant to the insulin signal even without the genetic knockdown of the metabolic enzymes involved in the use of glucose.

In both cases, glucose accumulation causes a wide range of cellular and tissue damage. For instance, excess glucose can bind to proteins (hemoglobin or antibodies, just to name a few), disrupting their function and altering the cellular morphology and, consequently, tissue function. As a result, diabetes causes affections such as atherosclerosis, hypertension, degradation of the nerves (including the optical ones, hence blindness), and fatal kidney damage. Furthermore, the surplus cellular glucose is converted into the alcohol sorbitol, which blocks the scavenging activity of myoinositol in removing free radicals from the cellular environment, resulting in additional damage to biomolecules. In addition, insulin depletion causes an abnormal breakdown of stored fat, releasing ketones (derivatives of lipids) into the bloodstream that can trigger mortal acidosis.

Starting from 1922, thanks to the work of Frederick Banting and Charles Bets, type I diabetic patients could be treated with insulin extracted from animal pancreas. Banting and Bets managed to isolate insulin from a dog's pancreas and subsequently injected it into the 14-year-old Leonard Thompson from Toronto, who was dying of diabetes. The little Thompson recovered quickly, and Banting and Bets won the Nobel Prize for their feat the following year. At the industrial scale, pigs and cows were used as sources of pancreas. The Danish company *Novo Nordisk* and the American *Eli Lilly* were the big players in the insulin business. Since the porcine and bovine insulin differ from the human protein by one and three amino acid substitutions, they function as the human's. However, about one in ten patients developed allergic reactions. In addition, back in the day, *Eli Lilly* alone needed to kill 56 million animals a year to sustain the demand for insulin. To overcome these pitfalls, *Eli Lilly* organized a meeting in Indianapolis where experts could discuss the future of the production of this hormone, essentially starting a race to produce synthetic insulin. The big players in this race were the company *Genentech*, the laboratory of Prof. Goodman at the University of California San Francisco (UCSF), and the laboratory of Prof. Gilbert at Harvard. The recent introduction of the so-called recombinant DNA technology was the key to victory.

Bacteria, like humans, can be infected by viruses. These bacterial viruses are known as bacteriophages (or phages for short). Bacteria have a repertoire of enzymes that can cleave foreign DNA sequences to defend themselves from infection. For instance, a bacterial "self" sequence might combine the nucleotides A, T, C, and G into a sequence such as ATGACATGACATGA. Each of these defensive molecules, known as restriction enzymes (or endonucleases), can cut a specific DNA sequence. Our example bacterium has an endonuclease that splits the sequence GGATAATAGG right in the middle and another endonuclease that cuts the sequence AGGAATTGC just before the C. The sequences GGATAATAGG and AGGAATTGC are not self; let us also assume that they are of viral origin. Once they are cut, the viral genome is compromised, and the virus becomes non-infectious; consequently, the bacterium can survive. While our example bacterium will have neither the sequence GGATAATAGG nor AGGAATTGC (otherwise, it would have killed itself), other bacteria or human genes might display these sequences by chance. For instance, the commensal bacterium (that is, a beneficial microbe present in our gut) *Escherichia coli* contains 43 sequences recognized by the restriction enzyme *SrfI* and four recognized by *FseI*. Obviously, *E. coli* expresses neither *SrfI* nor *FseI*, but other bacteria do (*Streptomyces* and *Frankia*, respectively), so it is possible to isolate these endonucleases from the different bacteria and cut *E. coli*'s genome. By extension, cutting the human genome to isolate precious genes is possible.

The idea behind the recombinant DNA technology is to use restriction enzymes to cut out specific parts of DNA molecules wrapped by sequence tags and then merge them with others using other enzymes known as ligases, which were initially isolated from phages. Afterward, DNA sequencing is required to ensure that the recombination has succeeded; otherwise, it would be like playing poker in a dark room without knowing if the served hand is a good one. Insulin comprises two chains: one short of 21 amino acids (chain A) and one long of 30 amino acids (chain B). The two chains are then joined by sulfur atoms (disulfide bonds). Both chains are enclosed between tags recognized by the endonucleases *EcoRI* and *BamHI*.

Genentech was founded in April 1976 by Herbert Boyer and Robert Swanson. Boyer, a professor at UCSF, was a leader in bacterial genetics and protein expression, whereas Swanson was a biochemist with a degree in management and worked as a venture capitalist. They met at the beginning of 1976 and agreed that the nascent field of recombinant DNA would open new opportunities that could be exploited commercially. Soon, with the contribution of the scientists Arthur Riggs, Keiichi Itakura, David Goedel, Daniel Yansura, and Roberto Crea, *Genentech* produced synthetic human somatostatin expressed in the commensal bacterium *E. coli*. At UCSF, Goodman, also a leader in recombinant DNA technology, teamed up with William Rutter. Instead, the University of Harvard hosted Walter Gilbert, another giant in DNA technology. Gilbert had worked with James Watson, who won the Nobel Prize with Francis Crick and Maurice Wilkins for the description of the DNA molecule. Gilbert perfected DNA sequencing with Allan Maxam in 1975 (both received the Nobel Prize in 1980 for their achievement), creating a technology still in use today that was pivotal, as mentioned, in the insulin race. In 1978, Gilbert founded *Biogen* to commercialize the insulin molecule he hoped to produce.

By working day and night and free from the red tape of academic institutions, the researchers at *Genentech* managed to express the human insulin in *E. coli* on the 24th of August 1978, winning the race. The next day, Boyer and Swanson signed a contract with *Eli Lilly* to supply the insulin required for a human trial. The trial included the first group of healthy volunteers in England and five diabetic people in Wichita. Richard Guthrie, a well-known authority in the field of diabetes, led the therapeutic trial. Despite his contribution to establishing the University of Kansas in Wichita, Guthrie remained down to earth. Being an admirer of the popular TV series *Star Trek*, he used to quote one of its characters, Dr. "Bones" McCoy, to describe himself: "I am just a simple country doctor." Guthrie, who died in 2020, is remembered for his kindness, compassion, and commitment to the care of children (he was the chairman of the Department of Pediatrics).

The first diabetic to receive the biosynthetic insulin was the 37-year-old Ms. Sandy Atherton. Given the trial's success, the bacterial-derived insulin developed by *Genentech* was approved by the Food and Drug Administration (FDA) in 1982 and marketed the following year. Since then, insulin has been more widely available and less expensive. More than that, the *Genentech* feat opened the path for producing many more drugs in bacteria, such as factor VIII for hemophiliacs and plasminogen for people who experienced a heart attack, but also vaccines, antivenoms, and antibodies.

1.2 1999, TOPEKA

In August, six of the eight Republican Kansas Board of Education members decided to alter the worldview of thousands of kids. The Board voted six to four (two Democrats also opposed) to modify the high-school science standard to include the teaching of "intelligent design" alongside evolution. Moreover, science was no longer to be limited to the investigation of physical phenomena. The chairman exulted by saying: "This is a great day for education … gets rid of a lot of dogma."

Evolution is a raw nerve in Kansas: two years later, when three members were replaced, the Board reversed the decision. Afterward, after another election, the teaching of intelligent design was reinstated in 2004. The Republican leader of the time and President of the USA, George W. Bush, praised the decision. The truth is, though, that teaching intelligent design is becoming an issue almost everywhere. Currently, it is endorsed in the US (apart from Kansas, it is taught in another dozen states) as well as in other traditionalist countries such as Turkey, Saudi Arabia, and South Korea.

According to Prof. Lawrence Lerner of the California State University, "evolution is intimately tied to ecology," altering young people's perceptions of the changing environment, specifically global warming. Hence, teaching intelligent design appears to be more of a political decision than an attempt to debunk dogmas. The choice between evolution or intelligent design is not the same as choosing between the "Big Crunch" and the "Big Rip" models about the universe's future, which depends on quantifying the total amount of energy in the cosmos. It is not like siding with either Newton or Leibniz for the paternity of the differentials, which is more a matter of personal taste. Choosing to teach intelligent design is akin to claiming that red is as much color as green, authorizing people to assert their right not to stop at a red light. It is easy to see how such a comparison can be hurtful.

Despite its smart sounding, intelligent design is simply a synonym for "creationism." The term "intelligent design" was introduced by the chemist Charles Thaxton in his seminal book *The Mystery of Life's Origin* (1984) and then adopted in the high-school textbook *Of Pandas and People* (1989) by Percival Davis and Dean H. Kenyon. The term was implemented to circumvent a ban against creationism introduced by a US Supreme Court decision in 1987. The substitution was so literal that, in the original manuscript *Of Pandas and People*, it is possible to read the word "cdesign proponentsists" (with "design proponents" not banned) instead of "creationists" (forbidden by the law). Thaxton is affiliated with the Discovery Institute, which founded the intelligent design movement's headquarters Center for Science and Culture. On the other hand, creationism has lived alongside evolution since the birth of natural philosophy.

By definition, religions assume that some deity has created human beings and all other things, hence the term "creationism." The ancient Greek philosophers spread another way of thinking in the West. Thales, the founder of natural philosophy who lived some 2500 years ago, was the first to teach that all things were composed of an eternal essence, namely water. About a century later, Democritus went even further by hypothesizing that matter was composed of discrete and eternal entities: atoms. Since things were not created but were eternal, such a concept went under the name of "materialism."

For us mortal beings, eternity is a concept difficult to grasp. Since we are accustomed to seeing babies being born and objects being made, it is logical to assume that somebody must have produced the universe. However, eternity is a more logical assumption than having a creator: since things have always been, there is no need for a creator. In other words, eternity is a more parsimonious concept than that of a creator. Moreover, eternity avoids the question implicit in a creator's existence: who created the creator? The classical answer to this question is that the creator is eternal. But such an answer duplicates the things (hence, it is not parsimonious): instead of having one eternal cosmos, there is (1) an eternal creator that created (2) the cosmos. Another answer to this question is the concept of "pantheism": the creator and the cosmos are the same thing. Not only do we live inside the creator, but we are made of the creator's bits. One of the champions of pantheism was the Dutch philosopher Baruch de Espinoza, who lived in the 17th century and elegantly theorized about an immanent God.

Moving on another path, the Greek philosopher Aristotle, who lived about 2300 years ago, introduced the concept of "causality." He noticed that all things were in motion and that some action caused such motion. A typical example is the billiard (invented at least a millennium after Aristotle's death): at the beginning of the game, the balls are set in motion by the cue ball, which moves because of a stroke by the cue stick, which the player swings. Logically, Aristotle thought there must be an entity that had set all things in motion, and he called that entity the "First Cause." In the Middle Ages, the founders of Christian theology tailored such a concept to that of the transcendent God of Jewish origin.

Conversely, the Greek philosopher Epicurus, who lived a couple of decades after Aristotle, reacted to such a model by renovating the materialist tenet. He did not exclude the existence of the gods altogether, but he expelled them from the natural world to get rid of superstition. During the Middle Ages, Christian theologians fiercely opposed the Epicurean doctrine. Then, in the 14th century, the English friar William of Ockham introduced the concept that neither senses nor experiments could demonstrate the existence of God. Thus, God belonged to the realm of faith and was not part of the scientific investigation. In addition, he introduced the principle of parsimony (Ockham's razor), which is still employed today in science and philosophy. Since then, religion and science have parted ways, each with its application area. The former is concerned with a moral quest to save people's souls; the latter is involved in describing nature's laws in more and more detail.

Two things are worth noting in this summary. First, theological ideas have a history. Unlike dogmas, mortal men introduce, adapt, and expand the concepts associated with deities and nature. They are not monoliths that appeared out of thin air. Second, the dissent is between creationism and materialism, not between creationism (or its sneaky rephrasing of intelligent design) and evolution. Nonetheless, intelligent design proponents elevated Charles as a champion of atheistic materialism and opposed the notion that nature requires a designer to account for its complexity. But doing so is double ironic.

First, Darwin was not an atheist, for he did not exclude the presence of an entity that defined the laws of the cosmos. He wrote that "probably all the organic beings … descended from some primordial form, into which life was first breathed by the Creator." Current science goes way beyond Darwin, dispensing for the creator altogether. One of the most accredited models regarding the origin of life postulates that life started in some abyssal thermal vents. Instead of a divine breath, it is more likely that, about 4.3 billion years ago, hot water squirting from the earth's bowels supplied energy to a ball of fat surrounding a still undefined mixture of organic molecules. That energy was enough to initiate a self-sustained chemical reaction, and the primordial proto-cell began to replicate and, more importantly, evolve. Such a model is not based on a scripture kindly provided by a god but is an educated guess based on chemical and geological experimentation. In addition, the model is being improved daily, whereas the scriptures have remained unchanged through the centuries. Thus, Darwin was not an atheist like Thales, Democritus, or even today's scientists, but he simply affirmed that species evolved according to natural selection.

The second element of irony is that Darwin introduced the modern meaning of intelligent design. In 1802, Archdeacon William Paley wrote an influential book (Natural Theology) that equated God to a clockmaker. He argued that the cosmos was too complicated to have raised on its own, just like a clock (the most complex machine at the time) could not have been built by randomly cramming gears and cogs together. In the centuries before, these similarities included sailing a ship and shooting arrows at bull's eyes. In 1861, Darwin wrote a letter to the physicist John Herschel, who inspired Darwin as a scientist but fell for Paley's argument, stating that "the point which you raise on Intelligent Design has perplexed me … one cannot look at this Universe with all living productions & man without believing that all has been intelligently designed; yet when I look to each organism, I can see no evidence of this." Such a passage defined intelligent design as the human feeling of sticking a creator to everything that seems too good to be true. However, such a practice hides a poor understanding of nature: looking at things this way would be like believing that the Giant's Causeway was built by the giant Fionn Mac Cumhaill rather than formed by erosion of basalt and chalk. While it is understandable that ancient people relied on myths to explain the world around them, it is no longer acceptable for a society that has painfully gathered evidence of the mechanisms of how nature works. Also, comparing things purposely built by men, such as clocks and boats, to natural elements, such as plants and animals, is like comparing apples with potatoes.

Darwin jeopardized the Christian tenet thrice by publishing his seminal books *On the Origin of Species* (1859) and *The Descent of Man* (1871). First, Darwin exposed the belief that species had been created once and remained as they were ever since. Second, he removed the causality implicit in the creationist view. Even assuming that species, once created, had evolved, they should have evolved according to a plan set by the creator, a concept known as "finalism." Darwin, instead, proclaimed that evolution was a stochastic affair. Third, he removed the halo of divinity surrounding human beings by demonstrating that men followed the same rules as all other living beings. After Darwin, Man was no longer God's beloved. It is understandable, therefore, how Darwin became the materialist icon and raised the hatred of the most conservative religious factions.

The term "Darwinism" was introduced a year after the publication of *On the Origin of Species* by Thomas Huxley. Huxley also revived the concept of a separation between religion and science, radicalizing the debate beyond Darwin's intentions. In 1864, Herbert Spencer coined the term "survival of the fittest," which he applied to endorse the draconian measures against the workers implemented by the British government at the time, establishing the political movement of "social Darwinism." A relative of Darwin, Francis Galton, proclaimed that evolution did not apply to civilized people and that it was required to use efforts to maintain the purity of the genes against the contamination imbued by inferior races, inaugurating the pseudo-science of "eugenics." Darwin's ideas have been improved over time by incorporating genetics and statistics, giving birth to what is now known as "neo-Darwinism" (or simply, "the theory of evolution"). Some scholars have expressed reservations about the model because of a few isolated exceptions that, however, can be interpreted within the framework of neo-Darwinism.

All in all, evolution is a well-established model that has passed the scrutiny of experimentation. Thus, it is a scientific model because it is characterized by being parsimonious, consistent, useful, empirical, correctable, and progressive. Conversely, creationism is not a science because it lacks all of these characteristics. As a result, when the Topeka guys likened evolution to creationism, it was like comparing apples to potatoes. Teaching creationism in the science syllabus was not about cracking a dogma but introducing one and calling it "science."

The European Council Committee on Culture published a report in 2007 that described creationism as a "blatant scientific fraud" and "intellectual deception" that "originates in forms of religious extremism closely linked to extreme right-wing political movements." The committee recommended, "resist presentation of creationist ideas in any discipline other than religion." The director of the Vatican Observatory, Father George Coyne, commented that "intelligent design isn't a science even though it pretends to be" that reduces "God to a mere engineer." Remarkably, while most religions have no problem incorporating evolution into their theologies, some minor churches charge headfirst evolution. The major of these (so to say) nominal churches is the American Christian Reconstructionism movement (founded by Rousas John Rushdoony in the mid-20th century), which is very popular in American conservative circles.

The diatribe that occurred in Topeka two decades ago and still reverberates today is, therefore, not about evolution and creationism but between progressive and reactionary forces. It is not so much about science but politics. Consequently, the Topeka affair reminds us of the importance of maintaining an open mind associated with science to avoid the rise of theocracies.

1.3 2011, STOCKHOLM

The 7th of July 2011 was a red-letter day for the Karolinska University Hospital in Stockholm, Sweden, and humanity. The Karolinska is not the usual highly rated institute: it is not just that it hosts some Nobel Prize winners, like other universities; it is the place where the Nobel Prizes for medicine are decided in the first place.

Thus, the announcement that the Karolinska was diffusing that day summoned extra attention: from that point on, it would have been possible to transplant new organs to recipients without donors. It was quite a step forward, for plenty of people need new organs, but very few donors are available, and even when a donor is found, chances are that the recipient will reject the transplanted organ. Karolinska's press release gave hope to thousands of suffering people who needed new organs to survive.

A month before, on the 9th of June, Andemarian Beyene underwent a game-changing surgical operation. Beyene, an Eritrean living in Iceland, had been admitted to the Landspital University Hospital in Reykjavik for an esophageal cancer that persisted despite several rounds of radiotherapy and that had wasted his trachea and lungs, making it difficult for him to breathe. A trachea (or windpipe) is a hollow cartilage tube covered with epithelial cells. Since the epithelial cells are mature cells with markers specific for each individual, their transplantation to another person will trigger the donor's immune response because the immune cell is specialized to remove everything that is not recognized as self.

A team led by the Italian surgeon Paolo Macchiarini implanted Beyene with a synthetic windpipe, where the cartilage was replaced by plastic material etched at the renowned University College in London (UCL) by Prof. Alexander Seifalian. The synthetic windpipe was then covered in stem cells harvested from Beyene's bone marrow through incubation on an ad hoc bioreactor made at the even more renowned Harvard Bioscience Institute. Stem cells are immature cells that can mature in virtually all types of mature cells (epithelial and immune cells but also bones, muscle, and so forth). Since these cells were derived directly from Beyene, his immune system recognized them as self—avoiding the risk of transplant rejection, one of the worst outcomes of transplantation. Thus, the possibility of implanting organs (or parts thereof) without the risk of rejection opened "new and promising therapeutic possibilities" in regenerative medicine and gave hope to "thousands of patients."

The experimental procedure was summarized in a scientific article published in the high-ranked journal *The Lancet*. The article was not any different from many other high-quality papers: it reported technical details of the bioreactor, plots of physiological markers, images of computed tomography scans of the trachea before and after the surgery, and a photograph of the bronchoscopy taken after the operation, showing how the implant was "integrated with surrounding tissues." The article highlighted that Beyene experienced bacterial pneumonia, which was successfully treated with antibiotics. Moreover, Beyene's condition improved within a week, and he was discharged to start rehabilitation. The paper stressed that "5 months after transplantation, the patient is asymptomatic, breaths normally, is tumor-free, and has an almost normal airway." Given the flying colors the operation displayed, the experimental procedure was applied to other patients. First, a British woman was operated on at the Royal Free Hospital in London, the University Hospital associated with the UCL; then, an American man; and then another six patients at the Karolinska.

The mastermind behind these miraculous operations was Paolo Macchiarini. He had been employed at the Karolinska from the end of 2010 due to his outstanding curriculum. That is to say, Karolinska's recruiters took at face value what Macchiarini had written in his CV. According to it, he obtained a Master's in Biostatistics at the University of Alabama, US; a specialization in surgery at the University of Pisa, Italy;

a Ph.D. in organ and tissue transplant, and another in life and health science both at the University of Besançon, France. Then he had a collection of professorships, for he was a tenured professor at the University of Paris-Sud, France; the Hannover Medical School, Germany; the University of Barcelona, Spain; and the Careggi University Hospital in Florence, Italy (a position allegedly sponsored by the Governor of the Tuscany region, Enrico Rossi of the Italian Democratic Party).

In June 2008, at the Hospital Clinic Barcelona, Macchiarini performed a seminal operation that was preliminary to that carried out 3 years later in Sweden. In this case, the trachea cartilage was not synthetic but derived from a corpse. The donor's epithelial cells were stripped away and replaced with the recipient's bone marrow cells. The donor was a 51-year-old man who died of a stroke in November 2007. In contrast, the recipient was a 30-year-old Colombian young woman living in Barcelona, Claudia Castillo Sanchez, who was suffering from inadequately treated tuberculosis. The donor's trachea was stripped of the original epithelial cells at the Carreggi Hospital in Florence, Italy, and the seeding was carried out at the University of Bristol under the supervision of Prof. Martin Birchall. The operation was hailed as the "dawn of the stem cell revolution." No wonder Karolinska's recruiters, including Prof. Urban Lendhal (who recommended the employment in the first instance), were transfixed by such a "super-surgeon," supported by solid recommendations such as Prof. Birchall, who moved to UCL.

The 2000s were Macchiarini's apex, not only professionally but also at a personal level. Although already married (and allegedly divorced) to Emanuela Pecchia in 1986, with whom he had two children, he planned to marry the beautiful and successful NBC producer Benita Alexander. The ceremony was supposed to take place in July 2014 and to be nothing short of what a Hollywood star might have conceived. The wedding party would have taken place at Castel Gandolfo, the personal summer residence of the Pope. Among the guests: the Putins, the Obamas, the Clintons, and the Sarkozys; the secretary-general of the United Nations, Kofi Annan; the world-renowned New Zealander actor Russel Crowe; the legendary English singer Elton John; and last but not least, the host himself: his Holiness Pope Francis I. Since the Pope was on the spot, he would not waste time and officiate the ceremony, also dispensing the Holy Communion to a gay couple (Macchiarini's tailor, Mr. Matthew Christopher, and his husband), an act hitherto unheard of within the Catholic world. The famous opera singer Andrea Bocelli was to sing during the gala.

Nevertheless, every apex is followed by a decline. First of all, the British and American patients died. The autopsy carried out on the British woman showed that the implant was unsuccessful: the synthetic trachea was not integrated at all with the surrounding tissues. Still, it was loose and obstructing the normal respiration. Chronic infections and a large amount of mucus also make respiration painful. In 2014, Beyene also died: he also suffered from a loose implant that had to be kept open by implanted forceps and experienced recurrent infections.

After Beyene's death, one of the members of Macchiarini's team, Dr. Karl-Henrik Grinnemo, and Dr. Thomas Fux, who assisted Beyene during the rehabilitation, smelled a rat. They reported to Karolinska's director that Macchiarini had falsified the data regarding the operation and grossly understated the actual conditions of Beyene. The vice-chancellor of the University, Dr. Anders Hamsten, authorized an external review carried out by Dr. Bengt Gerdin, general surgeon and professor emeritus at Uppsala University (Sweden). Gerdin, in May 2015, declared that there had been, indeed, scientific misconduct. However, Hamsten rejected Gerdin's verdict based on data not provided to Gerdin, and Macchiarini was deemed not guilty.

That was not the end of it. In January 2015, the Swedish television SVT broadcast a documentary (*The Experiment*) produced by journalist Bosse Lindquist that clearly showed the miserable conditions of Beyene after the surgery. It could be seen that the operation had been a complete failure, despite Macchiarini's flashy claims, and essentially the message delivered was that the "super-surgeon" was using people as guinea pigs. Lindquist's documentary raises a nationwide scandal. Hamsten had to resign, as did Urban Lendahl, who was a member of the Nobel Committee. Eventually, the dean Hans-Gustaf Ljunggren, the chairman Lars Leijonborg, and the vice-chancellor Jan Carlstedt-Duke of the Karolinska also had to quit. It turned out that Macchiarini did not even have a proper permit from the ethical committee. An expert in the field, Prof. Pierre Delaere from the Katholieke Universiteit Leuven, Belgium, declared that Macchiarini's procedure was theoretically impossible since it is well established in the field of staminal cells that bone marrow's stem cells do not adequately differentiate. The misunderstanding that stem cells could differentiate in, say, epithelial, blood, or bone cells was based on old studies that had been disproved. Thus, Macchiarini did not know even the bases of his trade, nor did Karolisnka's cadres, who relied on him for such expertise. Delaere dubbed Macchiarini's procedure "one of the biggest lies in medical history."

At the end of August 2016, Kjell Asplund (professor emeritus in medicine at Umeå University in Sweden, chairman of the Swedish National Council on Medical Ethics in Stockholm) deposited his investigation on Macchiarini's case, where he not only criticized the work of the "super-surgeon" but also the negligence of the Karolinska University. In September, the Central Ethical Review Board found Macchiarini and his co-workers guilty of scientific misconduct. In January 2018, the new vice-chancellor, Ole Petter Ottersen, declared Macchiarini and seven team members guilty of scientific misconduct. Six of the articles published by Macchiarini were retracted. Macchiarini was dismissed from the Karolinska in mid-2017.

Also, the Castillo's case was muddied by bitter controversies. Castillo's surgery is, to this date, supposed to be Macchiarini's stronghold since the studies published in *The Lancet* regarding this operation have not been retracted or questioned, and, more importantly, Ms. Castillo is still alive. Nevertheless, the testimony of a woman, Paloma Cabeza, recruited for the experimental surgery back in Barcelona, tells a different story. Ms. Cabeza deposited a statement to the Swedish authorities where she accounted how she was supposed to be the second patient to receive a trachea retrieved from a cadaver and how she met Ms. Castillo at the Hospital Clinic Barcelona. Since the two women became friends, Paloma claimed that Castillo advised her to withdraw from the experimental surgery. The reason is that Castillo's surgery did not go well at all. Although a follow-up paper in *The Lancet*, written five years after the surgery, claimed that "aside from intermittent bronchoscopic interventions, the patient had a normal social and working life" and that she "was not on immunosuppressive drugs," Castillo suffered from multiple lung infections and needed to get massive amounts of corticoids and antibiotics. Moreover, Ms. Cabeza claimed that Macchiarini falsified her CT scan and performed damaging preliminary interventions to demonstrate to the hospital's authorities a life-threatening case where there was no one.

According to Ms. Cabeza, while *The Lancet* paper reported that "the tissue-engineered trachea itself remained open over its entire length," Castillo's implant did not vascularize at all and ended up putrefied, causing infections. Albeit the press release in 2008 by Bristol University talked of "transplanting a trachea," what was done was to use "tubular tracheal matrices … to replace the recipient's left main bronchus." However, Castillo's left lung was removed in 2016, and before that, the rotting bronchus needed to be kept open using a silicon pipe known as "Dumon." Since the loss of a lung is not fatal, it explains why Ms. Castello could be heralded as the living proof of Macchiarini's might. In other words, she counts as a success because Ms. Castillo could not be added to the long list of patients (seven out of nine) who succumbed to Macchiarini's revolutionary promise. The fact that Ms. Castillo did not report Macchiarini was, for Ms. Cabeza, due to the fact that Macchiarini paid her a good deal of money while threatening her at the same time. Furthermore, according to the blogger Leonid Schneider (who lost litigation against Macchiarini), Birchall's Department in Bristol was not a hospital but a veterinary unit, the School of Veterinary Sciences in Langford, which is not accredited for working with human tissues.

Even the fabulous wedding ended up spoiled. First, Ms. Alexander discovered that the Pope had already organized a trip to South America for July 2014. When she asked the vowed groom about it, he replied that the Vatican had made a mess, and the Pope would simply have to cut the trip short. Then, the Vatican Public Relations Office director denied that the Pope even knew of one Paolo Macchiarini. Moreover, Ms. Alexander hired a private investigator who reported that Macchiarini never divorced Ms. Pecchia (and it is said that he also has an additional family in Spain). To make matters worse, it is known in the musical world that Andrea Bocelli does not do wedding parties. Simply put, the whole thing had been a lie, and Ms. Alexander had passed from living a Hollywood dream to re-enacting a Hollywood plot.

Nevertheless, as in all Hollywood stories, this melodrama has a happy ending, at least for the "super-surgeon." In 2019, an investigation from the Attorney General in Florence for extortion ended in 16 months of imprisonment due to forgery and office abuse, but Macchiarini was discharged on parole. In June 2022, the Solna District Court found Macchiarini guilty only of causing bodily harm, a simple felony. Not bad for somebody who had been implicated in the death of half a dozen people. Macchiarini continued to perform transplantation of synthetic organs, this time esophagi, at the Kazan Federal University in Russia. However, even his position there has not been extended. He even published his research in another high-end journal (*Journal of Biomedical Materials Research*), whose editor claimed to be unaware of Macchiarini's background. It is said that he now lives—in a villa, of course—in Cabrils, a suburb of Barcelona.

1.4 NEW YORK, THE PRESENT DAY

Wichita, Topeka, Stockholm—what do they have in common? This book will answer New York. Such an answer is not driven by the fact that New York is located at the crossroads of these cities but because it hosted summer boot camps that were involved, albeit not directly, in all of the events described in the preceding paragraphs.

New York is home to the Cold Spring Harbor Laboratory, established in 1890 by Eugene Blackford and Franklin Hooper. Over the years, this laboratory has provided compelling insights into agriculture, antibiotics, genetics, and oncogenesis. The Cold Spring Harbor Laboratory is home to eight Nobel laureates, including James Watson (1962), Alfred Hershey, Max Delbrück, and Salvador Luria (1969), Barbara McClintock (1983), and, most recently, Carol Greider (2009). Between 1945 and 1970, the Cold Spring Harbor Laboratory summer course on phage biology hosted attendees who became involved in the characterization of bacterial gene expression. Thus, these researchers collectively built the knowledge to produce insulin in bacteria. For instance, Esther Lederberg (née Zimmer) characterized factor *F* as an additional chromosome (a plasmid) present in certain bacteria. By doing that, she advanced our understanding of how bacteria encode and transmit genes. Joshua Lederberg (Esther's husband), Nobel laureate in 1958, further characterized the process of conjugation, which is the process where plasmids are mobilized inside a bacterial cell. Plasmids are essential in recombinant DNA technology because they contain the information required to produce the desired proteins. Plasmids are useless unless they are inserted into a producer bacterium. There is, therefore, an ideal link between the seminars held at the Cold Spring Harbor Laboratory and modern molecular biology, which includes the field of expressing proteins in bacteria.

We can imagine that love for science was the driving force for all these people. The eagerness to discover made them dedicate their lives to unraveling the secrets of the genetic code and microbial biology rather than the greed to gain money. The mindset of these scientists was based on the antipodes of Macchiarini's attitude. Macchiarini's case is by no means the only scientific fraud: let us think, for instance, of Elizabeth Holmes, who raised nine billion dollars in research grants for a revolutionary diagnostic device that was never functional. Or let us remember Hwang Woo-suk, who had been officially appointed "supreme scientist" by the South Korean authorities thanks to his high-ranking research, which included publications in the world-renowned *Nature Journal*, on human cloning and who was later revealed to have doctored his data and acted under fraudulent bioethical permits.

Nonetheless, the brutality and the halo of death surrounding Macchiarini's case set it aside from many other scientific misconduct cases. Macchiarini's deeds are so brutal that they make him resemble a character out of fiction, like the evil Dr. Wassory in Gustav Meyrink's novel *The Golem*. While Dr. Wassory was alone in misdiagnosing his patients to exploit them with treatments that left them disfigured, Macchiarini's con was not done in isolation: his and other fraudulent cases point to academic institutions warped for lucrative purposes rather than for the love of knowledge. Science is, in modern times, entangled with business, as the race for insulin clearly showed. Nonetheless, seeking a monetary return does not require the inhuman exploitation of patients. The trajectory taken by the Phage Group tells us that another way is possible.

The path connecting Topeka to New York also passes through Darwin (the person, not the city in California). Since Intelligent Design proponents use Darwinism as a proxy for materialism, the path passes through the quarrel between materialism and creationism. At the end of the day, knowing whether living beings have been created by a god or evolved on their own from chemical compounds does not change anything. Such knowledge neither pays the bills nor solves international conflicts. Nonetheless, science is about discovering the truth; thus, knowing whether orange paint is made by mixing green and yellow or red and green means a great deal to scientists—and painters. While some creationists might say that a deity started the world by allowing the Big Bang, some sentences of creationism are no longer defendable. Isaac Newton, for instance, claimed that God had to modify, from time to time, the path of the planets to keep them on the right track. Nowadays, hardly anybody backs such a statement.

Animals and plants evolve at a rate too low for humans to notice. Therefore, for the generations living before the invention of the microscope in the 17th century, it was commonplace to assume that all living species remained unchanged. It was once thought (and still is a creationist's tenet) that God created all of the species that inhabit our planet all at once. The Germ Theory and the field of microbiology changed that belief because microbes evolve at a rate high enough for people to notice. Science tells us that the immunodeficiency virus was not present before the 1930s; there was no SARS-coronavirus-2 (SARS-CoV-2) before 2015. One might object that viruses are not living beings, but can bacteria be ignored? Recent reports show that new species of bacteria are emerging that can degrade plastic to generate their sustenance. Since God did not create plastic, why would such bacteria be around? And since they were not here a 100 years ago, who created them? Such logical inconsistencies make creationism hard to digest. Would it not be more accessible to believe these plastic-eating bacteria have evolved due to human pollution?

Aside from logical consistency, there is another reason for defending Darwinism that makes a difference in our daily lives. History gives us countless examples of dogmatic societies where discrimination is endorsed. Creationism is, by definition, a dogma and is currently heralded by the political far right. Fearing that such political factions will gain power shortly is not science fiction.

As an example of the direction dogmatism can take us, in 1826, the reverend Thomas Malthus, one of the first economists and ardent believers in divine benevolence, wrote that God intentionally designed poverty to keep the human population low. He stated that if somebody "cannot get subsistence from his parents on whom he has a just demand, and if the society does not want his labor, has no claim of right to the smallest portion of food." Such a person should be considered an unwanted burden and gotten rid of. Thus, "we should facilitate, instead of foolishly and vainly endeavoring to impede, the operations of nature in producing this mortality." If famine is a too "horrid form" of disposal for our taste, "we should sedulously encourage the other forms of destruction, which we compel nature to use. Instead of recommending cleanliness to the poor, we should encourage contrary habits." Consequently, "we should reprobate specific remedies for ravaging diseases."

Malthus' ideas are a typical example of promoting the status quo of the time—and in the millennia before, the rich were rich because of God's will, and the poor were blamed for their poverty. No form of welfare state should be established. Fortunately for us, views have changed during the last two centuries, and poverty is now considered a plague to be eradicated. Many international agencies, such as the World Health Organization, aim to "ensure healthy lives and promote well-being for all." Such a change of heart is due to scientific progress in many fields of human knowledge, such as medicine, epidemiology, microbiology, and sociology. However, the creationists' anti-scientific attacks of today risk undermining such progress and re-establishing the status quo ante. Therefore, the defense of Darwinism, establishing its correctness, is just the tip of the iceberg of a much more critical social issue.

1.5 THE PURPOSE OF THIS BOOK

There are other excellent books regarding the Phage Group. What was the purpose of writing another? This book was written neither by Delbrück's friends nor Luria's colleagues; it was conceived neither by world-renowned professors nor Nobel laureates. On the contrary, it was drafted over two generations after Delbrück, Luria, and the other fellows of the Phage Group. The motivation underlying this endeavor was not personal affection but rather academic admiration for the past in scientific endeavors. The Phage Group has profoundly altered our world, propelled by a sake of knowledge that the commercialization of scientific research has diluted.

Furthermore, the discoveries directly made by the Phage Group or stimulated by it are nowadays taken for granted. The field of biology is now dominated by kits that routinely do tasks that were state-of-the-art endeavors just a few decades ago. The entire human genome has been mapped, and we are now at the point where 1000 genomes will be mapped to understand the variability between people. The applications of such mapping are expanding. Today, it is possible, for example, to determine if somebody is at risk of cancer simply by scanning their DNA. Cancer treatment can be tailored by using the specific fingerprint of each patient.

However, since the history of science is not part of the standard curricula in molecular biology or microbiology, the process of discovering the inner features of living beings is falling into oblivion. We are left only with the outcomes of such a process. The investigative quest that uncovered such a body of knowledge is lost, which is a pity because it could teach today's scientists how to approach the many unknowns of modern research. Thus, the goal of this book is to preserve such a memory. This book describes how the researchers involved in the development of molecular biology proceeded to understand how cells work. It is the chronicle of how ideas that changed our world were brought up.

Moreover, this book depicts the lives of these brilliant pioneers. They were ordinary but dedicated people, humbly focused on decrypting the secrets of life. They did so during highly upheaval times: the quantum-mechanic revolution, the Russian revolution, the rise of the Nazis, the Second World War, and the boom of post-war society. These events have shaped the world we are living in, yet they are now part of a past that is rapidly fading away. This book wants to keep the remembrance of those days.

In conclusion, the story of the Phage Group (or Phage Church, as was sometimes called) is not just a chronology of discoveries or an overview of viral biology. This story is a fusion of scientific concepts and historical events that are closely intertwined. It is a look back into a glorious past of great scientific discoveries and the development of human society.

The opera is conceived as a journey in space and time that will take us from the shores of a decadent Europe to the new frontier of America, from the darkest hour of human history to the dawn of our current age.

2 London and Paris

2.1 UNDERFUNDED

"Curioser and curioser," Frederick Twort must have thought as he turned the dish in his fingers to get a better illumination from the light passing through the window. Outside, in the yard, the neighing and pawing of the horses mixed with the bleating of sheep and goats. Twort's laboratory was located within the Brown Animal Sanatory Institution in the south of London. The Institute was born as part of the University of London thanks to a donation from Thomas Brown, a wealthy donor, and inaugurated in 1871 to provide veterinary care for "quadrupeds or birds useful to man," as Brown expressed in his testament. However, in the following 40 years, the Institute has become a free veterinary hospital for poor people. No fees were imposed, only a contribution to the expenses. No wonder its status had decayed during those years. Twort was the eighth Institute's superintendent, appointed in 1909, and struggled for funding like his predecessors. Still, the laboratory was on the front line of microbiological research, and he kept doing lab work despite being the director. Six years after being appointed to the Institute, Twort made a pretty unusual discovery while trying to grow a virus in a test tube.

Frederick William Twort—son of William Henry, a local GP with a bad temper who did not get along well with his 11 children—was born on the 22nd of October 1877 in Surrey, south of England. At 17, he enrolled at the University of London to train in medicine at the St Thomas' Hospital. Despite his intelligence and curiosity, he barely obtained his degree, which was granted in 1900. However, he compensated with a cheerful spirit that made him many friends and warranted the moniker "Peter" after Tsar Peter the Great of Russia. In 1901, he started his career as a bacteriologist, becoming a laboratory assistant first to Dr. Louis Jenner—not be mistaken with the physician Edward Jenner, who introduced the practice of vaccination in 1796—and then to Dr. William Bulloch, who became Twort's mentor. Twort and Bulloch worked happily together for seven years, during which the former distinguished himself for his hard work, attention to detail, and stubbornness. In 1908, Twort helped his father contain an outbreak of diphtheria at the Royal Albert Orphanage in Camberley, Surrey, during which he also studied its etiological cause and mode of transmission of the disease. The following year, he was appointed as the Brown Institute's director.

His first contribution to the field of microbiological research was on bacterial fermentation. About a decade earlier, Alfred Theodore MacConkey, an English bacteriologist working in Liverpool, had developed a medium that inhibited the growth of harmless bacteria but enhanced that of enteric pathogens. The trick was to add bile—an organic salt naturally present in the intestine and necessary to digest fats—to the formulation: only bacteria that evolved to live in the intestine could sustain bile's disruptive action on the bacterial cell. That medium—now known as "MacConkey"—was called "selective" because it selected a specific class of bacteria, such as *Escherichia coli* and *Salmonella enterica*, which can cause gastroenteritis.

DOI: 10.1201/9781003657712-2

Dr. MacConkey reported that it was possible to classify bacteria isolated with his medium by how they fermented selected sugars. In this biochemical characterization, a bacterium could ferment one type of sugar but not another. Twort, on the contrary, showed that the same bacterium could ferment different types of sugars in distinct conditions. MacConkey's typing system, thus, was not as clear-cut as hitherto thought.

Twort even suggested that harmless environmental bacteria could change their metabolism in the animal body and become pathogenic. After all, was that not what happened to the Koch's bacillus? *Mycobacterium tuberculosis*—as the bacterium was officially named—originated 150 million years ago as a simple soil bacterium—an ordinary earth organism. Then, three million years ago, it managed to infect primitive humans, and about 20,000 years ago, the terrible etiological agent of tuberculosis—also known as consumption, phthisis, or the White Plague—surged. At the time, tuberculosis was killing about one-quarter of the adult human population. Just in Britain, in the 60 years before Twort's employment as director of the Brown Institute, consumption had killed four million people. One century later, tuberculosis still kills about two million people a year. Back in the Middle Ages, the "Black Death"—caused by the bacterium *Yersinia pestis*, isolated by the Swiss-born French physician Alexandre Yersin in 1894—had wiped out about one-third of the European population, some 25 million people, and hence the tuberculosis' nickname "White Plague."

Twort was confident that his discovery would allow him to make a name for himself internationally. Instead, his paper, published in 1907, went virtually unnoticed. The same destiny doomed his second, published in 1908. He reported that ericolin, a bearberry sugar, could inhibit many bacteria but not Koch's bacillus. He tried to use ericolin to isolate not the bacillus of tuberculosis but that of leprosy—a related bacterium known as *M. leprae*, isolated in 1873 by the Norwegian physician Gerhard Armauer Hansen. However, he failed to isolate *M. leprae*. Because *M. tuberculosis* could grow in ericolin-containing media, he attempted to grow both mycobacteria together, hoping that *M. tuberculosis* would produce some "essential substance" to aid *M. leprae* growth. This time, he succeeded. Adding dead *M. tuberculosis* bacilli to the medium allowed the cultivation of *M. leprae*. It took a quarter of a century to assess that the "essential substance," the growth factor, was vitamin K. Yet, back in the day, his work went through the bacteriology community as water on glass.

Things went differently on his third strike. In 1912, George Ingram and Twort isolated Jonhe's bacillus, *M. paratuberculosis*, which induced chronic gastritis in cattle and a substantial economic loss. Chronic gastritis does not sound like much, but it caused Britain to lose one million pounds (about 150,000,000 € in today's terms) a year. Even today, it is relevant not only in veterinary terms but also clinically since it can cause deadly zoonoses. *M. paratuberculosis* is also fastidious, meaning it was challenging to cultivate. Ingram and Twort applied the approach of using a supporting bacterium and managed to isolate the bacillus. The international community praised them for this feat, and Twort was finally recognized as a first-rate bacteriologist. Twort was planning great things for his career, but his dreams of glory were shattered even this time. In fact, despite the recognition from the Royal

Society, he still did not get enough funding to pursue his discovery in full. Soon after being appointed director, he remembered when he asked the University of London for a grant of 3000 £ (460,000 €) to refurbish the Institute and establish a library. In return, the Dean has reduced his salary to 300 £ (46,000 €) annually. And how much paperwork did he fill out to beg for a 500 £ grant (76,000 €) that was promptly denied?

Twort was shocked that the Board of Agriculture was willing to pay for somebody else's research on Jonhe's bacillus—the "exploiters"—but not the guy who first isolated it—meaning him. That was his doom: lack of funds. He always struggled to meet ends, always short of resources to follow his lead until the end. If only he could have access to better equipment, if only he could buy finer chemicals, if only Twort could employ an additional assistant, God knows what he could have achieved by now. On top of it all, Ingram died two years after their work was published, leaving Twort essentially alone in characterizing *M. paratuberculosis*.

Since the avenue of research of Jonhe's bacillus was not going any further, Twort decided to experiment with the cultivation of viruses because that was the hot topic of the day. Maybe he would finally break through and get the much-needed grant to set up a proper laboratory of bacteriology. Fortune favors the brave, after all. And so, after three years of painful trials and with the help of his brother Charlie (Charles Claud), Twort had found something. Not what he was looking for, though.

Viruses were a tricky novelty at the time. Ever since the Dutch glassmaker Antoni van Leeuwenhoek showed the existence of microscopic animals more than two centuries ago, it was understood by the medical community that these "animalcules"—Latin for "tiny animals"—could cause illnesses. Some of the scariest diseases at the time, such as typhus, cholera, and tuberculosis, were due to microscopic organisms. About 70 years before Twort's discovery, in 1828, the German naturalist Christian Gottfried Ehrenberg had called these animalcules "bacteria," from the Greek for "rods." Twenty years later, the French chemist Louis Pasteur demonstrated that these "microbes"—a term he coined in 1878—did not arise spontaneously from the decay of organic matter but developed from pre-existing microorganisms. Heinrich Hermann Robert Koch, a German physician, was able to isolate the bacterium that caused tuberculosis and then cause the disease in a healthy animal by injecting the purified microorganism. Although Koch died four years earlier, in 1910, his imprint in the realm of bacteriology had been profound. Apart from the isolation of the most feared bacterium of the late 19th and 20th centuries, he also developed the vast majority of the techniques of the bacteriological trade. Even the dish Twort held in his hand had been invented by one of his associates, Julius Richard Petri, who perfected one of Koch's contraptions.

Bacteria are so small that one needs a microscope to see them—precisely as van Leeuwenhoek did. Then, about a decade before that day, three botanists—of all people—changed the field of medicine for good. It all started with a German working in the Netherlands, Adolf Mayer. Mayer described a disease of tobacco plants, which he called "mosaic," and revealed that an infusion of affected plants could cause the same illness in healthy plants. A few years earlier, Koch had put forward a framework to demonstrate the infectivity of a bacterium, a logical sequence that one of Koch's pupils, Friederich Löffler, dubbed "Koch's postulates." Such a "trinity of experimentation" went something like this: isolate a bacterium common to all cases of a disease, let's say tuberculosis; inoculate that purified bacterium in a naïve animal, and it will cause tuberculosis in it; the same bacterium must be present in the diseased testing animal. Even if Mayer demonstrated the agent's infectivity that caused mosaic, he could not isolate any bacterium. Thus, the postulates were not fulfilled.

The Russian botanist Dmitri Iosifovich Ivanofsky worked along that line of research. The Russian Department of Agriculture commissioned him to find a cure for mosaic, which was devastating tobacco cultures in Ukraine. He repeated Mayer's experiments but added the step of passing the infusion through a Pasteur-Chamberland filter, a porcelain tube Louis Pasteur and his assistant Charles Edouard Chamberland patented in 1893 to obtain drinkable fresh water by physically removing bacteria from the liquid. The filtrate was still capable of infecting naïve plants. The Dutchman Martinus Beijerinck, who had worked with Mayer, independently repeated Mayer's experiments—and he, too, passed the infusions through a Chamberland filter, obtaining the same results as Ivanofsky. Thus, two researchers, working independently, had found the same result: there was an infectious disease caused by a "virus" much smaller than a bacterium. Beijerinck had called that something "*contagium vivum fluidum*" ("contagious living liquid"). "Virus" was a general term, just a bit more focused than calling a microbe a "thing." When someone did not know what microbe caused a disease, it was customary to call it a "virus." The ancient Romans used the term "virus" to describe a poison like the venom of a snake or the drool of a rabid dog—both could kill a person, albeit in very different ways. At the beginning of the 20th century, the term was applied more and more to describe these elusive tiny microbes introduced by Mayer. Twort called them "filter-passing viruses" or "ultra-microscopic viruses." Today, the adjectives "filter-passing" and "ultra-microscopic" are dropped altogether.

After these three botanists' work, violent discussions ripped the microbiology's world apart. What was the virus—now known as tobacco mosaic virus (TMV)—that caused mosaic? A tiny bacterium? Or something different, maybe a toxin, like Ivanofsky suggested? However, if TMV were a chemical, one would expect that the dilutions of the filtrate would become less and less effective in causing mosaic, which was not the case. It was like TMV, once immensely diluted, could regain its strength, which is why Beijerinck called it "vivum." If that "thing" was alive, it must be possible to isolate it and cultivate it, which was Twort's specialty.

Twort based his hunt for viruses on the hints he gained in his previous research: there are many more harmless microbes than harmful ones; the former are more easily cultivable than the latter. Moreover, harmless microbes can easily be isolated from the environment. Thus, he soaked specimens of soil, dung, grass, and pond water in sterile liquid, then passed the fluid through Chamberland filters—and their German versions, the Berkefeld filters—and inoculated the filtrate in hundreds of different types of growing media. He also inoculated the filtrates into animals and even rubbed them on the shaved skin of a volunteer. He did not find anything. None of the test subjects developed any abnormal symptoms. On the contrary, his work was spoiled by false results due to bacterial spores passing through the filters.

Frustrated, Twort tried another avenue. He inoculated agar dishes with fluid obtained from the calf's lymph nodes. Agar was another of Koch's innovations. Agar is a sugar composed of agarose and agaropectin, and it is extracted from algae. Commonly, it is used to make puddings, but Koch found that it could solidify a liquid growing medium, allowing bacteriologists to work with solid surfaces, which have the advantage of fixing an individual colony on the spot, improving a lot of the task of isolating bacteria. Calf's lymph nodes, instead, have been used since Edward Jenner's time to produce smallpox vaccines. More than a century earlier, Jenner noticed that maidens milking cows were less susceptible to smallpox.

Smallpox was another of humanity's momentous plagues: before the introduction of vaccination, about one in ten people worldwide died of this infection. It turned out that cows were infected with cowpox rather than smallpox, but the two viruses were similar enough to produce cross-reactive immunity. Exposing people to cowpox would induce a milder disease than smallpox, but the resulting immune response would protect them from subsequent infection with smallpox, saving their lives. Calf's lymph nodes were filled with cowpox viruses; thus, they were the basis for vaccine production. Nonetheless, these lymph nodes also contained tiny round bacteria, known as micrococci, that would pass through the filters as contaminants. Micrococci were, by definition, "filter-passing viruses," but not the kind of viruses Twort was after because they were established to be bacteria.

Therefore, at the time, Twort's laboratory was crammed with dishes containing micrococcal cultures, a slew of false-positive results. Then, one morning, Twort found something odd. In theory, the dishes should show a homogeneous patina of white bacteria. Instead, to his surprise, there were "watery-looking areas" all over. The agar on the Petri dishes showed transparent spots on an otherwise creamy surface.

Turning the dish with his fingers to better see the transparent colonies, he considered what might be the cause of this "glassy material." He had started a new colony of cocci, which had grown in its usual way. The previous day, he had filtered one of those glassy colonies through a Chamberland, diluted it one million times in sterile saline, and inoculated the fresh colony. That morning, the new colony had turned transparent. Whatever was causing the bacteria to become glassy could be transmitted. The cause of this transformation was not only transmissible but also akin to the viruses found by Mayer, Beijerinck, and Ivanofsky. God willing, he was on the right track!

What was the "virus" he was after? A tiny bacterium that could slip through the filters? A kind of fastidious micrococcus? A spore, perhaps? A minuscule amoeba? Putting the dish on the bench before him, Twort must have pondered whether viruses might be even simpler organisms than amoebae, maybe some protoplast. Or was the cause of this glassy phenomenon derived from the cells altogether? Some kind of enzyme? A toxin, as Ivanofsky hypothesized?

Twort will have to wait a long time for his answer. He continued to experiment with his glassy cultures in the weeks that followed that morning, constantly struggling with money. He observed that if he spread even one drop of these transparent colonies on the surface of the agar, no cocci would grow there anymore. He noticed that bacilli and yeast were immune to this virus and that not all cocci were equally susceptible. For instance, Staphylococci—which usually causes furuncles—were transformed much more quickly than Streptococci—which predominantly provokes laryngitis—or bacilli. He recorded that the transparent spots could spontaneously generate if he kept a culture running long enough, and the newly developed glassy colonies could be transmitted in the ways he had observed. He argued that the transforming factor must have been something inherently present in the cocci, albeit he could not exclude an infectious agent resistant to his attempts to purification, perhaps dormant until certain conditions were met. But despite his efforts, he could not finish his research.

The First World War broke out in July 1914, just weeks before that faithful day. A few weeks later, he volunteered in the Royal Army Medical Corps and was dispatched, with the rank of captain, to Thessaloniki, Greece, to carry out an intensive diagnostic routine on dysentery—another bacterial disease that afflicted the entire world at the time, killing thousands of people each year. He will continue to struggle to find funding and time to study his "glassy effect." He was well aware that even in this case, when examining *M. paratuberculosis*, he had lost the race. In the short report on his finding on the glassy phenomenon—only six columns, released in 1915, in the journal published by the Royal Society—he wrote the epitaph to his investigation:

> I regret that financial considerations have prevented my carrying these researches to a definite conclusion, but I have indicated the lines along which others more fortunately situated can proceed.

These words are the nightmare of every researcher back in the day as today: not gathering enough money to carry out experiments and watching others do them instead. In the academic environment, such an ordeal even has a name: the "Matthew effect"—scientists who have won large grants will easily win even larger ones; those who don't stay out of the golden circle of supported research. Twort was right: the delay in obtaining funds ended up fateful. Somebody else was about to answer his questions before he could.

2.2 INSTRUMENT OF RECOVERY

"How wondrous," thought Félix d'Hérelle, comparing the two test tubes he was holding. One contained a turbid suspension, the other a clear liquid. They could not have been more different, and that was amazing. Both tubes started as liquid cultures of *Shigella*; in the cloudy tube, the bacterium grew as usual, causing turbidity, but in the second, the bacterium dissolved "like sugar in water." Félix d'Hérelle remembered with delight the path that brought him this morning and how faith had worked to crown his success.

After the First World War started, he volunteered at the Pasteur Institute (PI) as part of the war effort. Within a few weeks, he became laboratory director—just an honorary title with very little economic output—in charge of producing 20 million doses of anti-dysentery vaccine for the French army. He worked shifts lasting from eight in the morning to one in the following morning. He was one of the first physicians to introduce animal models to test the pathogenicity of microbes. He also introduced, with the blessing of the director of the PI at the time, Emile Roux, the use of mustard to kill *Salmonella typhimurium* and *Shigella dysenteriae*. Mustard was relatively inexpensive and readily available; since it could kill bacteria more effectively than other methods, it was optimal for developing vaccines that used dead bacteria to raise immunity in people.

In the summer of 1915, an outbreak of dysentery occurred among a cavalry regiment deployed in Maison-Lafitte village, located on the outskirts of Paris and approximately 80 km from the front lines. Dysentery is not just a form of diarrhea: the bacteria attack the intestinal tissues, causing necrosis, and the disease is often fatal. The army doctor of the regiment, George Bertillon, quickly realized that the outbreak was atypical and contacted the PI for help; d'Hérelle was instructed to provide such support and he quickly isolated the cause of *Shigella*'s epidemic (shigellosis), confirming that it was a novel and atypical strain.

It was during this time that he made a remarkable discovery. He plated *Shigella* on Petri dishes and added filtrates from the patients' stools. He noticed that in some of the cultures, the creamy patina of bacteria was scattered with perfectly round spots—"taches perfaitment circular," which later he called "taches vierges" or simply "taches" or "plaques," a term still in use to this day. It was a rediscovery, for he had noticed the same phenomenon about three years earlier in Mexico when fighting locusts. At the time, he did not know what to make of such observation, but when the plaques reappeared, he "finally had the proof that the phenomenon of the clear spots was not limited to the coccobacilli of locusts but that bacteria to man were equally susceptible to it." While remembering how all this came to be, he tried to make himself more conformable on his stool.

Born in Montreal, Canada, on the 25th of April 1873 but raised in Paris, Félix Hubert d'Hérelle has always been a dynamic chap, almost an adventurer. For his 16th birthday, his mother, Augustine, gave him a hefty sum of money to travel around Europe by bicycle. During the holiday, in the town of St. Hubert, he met a family whose son had been infected by the rabies virus. That encounter sparked a life-long interest in microbiology. After graduating from high school the following year, his mother gave him an even heftier sum that he used to travel around South America for three months. On his way back to Paris via transatlantic cruiser, the Royal Mail, he was caught in an outbreak of yellow fever. Despite 20 passengers dying from the disease, he never panicked. On the contrary, he clinched his commitment to microbiology at that moment. His journey throughout microbiology, though, did not start straight away.

After attending some classes at the University of Bonn for a few months, he left for Turkey in the Spring of 1893 for another long holiday. There, he met the daughter of the French consul, Adèle Caire, and they married in the summer of the same year. The new couple went to Athens for the honeymoon and then moved to Montreal. In 1897, d'Hérelle established his laboratory in his house, paying for all the equipment and reagents. That was a notable feature by Canadian standards at the time, as there was only another bacteriologist in the country, Dr. Bernier, who later became a professor of bacteriology at the University of Montreal.

About a year later, a family friend, Sir Henri Gustave Joly de Lotbinière, asked him to find a way to ferment the excess maple syrup, which Canada was literally throwing away, to produce whiskey for the US market. To allure d'Hérelle to the task, Sir Henri compared him to the most outstanding microbiological researcher of the time, Louis Pasteur. "Pasteur," de Lotbinière said, "made a good beginning by studying fermentation." About 40 years earlier, a winemaker in Lille, Mr. Bigot, had asked Pasteur to solve a similar problem: ferment beetroot to make an alcoholic beverage. Needless to say, d'Hérelle accepted. He and Adèle moved to Bauceville, near Québec City, and within a year, d'Hérelle had isolated a yeast that could do the job. Unfortunately, by then, the American market was already flooded with whiskey, so there was no place left for maple surrogates.

D'Hérelle did not lose heart. One year did not pass, and he was involved in another endeavor, this time with the involvement of his brother Daniel: a chocolate factory. He and his brother invested their mother's fortune to build a factory in Longueuil, on the outskirts of Montreal. However, already in 1900, Félix had a change of heart and sold his shares to his brother. That was good foresight, for the firm went bankrupt 1 year later, wiping out the family's fortune. He joined as a physician in a gold-seeking exploration in Labrador, but the expedition went uneventfully.

D'Hérelle's fortunes changed in the Spring when he was hired as a microbiologist in Guatemala City. He shipped his personal laboratory equipment and relocated his family there for good. If Canada had only two bacteriologists, Guatemala had only two scientists: apart from the newly appointed d'Hérelle, the other physician was René Guérin. Those were delightful times for d'Hérelle, with little work, no stress, and surrounded by exotic beauty. One day in 1912, knowing d'Hérelle's previous experience fermenting maple syrup, the agriculture minister asked him to ferment bananas. The minister, too, used the Pasteur parallel to entice d'Hérelle, and the proposal was accepted once more. D'Hérelle moved to Sao Tomàs and, within 1 year, had found a yeast that could do the trick. The resulting liquor was marketed as Mon banana whiskey and tasted like "a good Armagnac."

In 1905, he was charged by the government of Guatemala to organize a first line of defense against yellow fever, but he contracted malaria. In 1906, the minister of health gave him another challenge. There was ongoing devastation of tobacco plants due to some sickness; d'Hérelle—like Ivanofsky before him—was to find a cure for it. Within a few weeks, d'Hérelle isolated the infectious agent: a fungus that he called *Phtora vastatrix* ("devastation devastator") that thrived in acid soil. He reduced the infection of the plants by alkalinizing the ground where the plants were growing. In addition, he also discovered that some of the tobacco plants suffered from opportunistic bacterial infections.

His reputation as a chemist spread throughout neighboring countries. In 1908, the Department of Agriculture of Mexico asked him to ferment agave (bagasse) to produce Mexican whiskey. D'Hérelle accepted the challenge, relocating to Mexico City, and in March 1909, a factory was inaugurated to put d'Hérelle's method into practice. The resulting liquor had an "anise" flavor. He then volunteered as a laboratory assistant at the PI in Paris in April, returning to Mexico in November.

In the summer of 1910, a swarm of locusts wreaked havoc on the Mexican countryside. He was tasked with getting rid of the locusts, which caused a scourge that is very much present even today. His approach to the problem was entirely biological: he looked for dead locusts and then investigated the cause of death. While most of the locusts he collected died because of infestation with larvae of the fly *Sarcophaga*, others died more peculiarly. A black fluid filled the gut of those locusts and, under microscopical examination, contained bacteria of the species *Coccobacillus acridiorum*. D'Hérelle isolated the bacterium and then used it to start an outbreak among the locusts by infecting some animals and releasing them in the field. Essentially, he invented biological pest control—which today employs mostly *Bacillus thuringiensis*. The hoard of locusts subsequently receded, saving Mexico from famine. During this period, he first noticed the phenomenon of plaques.

In the Spring of 1911, he and his family moved to Paris for another turn as a laboratory assistant at the PI, where he presented his results on the fight against locusts. His seminar was reported in the local press and caught the attention of the Argentinian ambassador Enrique Rodrìguez Larretta. Larretta asked d'Hérelle for help in containing an invasion of locusts affecting Argentina; the latter accepted, and d'Hérelles moved to Buenos Aires, which at the time was competing with Paris and New York as the fanciest metropolis in the world. D'Hérelle's method worked and the swarm subsided, and his bacterial strain was subsequently employed in Colombia, Cyprus, and Algeria. However, d'Hérelle's success sparked a wave of chauvinism within the Argentinian Department of Agriculture, prompting the formation of a committee of inquiry to assess the efficacy of his method. The committee, formed only by Argentinian doctors, concluded that the bacterium isolated by d'Hérelle was not effective and he had used a flawed methodology. D'Hérelle quickly resigned, offended that his assistance had been met with haste rather than grateful recognition. This happening inspired a satirical novel (*El coccobacilo de Herrlin* by Arturo Cancela, published in 1923) about political dullness and opportunism. Despite the Argentinian diatribe, d'Hérelle's method was highly requested abroad: he spent most of 1913 in anti-locust campaigns in Algeria, Turkey, Tunisia, and Corsica. With the onset of the First World War, he returned to Paris, where he volunteered as a laboratory assistant—still an unpaid position—at the PI as part of the war effort. Soon, he became laboratory director—just an honorary title with very little economic output—in charge of producing vaccines.

So here he was, about a year later, with the results of his second investigative phase. He discovered that the plaques formed during the patient's recovery, implying that whatever caused the inhibition of the bacterial growth was involved in the infection's clearance. He immediately realized that the agent counteracting *Shigella* could be used as an "instrument of recovery." He subsequently demonstrated that the agent was filterable. However, it could not be an enzyme or a toxin; otherwise, it would have diffused throughout the agar, and the inhibition would have been complete. Only a sizable corpuscular body could cause the phenomenon of the spots: from a single point, the agent could diffuse radially only within a limited range. The agar's porosity that assured the agent's diffusion also acted as a filter, limiting its movement. His trials with filters of different pore sizes showed him that the agent was about the size of human albumin—about 8–22 millionths of a millimeter (nanometer, shortened to "nm").

He also found that the agent was transmissible, for a drop of a suspension made by inserting the tip of a needle into a spot and stirring it into sterile water was capable of generating the exact spots of inhibition. It was a specific agent because the inhibition phenomenon did not occur when he used other bacterial species or even other strains of *Shigella*. Finally, the agent must have been a bacterial parasite since he could not grow it without bacteria. In addition, he determined whether the inhibitor was intrinsic or extrinsic to the bacteria. In the former case, the inhibition should be proportional to the number of bacteria placed in a Petri dish. In the latter case, the inhibition would be weakened by increasing the dilution of the extracts derived from the plaques. So, d'Hérelle prepared serial dilutions of *Shigella* and added the same amount of inhibitory filtrates to them. His results were in line with the second hypothesis. He later suggested the name *Bacteriophagium intestinale* or "bacteriophage" in vernacular. Today, the term "phage" is widely employed.

D'Hérelle determined that the factor inducing the inhibition of *Shigella* was external to the bacteria with the following experiment (Figure 2.1). He prepared a serial dilution of bacteria—that is, a series of test tubes with increasingly less concentrated bacteria suspended in culture media. In each tube, he added the same amount of phages. If phages were associated with the bacteria, there should be a number of plaques proportional to the concentration of the bacteria. If phages were independent entities, the number of plaques would be proportional to the concentration of these elusive "filterable ultraviruses." His results were consistent with the latter scenario.

The outbreak of shigellosis did not go away, but it did not bring Paris to its knees either. A woman had been admitted to the hospital affiliated with the PI a few weeks before that faithful morning. He had closely followed her case, each night seeding filtrates of her stools. The test tubes he held had grown evenly over the first three days, but one had completely clarified that morning. Excited as he had been only a few times before, he carefully placed the tubes in their racks and almost dashed toward the hospital. There, he discovered that the lady had recovered from the fever that had consumed her. "I have never seen such a grave case of dysentery clear up so quickly," the nurse attending the lady told him. D'Hérelle was overjoyed: the lady's recovery was more than just good news for her. It meant phages had medical potential, which he was eager to investigate and apply.

2.3 A CORNER OF A TABLE

In the time that followed that primeval discovery, d'Hérelle found out that the phage-killed *Shigella* could not cause disease in rabbits but would maintain their immunogenic power, allowing for the implementation of an effective vaccine. Typically, rabbits infected with *Shigella* die too fast to raise an immune response. Immunizing rabbits with phage-killed bacteria permitted them to stay alive for enough time to establish an immune response and harvest the antibodies. One of his colleagues, Tamezo Kabeshima, exploited this feature and became the first employer of phages in medicine apart from d'Hérelle. However, since phages were stable in temperatures up to 70°C, Kabeshima concluded, unlike d'Hérelle, that they were "not a living being" but "nothing but a sort of catalyzer." Kabeshima published his results in 1919, but he clashed with d'Hérelle, who was developing a model that considered phages independent living organisms. "Enzymes," d'Hérelle argued, "do not attack living matter."

During an outbreak of dysentery in 1918, d'Hérelle quantified the number of phages during the infection in 34 patients. He observed that phages were present

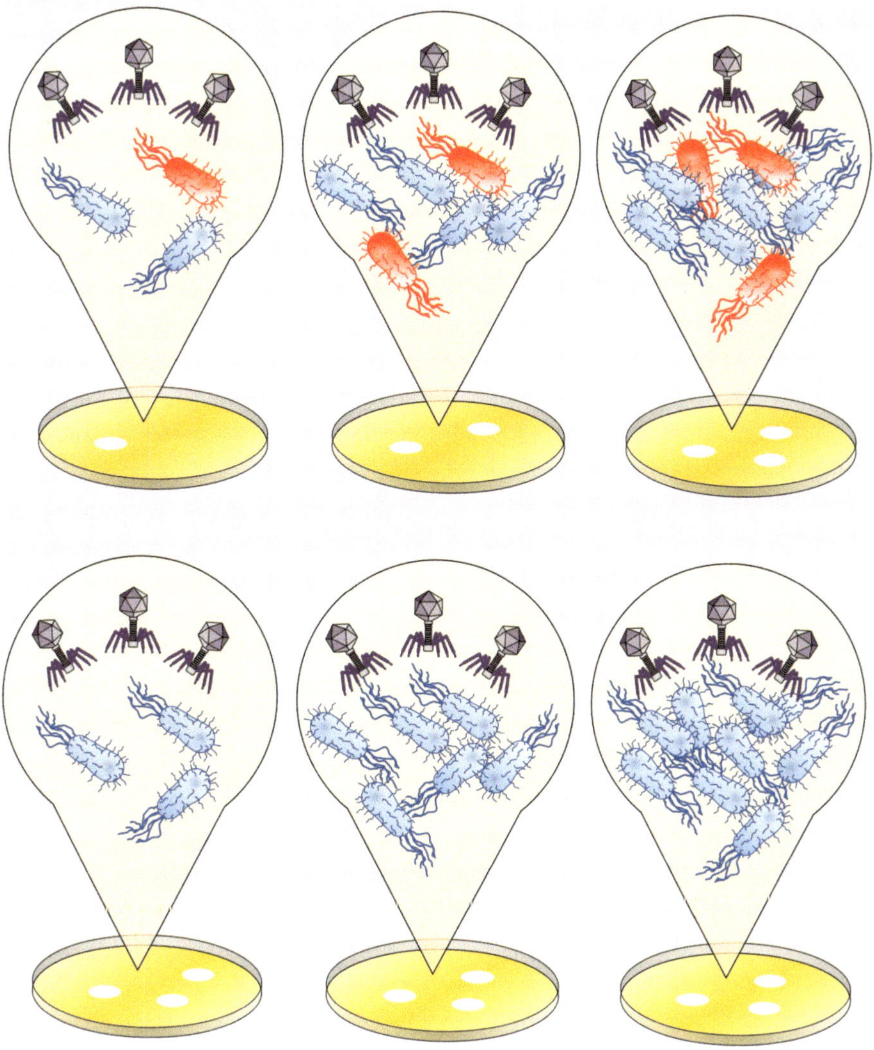

FIGURE 2.1 *Demonstration of the origin of bacteriolysis.* The factor triggering the observed lysis of bacteria associated with the plaques might reside on either the bacteria themselves (as stated by Bordet) or an independent element (as declared by d'Hérelle). In the former case (upper panel), the plaques would be proportional to the number of bacteria present in a plate. Assume that a certain proportion of bacteria is susceptible to spontaneous lysis (in red): the number of these lytic bacteria would increase with bacterial concentration (from one to three in the example). The number of plaques should be proportional to the number of lytic bacteria after mixing the different bacterial concentrations with an equal amount of phages. In the latter case (lower panel), conversely, the plaques would be proportional to the number of phages. Because the phages are added at a constant concentration, the plaques would look the same no matter how many bacteria were added. The experimental results supported the latter hypothesis, proving that d'Hérelle was correct: bacteriolysis was caused by a factor autonomous from the bacteria.

even in healthy people; thus, he believed they were part of the normal intestinal flora and could adapt to invading pathogenic bacteria. In the Lamarckian mindset typical of pastorians at the time, d'Hérelle thought that the same organisms could adapt in countless ways according to the environmental conditions. The adaptability of phages was a crucial point that will be covered in detail later by the fellows of the Phage Group.

Nonetheless, d'Hérelle left for Saigon—the former capital of Indochina, currently known as Ho Chi Minh City, in South Vietnam—in March 1920 to work with Alexandre Yersin and put the vicissitude aside. Saigon, d'Hérelle remembered, was a microbiologist's land of "dreams: one meets cholera, plague, and various epizoonotics." While he was away from Paris, things at the PI changed dramatically for him. D'Hérelle's mentor, Emile Roux, fell ill and was replaced by Albert Calmette, who previously developed vaccines against smallpox and rabies and introduced the first antivenoms for snake bites. In collaboration with Yersin, Calmette produced antisera against the plague. Around 1920, in partnership with the French immunologist Jean-Marie Camille Guérin, Calmette isolated an attenuated subtype of *M. tuberculosis*, the "BCG" strain. Calmette convinced the board of the PI to apply this strain for vaccination purposes. As head of vaccine production, d'Hérelle was particularly interested in the subject: he realized that attenuated vaccines had an inherent problem.

A microbe can be considered as a collection of compounds—known as "antigens"—that can induce an immune response in the mammalian body, including people and farm animals. There are two types of adaptive immune response. The primary response is slow, taking over two weeks to mount, and relatively mild. The secondary response is faster, taking less than a week, and very robust. By exposing the body to selected antigens in a controlled way, that is, during "vaccination"—which technically goes under the name of "artificial active immunization"—it is possible to kick-start the primary response. A pathogen infecting a naïve person or animal can also induce a primary response—technically "natural active immunization"—but he microbe will grow faster (usually in a couple of days) than immunity, leading to severe pathogenesis. Immunization causes a pathogen to face a secondary response immediately.

Consequently, the immunized person or animal might experience either a mild disease or no symptoms when infected with the actual pathogen. That is why people who have been vaccinated or recovered from an infectious disease are immune to a specific pathogen. By comparison, subjects who are not immunized might even die.

Back then, there were just two types of vaccines: those with inactivated microbes and those with attenuated microbes. The former is based on killed microbes that are entirely inert. The latter is established on still-alive microbes that induce fewer compelling symptoms than those of the original strain. It is said that these vaccine-grade microbes are "attenuated." Usually, attenuation is obtained by growing the microbes in suitable culture media. For instance, Calmette and Guérin isolated the BCG by cultivating *M. tuberculosis* on bile-containing media for years.

The process of "attenuation" can be summarized as follows. In the natural environment, such as the human body—that is, *in vivo*—a pathogen, such as *M. tuberculosis*, must compete for metabolic resources against a profusion of other species. Every element that can give an edge will increase the pathogen's survival. These elements are known as "virulence" factors. For instance, *M. tuberculosis* produces a sugar (phosphatidylinositol mannoside) that assembles as an envelope around the bacterial cell and facilitates macrophages' adsorption. The bacterium then grows inside the macrophages, part of the immune system, shielded from the immune response. There is a catch: the virulence factors require metabolic energy. In natural conditions, the biochemical expenses to maintain an additional piece of DNA and associated proteins are offset by obtaining an edge over microbial competitors. Still, in a controlled environment such as a Petri dish—that is, *in vitro*—where the microbe grows alone, a virulence factor is simply a burden. In this context, a microbe (let's say strain α) competes against itself. A mutant (let's say strain β) that does not produce a virulence factor will appear with enough time. Strain β can use the energy needed to make the virulence factor grow a little faster than α, and within a few generations, β will be the predominant strain in the culture. However, strain β will lose the struggle against the resident bacteria when introduced back into a body. Nonetheless, strain β will endure enough to trigger the immune response.

Attenuated strains lose their ability to produce virulence factors in a variety of ways: random mutations can affect a gene that codes for a virulence factor; the gene encoding the virulence factor may be present on a plasmid and thus be lost entirely; or the encoding is switched off by epigenetic means while the gene itself remains intact. Nonetheless, there is a risk of "reversion": an attenuated microbe can become highly pathogenic again by re-encoding the virulence factor lost *in vitro*. Attenuated vaccines are no longer popular precisely because of the risk of reversion; nowadays, vaccines are mainly based on a single antigen rather than the whole microbe. Reversion is not a theoretical issue. For example, the vaccine against poliovirus, made with attenuated viruses, reverted several times. More recently, it has been reported that even a vaccine strain of SARS-CoV-2 produced during the COVID-19 pandemic, obtained by removing the envelope gene from the wild-type virus, reverted to full pathogenicity.

Back in 1920, while in the mess room of the PI, d'Hérelle expressed concerns about using BCG, which Calmette was planning to use on children. D'Hérelle's colleagues immediately reported his words to Calmette, who was taken aback. Calmette summoned d'Hérelle upon his return from Saigon to discuss the rumor, which d'Hérelle confirmed. From that day on, Calmette became d'Hérelle's "declared enemy." The BCG vaccine was administered to children just as planned, but in 1930, 251 infants were vaccinated with a virulent strain of tuberculosis in Lübeck, Germany, not because of reversion but because of stock contamination; 67 of them died. Calmette might have been a proud cadre but not an insensitive bureaucrat: the episode startled him, affecting his health; he died 1 year later. In 1920, however, Calmette acted just like a proud bureaucrat: in retaliation to d'Hérelle's legitimate concerns, he reassigned Félix's laboratory. Despite being still the head of vaccine production, d'Hérelle could remain at the PI only because he could "use of a corner of a table" in the "minuscule laboratory" of his colleague and friend Edouard Alexandre Pozerski.

Despite these drawbacks, d'Hérelle proceeded steadily in the characterization of phages, publishing a first monography in 1921 (*The bacteriophage, its role in immunity*). D'Hérelle benefited from a vacation taken by Calmette to propose his book to Roux, obtain his patronage, and have it published as part of the *Annals* of the PI. D'Hérelle assessed that phages were very specific because they could inhibit the strains of *Shigella* that he isolated but not those isolated by Hiss and Flexner. D'Hérelle also recorded a reconstitution phenomenon similar to that reported by Beijerinck over three decades earlier: a significantly diluted suspension of phages could inhibit a whole culture of bacteria. Such behavior was difficult to explain in the light of a soluble enzyme. Instead, he thought of a living entity as a "filterable ultravirus" in the league of smallpox virus or TMV. In addition, the inhibition was due to the lysis of the bacteria, not to their slower growth.

In 1921, he also defended his thesis against the pastorian Alessandro Salimbeni, who claimed that bacteriolysis was due to the action of a slime mold (myxoamoeba). Not only could d'Hérelle not recover any amoeba in his preparations but also visually described the phenomenon of bacteriolysis due to phages—which he described as "very minute brilliant points." Equipped just with an "ultramicroscope"—powerful for the time, but it was, for modern standards, just an ordinary optical microscope—he recorded that:

> At the beginning the bacilli present a normal appearance. After forty-five to sixty minutes fine granules are seen, ever becoming more and more abundant, within the interior of the bacterial cell. The number of bacterial cells containing granules also rapidly increases with a corresponding diminution in the number of normal bacilli … The bacilli begin to swell … in the place of the spherical body there remains a slight cloudy floccule that slowly dissolves, thus liberating the fine granules … There can be no question concerning the nature of these bodies; they are bacilli which, operated by a force which can only be internal, take a first a globoid form and then rupture … This observation provides direct proof that the ultramicrobe develops and exerts its action within the bacterial cell.

"What do the fine granules that can be seen under the microscope represent?" d'Hérelle asked himself. "While nothing can be affirmed with absolute assurance there is nothing to preclude the supposition that they represent the ultramicrobes," he answered. "Bacteriophagy always takes place in the same manner," he noted:

> The bacteriophage particle must invariably become fixed to the bacterium to exercise its action. Destruction of the bacterium is always accomplished by bursting. The bacteriophage corpuscles always multiply within the bacterial cell and are always liberated with the rupture of this cell.

Furthermore, he noted that "the multiplication of the ultramicrobes is extremely rapid and, what is most remarkable, it is associated with successive jumps" (Figure 2.2). The Phage Group will replicate these experiments with more sensitive equipment about two decades later.

In the Spring of 1919, d'Hérelle introduced phage therapy for veterinary purposes. He isolated the bacterium *Salmonella gallinarum* during an outbreak of fowl typhoid—or "avian typhosis," as d'Hérelle named the disease—in the village of Pougy that was decimating the chicken's production. He then isolated phages against this pathogen and managed to get rid of the epidemics after treating the chickens. Afterward, he partnered with the renowned pediatrician Victor-Henri Hutinel, who was working at the Hôpital des Enfants-Malades in Paris, to find patients to test the clinical value of phages. In typical 19th-century spirit—exemplified by Dr. Jekyll drinking his potion

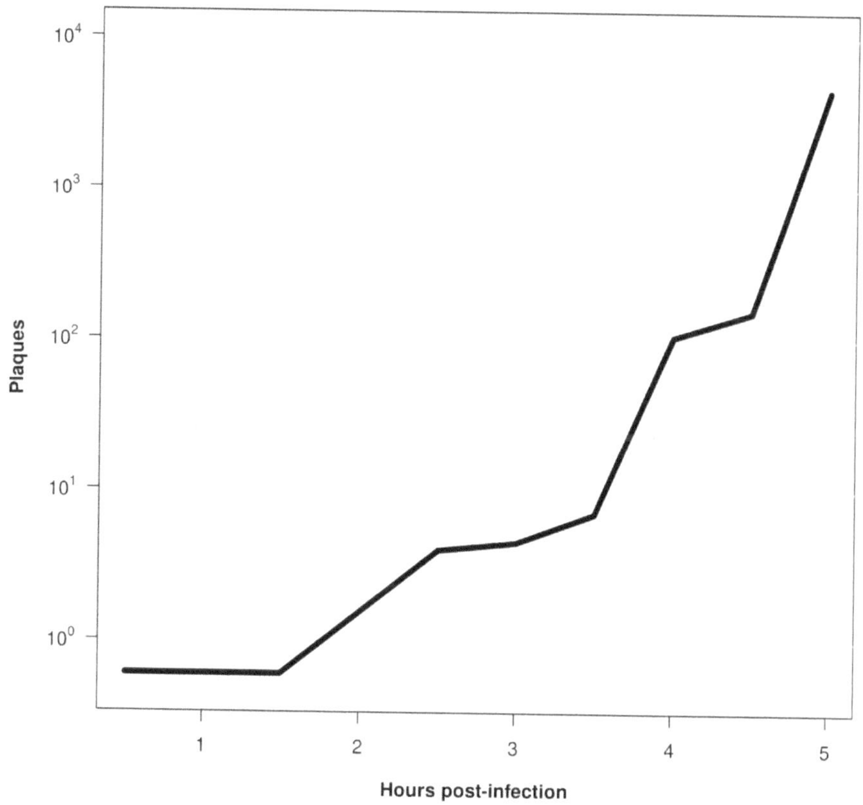

FIGURE 2.2 *Growing profiles of phages.* Plot of the number of plaques obtained from a series of bacterial cultures over time. Notice how there are almost no plaques at the start of the exposure; this period with an absence of phages is known as the "eclipse." Afterward, there is a ten-fold increase in plaque-forming units (PFU) in a short time (about an hour in the figure), which takes the name of "rise," followed by another dormant period and another ten-fold PFU increase. Such a profile is better known as "one-step growth curve of bacteriophage."

instead of going through a randomized, placebo-controlled, double-blinded clinical trial—d'Hérelle tried phage preparations, experiencing no side effects. In fact, before accepting the partnership, Hutinel requested that d'Hérelle should try a dose of phages 100 times more concentrated than the one he intended to give to the sick children. D'Hérelle agreed, and Hutinel followed suit; they both acknowledged that "the flavor, if not delicious, was not too disagreeable." None experienced any side effects. In August 1919, a 12-year-old child was hospitalized with dysentery, given a dose of phages, and recovered the following morning. In September, three brothers of 12, seven, and three years also recovered from dysentery within one day of the phage treatment. Such recoveries were hitherto unheard of.

After these successes, d'Hérelle switched to fighting another pressing infectious disease of the time: bubonic plague. In 1920, while in Saigon visiting Alexandre Yersin's laboratory, he isolated the first phage against *Ya pestis* in 1920, which he tested successfully in 1925. Meanwhile, since his position at the PI was compromised—an early case of "mobbing"—d'Hérelle gladly accepted a position at the University of Leiden in the Netherlands (from 1922 to 1924), offered by Prof. Paul Christian Flu, and then at the Quarantine Service in Alexandria, Egypt (from 1924 to 1928). In July 1925, he identified four cases of plague in ships approaching the city; even in this case, phage treatment resulted in healing infected people who would have died otherwise. This feat reverberated within the British medical community—since Egypt was an English colony back then. In 1926, the British government charged the Haffkine Institute in Mumbai, India, with a full-scale assessment of the efficacy of phage therapy. A first study showed a 100% recovery rate in 16 patients treated with phages, whereas 13 out of 33 patients treated with the standard methods died. A second study showed that the mortality rate in the phage-treated group was 8%, whereas, in the control group, it was 63%. The director of the Haffkine Institute, Lieutenant Colonel John Morison, was so amazed by the results that he authorized the use of phages as prophylaxis to cholera in the region of Naogaon. The study involved hundreds of thousands of participants, but during its implementation (1930–1935), there were no cholera epidemics in this region. In contrast, over 400 people died from cholera in the neighboring region of Habiganj, which shared many similarities with Naogaon and was used as a control for phage treatment. The British government decided that the protective power of phages was well proven, and phages were also applied as prophylaxis for cholera in Habiganj. From then on, phage therapy started to be used worldwide.

Nonetheless, d'Hérelle sometimes overstretched. Due a bit to his autodidact knowledge—thus missing a formal formation in biology—and a bit to his self-confident personality, d'Hérelle believed that the bacteriolysis that wiped out pathogenic bacteria in the human body was entirely due to phages. In other words, he did not believe in the existence of antibodies. On the contrary, his tenet was that a single type of phage would adapt itself—in fully Lamarckian terms—to feast on the first pathogen attacking the body. D'Hérelle's theory of the non-existence of antibodies reached the hears of Jules Jean-Baptiste Vincent Bordet in the winter of 1920. Bordet, director of the PI in Brussels, had won the Nobel prize in 1919 precisely for the characterization of the antibodies and their role in bacteriolysis. He decided to get involved in phage research solely to dismount d'Hérelle's theory and strengthen his own model.

Bordet became d'Hérelle's nemesis, albeit their dispute remained on more academic grounds than the feud with Calmette. Nonetheless, Bordet assembled a group of high-caliber scientists, including Nobel laureate John Howard Northrop, that became known as the "Belgian Group." Bordet discovered that bacteria exposed to animal serum lysed, dubbing these bacteria "lysogenic" ("formative of lysis"). He believed that the immune system was starting a modification in the bacteria that triggered the lysis even outside the mammal body. Once activated, these lysogenic bacteria would remain susceptible to lysis even if they did not come into contact with bodily fluids afterward. One of Bordet's collaborators, Mihai Ciuca, contextualized such bacterial modification in metabolic terms. He theorized that the immune system could alter the metabolism of the bacteria, causing them to become lysogenic. Ciuca named this process "nutritive vitiation." D'Hérelle responded to Bordet with the assistance of his new friend Georgi Eliava. Eliava also worked in Pozerski's laboratory and became the greatest of d'Hérelle's apostles. D'Hérelle and Eliava demonstrated that phages could lyse bacteria without needing mammalian tissues, including blood. In other words, phage-driven bacteriolysis could occur entirely *in vitro*. If the immune system did not activate the bacteria, then the whole theory of "vitiation" was flawed.

The size of the lytic factor was the second broadside of the Belgian Group against d'Hérelle. André Gratia, a Belgian working in the Department of Bacteriology at the University of Brussels under Jules Bordet, proposed that the lytic factor—defined as a "soluble ferment"—would diffuse from one bacterium as a source, forming a halo around the dead bacterium that would otherwise have grown to be a colony. D'Hérelle had already rejected this possibility but proved Gratia wrong with the following experiment (Figure 2.3). He noticed that the phages were active at a dilution of one in ten billion (10^{-10}) but not at a dilution of one in one hundred billion (10^{-11}). The lytic factor, therefore, lost its power between these two limits. He added 1 mL of the 10^{-10} dilutions to each of the ten tubes containing 9 mL of bacterial suspension, generating dilutions of 10^{-11} of the phages. If the lytic factor was soluble, as Gratia claimed, the tubes should be filled with fully grown bacteria after a few days of incubation because the lytic factor would have been too diluted to work. Instead, he discovered that one-third of the test tubes lacked bacteria. Lysis had occurred, but only in a few cases. Only a large, sizable entity could do that. It was a little like throwing a ping pong ball on a table filled with cups at an amusement park. D'Hérelle

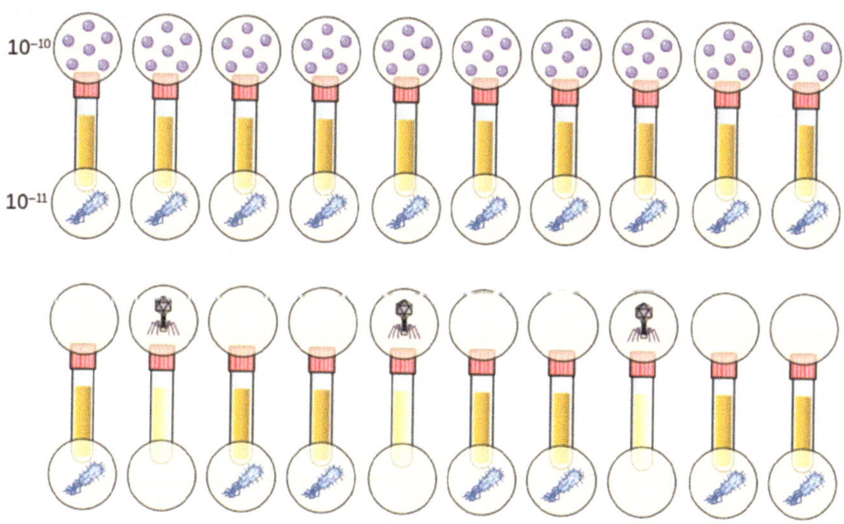

FIGURE 2.3 *Demonstration of the corpuscular nature of phages.* The bacteriolytic factor might be either a soluble element (as stated by Bordet) or a corpuscular entity (as claimed by d'Hérelle). In the former case (upper panel), the lytic factor would be homogeneously distributed in a series of tubes; in the latter case (lower panel), only a subset of vials would contain the lytic factor. The lytic factor was known to be active at a dilution of 10^{-10} of the original phage isolate, but lost its lytic power with a subsequent ten-fold dilution. If the lytic factor was soluble, no ten-fold dilutions of an active preparation would cause bacteriolysis, and the bacteria would continue to grow unaffected. Bacteriolysis would occur in a subset of tubes (lighter color) if the lytic factor was corpuscular, and bacteria would grow only in the vials where the phages did not end up. The experimental evidence was consistent with the latter hypothesis, proving that d'Hérelle was correct: phages were corpuscular entities.

had his hypothesis confirmed by none other than Albert Einstein, who was employed at the University of Leiden in 1920. Such an experiment was central to understanding the nature of phages and will be repeated by Delbrück and Luria a few decades later.

The Belgian Group's third strike against d'Hérelle was intended to strip him of the phages' paternity. Bordet and Ciuca re-read Frederick Twort's 1914 article at the Belgian Society of Biology Congress in March 1921. Twort was bestowed as the true phage's discoverer even though, unlike d'Hérelle, he did not offer a consistent theory about the origin of the lytic phenomenon he had observed. Such an attempt to discredit d'Hérelle was a significant setback for the Canadian scientist. Still, by that point, he had established himself as the authority on phage biology and phage therapy. The paternity of phages was no longer a big deal. D'Hérelle and Twort met for the first time at the Congress of the British Medical Association in Glasgow (1922). While d'Hérelle presented his phage biology model, Twort had little to offer over his original paper. At the meeting, Gratia claimed that "the Twort phenomenon and the d'Hérelle phenomenon are identical." From then on, the phage-driven bacteriolysis became known as the "Twort-d'Hérelle phenomenon."

In 1928, d'Hérelle was invited by Edwin W. Schultz to join the School of Medicine of Yale University, directed by Milton C. Winternitz. D'Hérelle promptly accepted, thus seeing his dream of recognition crowned. Such a dream started to come true when the American writer Sinclair Lewis wrote the novel *Arrowsmith* in 1925. Being the inspiration for a novel that even won the Pulitzer Prize—Lewis won the Nobel prize in 1930, the first American writer to obtain such an honor—meant that he was widely recognized as an international public figure. His fame extended also to the academic level. The previous year, he received an honorary degree in medicine from the University of Leiden and, in 1925, a Leeuwenhoek Medal from the Royal Academy of Sciences of Amsterdam. He was particularly proud of the medal because Louis Pasteur was the only other non-Dutch who won it. Like his predecessor, d'Hérelle was now a world authority in the realm of microbiology. Nonetheless, d'Hérelle was never awarded the Nobel Prize, most likely because the high echelons of the PI cast him in a negative light with the Karolinska Institute's selection committee.

2.4 FROM THE LAB-BENCH TO THE BEDSIDE

The work of Twort and d'Hérelle had fixed the family tree of viruses. Together with the viruses infecting plants (TMV), animals (foot-and-mouth virus, isolated in 1898 by Friedrich Löffler and Paul Frosch), and humans (yellow fever virus, isolated in 1932 by Max Theiler), it became established that viruses could infect all forms of life. New species were discovered or evolved—for instance, SARS-CoV emerged around the year 2000 and SARS-CoV-2 about two decades later—but the tree remained essentially unaltered for a century. The picture changed in 2008 when viruses infecting viruses were discovered. These novel viruses, named virophages and whose prototype is the Sputnik virus, infect giant viruses that are also a novel addition to the family tree. The first giant virus (*Acanthamoeba polyphaga mimivirus*) was isolated in 2003 from amoebae, demonstrating that viruses could be nearly as large and complex as small bacteria.

Phages were unknown when Max Delbrück was born in 1906, but they had been intensively studied by the time he finished his Ph.D. in 1930. Furthermore, they were widely used in the medical field. If Delbrück had walked into a pharmacy, he might have noticed phagial preparations like *Bacté-coli-phage* and *Bacté-staphy-phage* produced by the *Laboratoire du Bactériophage*, a firm that d'Hérelle founded in 1920. Later on, *Synthélabo France*, a subsidiary of *L'Oréal*, purchased the company, and its products remained on the market well into the 1970s. Undoubtedly, d'Hérelle's influence in applied microbiology can be felt even today. His legacy, however, is passed down to the ex-Soviet block rather than the West. The explanation for Eastern countries adopting Western medical methodologies is an amalgam of personal and political reasons.

Even if the professorship at Yale University had been a personal coronation of his status as a first-class microbiologist, he did not fit well into the academic establishment. Quickly, he developed a personal conflict with Winternitz, who did not quite like that d'Hérelle would spend much of the academic year either traveling for paid conferences or at his firm in Paris. Thus, d'Hérelle promptly accepted the invitation given by his friend Eliava.

Georgiy Grigor'yevich Eliava, born in 1892, worked as an army bacteriologist during the First World War. In 1918, he visited Pozerski's laboratory at the PI, where he befriended d'Hérelle. He returned to the USSR in 1921, obtaining the Chair of Hygiene in 1927 at the University of Tbilisi. In 1929, he co-founded the Bacteriology Institute in Tbilisi and became Professor of Bacteriology in 1929. He introduced phage therapy into the Institute, obtaining great medical success. In 1933, Eliava proposed to d'Hérelle to join him in Tbilisi to further work with phage therapy.

Eliava had at his disposal the resources not merely of a research grant but of a whole country; thus, he could offer things that the US—which was experiencing the "Great Depression" following the market crash of 1929—could not: a fully equipped laboratory with dedicated personnel, a hefty salary, academic independence, the opportunity to work half a year at his *Laboratoire du Bactériphage* in Paris, a villa, and even a chauffeur. At the time, the Soviet Union was in a phase of intense modernization of its infrastructure. Therefore, there was considerable interest in importing the most effective medical practices of the West. Having a world-renowned American professor move to the USSR would have been a point of honor for this young country.

Vladimir Ilyich Ulyanov, known as "Lenin," gained power in one of the world's most influential countries: Tsarist Russia. Before the First World War, Imperial Russia accounted for one-third of the European population (498 million). However, Russia was still a medieval country that could not compete with the European countries of advanced capitalism (United Kingdom, Germany, and France). While the Tsars and their court lived lavishly, millions of Russians were starving to death every year. The Tsars compensated for administrative incompetence with ruthless state police. When the Tsar of the time, Nicholas II, sided against Germany during the war, he went a step too far: within a couple of years, the country was on its knees and was ripped apart by riots. The communist party led the anti-Tsarist protest but was divided into two caucuses. Lenin was the leader of the minority faction—"Bolshevik" in Russian. In February 1917, a revolution caused the Tsar to allow the instantiation of a provisional government, which continued the war against Germany.

In November, Lenin led a coup—probably facilitated by the Germans—against the government and took control of the whole of Russia, funding the Union of the Soviet Socialist Republic (USSR), which promptly signed a peace agreement with Germany (treaty of Brest-Litovsk).

After the war, though, the capitalist countries of the West tried to reinstate the Tsar by funding royalist troops, and a civil war erupted, ending in 1922 with the victory of the Bolsheviks. While the US recognized the USSR only in 1933, the European countries found it more convenient to engage in commercial activities with this young state. Lenin supported such a trade, intending to revive a bled-out USSR, inaugurating the "New Economic Policy." This new phase of socialism, anointed "state capitalism" by its Bolshevist detractors, had the benefit of raising the standard of living of the Soviet society.

Thus, in the late 1920s, the USSR was a liberal and sparkling country actively seeking a new type of society. In stark contrast, the West entered a new phase of economic recession after the market crash of 1929. While the First World War was viewed as a massacre perpetrated by imperialist countries to increase their profits, the socialist society desired to be one founded on logic and fraternity rather than greed. Flocks of Western intellectuals moved to the USSR to be part of this new human experience. Being a liberal, d'Hérelle was not deaf to such a call; thus, there was a more personal benefit in joining Eliava's group than mere academic opportunity.

Furthermore, there must have been a more intellectual reason. D'Hérelle, who was of French origin and a pastorian, was a Lamarckist. One could argue that Jean-Baptiste Pierre Antoine de Monet, Chevalier de Lamarck, who passed away about 50 years before d'Hérelle was born, was Darwin's opposite. While Darwin was a scientist who based his work on empirical fieldwork, Lamarck has been defined as "the last of the Greek philosophers of nature." As an Appointed Academy of Science member from Louis XVI's reign to that of Napoleon throughout the French Revolution, Lamarck was a fervent follower of Heraclitus, believing that everything was in continuous mutation. He claimed that even crystals were bound to change their composition—so much for zircon, a mineral made by zirconium, silicon, and oxygen that can be as old as four billion years. Lamarck did not believe in species, considering them human constructs without biological relevance. Living beings showed a continuum of characters that did not allow them to separate from one another.

Both Lamarck and Darwin, however, assumed that species could evolve, even if Lamarck did not believe in the transmutation of the species until 1802–1809. Each species' characteristics (such as the number of legs or the shape of a bacterial cell) were encoded in what we now call "genes." Both men understood that organisms could adapt to their environment (for example, by switching off a gene or producing a modified protein). Lamarck believed that the environment could change the genes, allowing the adaptation to be passed down to offspring. Darwin, instead, thought that the genes were already altered, even before the environmental transformation. Therefore, the flux of genetic information was the contrary between the two scientists. D'Hérelle must have felt uncomfortable in the Anglo-Saxon academic environment of Yale University, strictly Darwinist but cozier in the Soviet worldview, which was shifting toward the Lamarckian side. The communist movement, based on the writing of the German philosopher Karl Marx, believed that the environment could determine how people behaved. Therefore, the Soviet experiment was directed precisely at modifying the environment (the working places, the administration, and so forth) to create a better society. Although officially Darwinism was incorporated into the Communist Party's tenet, there was a tendency to side with Lamarckism instead. The champion of Soviet Larmarkism was the agronomist Trofim Denisovich Lysenko.

In 1929, Lysenko conducted experiments in the Lamarckian framework to improve crop yields that were deemed very successful. Cereals come in two varieties: Spring and winter. Spring wheat is planted in the Spring and harvested in the autumn. Winter wheat, the most common, is planted in the autumn, grows partially, goes dormant during the winter, resumes growth in Spring, and is harvested in early summer. Spring wheat is grown in the colder northern areas, and winter wheat is grown in more temperate regions. Lysenko persuaded his father, a Ukrainian peasant, to wash winter wheat seeds, bury them in the snow, and sow them in the Spring like spring wheat. Lysenko managed to speed up winter wheat germination; apparently, his father's harvest was the most abundant in the neighborhood. Lysenko named his procedure "vernalization" (from the Latin "vernum," Spring) and assured him a position at the Ukrainian Institute for Selection and Genetics at Odessa. Vernalization did not become a widespread practice after all, but it appealed to the cadres of the Communist Party. Lysenko went even further, assuring that he transformed winter plants into spring plants simply by manipulating seeds. Plants are improved in traditional agriculture through grafting, which requires the physical combination of one donor (scion) and one recipient (stock) to generate a hybrid between the two. Lysenko's breakthrough, which eliminated the need for the tricky grafting procedure, was truly remarkable, placing Soviet science at the forefront of worldwide research and offering new life to Lamarckism.

Another reason for d'Hérelle relocating to the USSR was that, in true pastorian fashion, d'Hérelle was more concerned with the clinical application of phages than with their biological characterization. The PI was deeply involved in applied sciences, with all its research aimed at improving diagnostics and treatment, summarized by the motto "from the lab bench to the bedside." Despite studying phage mutations, he was not completely enamored with this line of research. "While in residence here," stated the official press release provided by Yale University upon his withdrawal, "Dr. d'Hérelle confined himself largely to the study of bacteriophages as a laboratory phenomenon. The control of infections by the use of bacteriophages has been a primary interest for him." Such a primary interest led him to leave the US in 1933 and relocate to Tbilisi. He worked hard alongside Eliava, and the Bacteriophage Institute accomplished many therapeutical successes, growing steadily. Those were happy years, but neither knew the Soviet dream would soon become a nightmare.

2.5 HAPPY YEARS

Despite the appearance of a modern and egalitarian society, the Bolshevist Republic was essentially a dictatorship, which conflicts with the open-mindedness required for academic advancement. Lenin declared that illiterate factory workers and peasants should control all intellectuals, and he enforced this view by exiling over 2000 scientists from the USSR. A new generation of academics was formed from "proletarians" (workers and peasants), who were, unsurprisingly, significantly less qualified than the old cadres. Lysenko, the son of a peasant, belonged to such a new generation of Russian scientists. But things were about to get worse. Lenin became increasingly ill beginning in 1921 and died in 1924. He left a testament to the Communist Party's upper echelons a year before his death, naming Leon Trotsky (or Trockij) as his successor and suggesting Stalin's ostracism. Who was this Stalin character capable of upsetting even one of history's most successful revolutionary?

Ioseb Besarionis dze Jughashvili, better known as "Joseph Stalin," was born in Gori, near Tbilisi, in 1878 and joined the communist party around 1901. In 1907, he participated with Lenin in the robbery of the Tbilisi Central Bank, which ended up with a haul of 341,000 rubles (about 2.3 million euros in today's terms) and over 40 casualties. In 1908, he was arrested and, in prison, organized the assassination of several anti-communist informants. After evading, Stalin became the director of Pravda, the newspaper published by the communist party. In May 1918, during the civil war, he was sent to Volgograd—later renamed Stalingrad—for food procurement, a directive he fulfilled even by burning down villages. He went beyond that order by commanding the revolutionary militia, known as the "Red Army," to battle against anti-Bolsheviks troops but incurred heavy losses. He also seized the local branch of the "Cheka"—the USSR's secret service at the time, which re-established the Tsarist secret police, the "Okhrana"—and ordered the assassination of many dissidents. In 1921, Stalin invaded Turkestan, forcing it into the USSR.

Lenin despised Stalin, finding him rude and stupid. Nonetheless, after Lenin's death, Stalin became his herald, giving several lectures on the "foundation of Leninism" at Moscow's Sverdlov University. Stalin formed a cabal with Lev Kamenev and Grigory Zinoviev to oppose Trotsky. Thus, in 1923, the USSR was under the guidance of the "triumvirate" Stalin-Kamenev-Zinoviev. The following year, however, Kamenev and Zinoviev sided with Trotsky against Stalin but the latter prevailed. The Communist Party, under Stalin's orders, expelled Trotsky, Kamenev, and Zinoviev. While Trotsky remained a staunch opponent of Stalin, even if in exile, Kamenev and Zinoviev apologized publicly for their treason and were readmitted to the party in 1927. They were expelled again in 1932 but reinstated the following year after another round of public pillory.

In 1928, five years before d'Hérelle moved to the USSR, Stalin started the forced collectivization of Soviet agriculture by incarcerating over ten million farmers who possessed their land ("kolkhozes"). However, contrary to expectations, the state-directed administration of large swaths of state-owned land did not improve over the individual enterprise of the kolkhozes. Soviet agriculture was disastrous and led to widespread famine. To overcome such a debacle, in 1931, Stalin simply ordered the Ministry of Agriculture to develop a new vigorous crop within five years. To grasp the herculean task, one should consider that it takes at least a decade to establish a new crop variety using conventional agriculture. In this context, Lysenko's ascent took place: he could fulfill Stalin's directive in just a couple of years while all other agronomists failed.

Afterward, however, Stalin's paranoia about possible plots against him reached a boiling point. On the 1st of December 1934, he ordered the Cheka—meanwhile renamed People's Commissariat for Internal Affairs, or "NKVD" for short—to murder Sergei Kirov, the second in command after Stalin, starting what became known as the "Great Purges." Zinoviev and Kamenev were arrested again and forced to confess in a public trial—that was nothing more than propaganda—their "moral complicity" in Kirov's assassination. Zinoviev and Kamenev were put on trial once more in the summer of 1936 and were forced to plead guilty in a show trial to more and more elaborate plots against Stalin. They were executed on the 25th of August 1936. Another 14 high-ranking party cadres were processed together with them and suffered the same fate. These high-ranking echelons were not the only ones persecuted during this period; 700,000 people were executed, and over 100,000 died in prison camps known as "Gulags."

Eliava and his wife, the renowned Polish singer Amelia Vol-Levitskaya, also succumbed to this carnage. It's unclear why Eliava, who wasn't politically active, was on the NKVD's death list. He may have had a personal feud with Lavrentiy Pavlovich Beria, who was the First Secretary of the Communist Party of Georgia at the time and later became the head of the NKVD. Eliava could also have been evicted because one of his relatives, Shal'va Zurabovich Eliava, was a fervent anti-Stalinist. Beria could also have a monthly quota of political opponents to present to Stalin, so he filled it freely. Whatever the reason, the Eliavas were seized on the 22nd of January 1937 and shot in July after months of harsh incarceration. Even their daughter, Hannah, who was 23 years old at the time, was imprisoned for nine years. The official excuse for the arrest was that Eliava was deliberately infecting children with his preparations; the Soviet press commented: "On the order of the bloody fascist Trotskyites, he was preparing bacteria to kill the Soviet people." The gratitude Eliava got for having saved thousands of lives with vaccines first and phages later was the epitaph: "This animal got the response of the Soviet people that he deserved."

While the Russian society was purged of anyone who opposed Stalin, "Stalinists" had the opportunity to advance their careers. Among the people who profited from the Great Purges was Lysenko, appointed director of the Lenin Academy of Agricultural Sciences in 1935. Soon afterward, he denounced "Mendelian genetics" as a "fascist science." It was an exaggeration: Mendelian genetics, and more in general genetics, was a research hot topic in Britain and England, both top capitalist countries—with their good share of social inequalities—but, unlike fascism, also strongholds of constitutional democracy.

Gregor Mendel, an Austrian friar, had described how pea plants' morphological characteristics ("phenotypes") were passed down from generation to generation in a book (*Experiments on Plant Hybridization*) published in 1866. His work went unnoticed for about three decades until Hugo de Vries (Netherlands), Carl Correns (Germany), and Erich von Tschermak (Austria) wrote scientific articles that cited Mendel's book. Later on, John Burdon Sanderson Haldane (England) published a report on the transmission of morphological characteristics in mice in 1904. Sir Ronald Aylmer Fisher (England) published a statistical analysis of phenotypic inheritance a decade later (*The Correlation Between Relatives on the Supposition of Mendelian Inheritance*). In the first half of the 20th century, the American Thomas Hunt Morgan established himself as the champion of this novel branch of research—"genetics"—by publishing a seminal book (*The Mechanism of Mendelian Heredity*) that summarized the state-of-the-art in the field. Genetics assumes that phenotypes are dictated by physical entities known as "genotypes," which are physically present in "genes."

The statistical analysis of the inheritance of phenotypes moved alongside a deep understanding of animal and plant physiology. The German August Friedrich Leopold Weismann demonstrated, at the end of the 19th century, that higher organisms, including humans, pass on morphological characteristics (for example, the shape of a flower or the color of the hair) via a subset of cells known as "gametes." Men's sperms, for example, mate with women's eggs to form a zygote that develops into a baby. What is inside the sperm and eggs will determine the baby's outlook. The rest of a plant's or animal's body is made up of "somatic cells." What happens to somatic cells during their lifetime is not passed down to their descendants.

On the contrary, any damage to the gametes is transferred to the offspring. As a result, biological information can travel from gametes to somatic cells but not vice versa ("Weismann barrier"). In 1895, the Canadian-Scottish anatomist George John Romanes coined the term "neo-Darwinism" to mark this turning point in biological understanding. While Darwin was willing to accept some of Lamarck's ideas, modern science rejected Lamarckism entirely. Ironically, Russian scientists were at the vanguard of defining the cellular biology of genetic inheritance. In 1903, Nikolai Kol'tsov described the cytoskeleton and, in 1927, the chromosomes. In particular, he illustrated how these giant molecules would join in pairs and use each other as a template for their replication. However, Kol'tsov was sentenced to death by the Bolsheviks soon after the revolution and pardoned because his friend Max Gorsky was a major contributor to the communist cause.

It was unknown what genes were made of; the Phage Group—another Anglo-American endeavor—will be instrumental in establishing the nature of genes. However, it was known how genes worked. Let us assume that gene G encodes a morphological characteristic, namely the color of peas: either dark (phenotype D) or light (phenotype L). This gene can take two flavors, known as "alleles": light and dark, represented by the symbols G^L and G^D. It was also recognized that not all alleles were equal: some could overwhelm others. Therefore, genes were (and still are) subdivided into "dominant" and "recessive." Let us assume that G^D is dominant over G^L. Because of dominant and recessive genes, geneticists working in the first decades of the 20th century understood, based on Mendel's work, that it was indeed the genotypes that determined the phenotype and not vice versa. For example, there could have been cases where dark parent pea plants produced an offspring of light peas. How could light peas come from dark peas? Geneticists discovered that genotypes were formed by combining two alleles, which determined the phenotype. In the example, phenotype D was generated by G^D/G^D and G^D/G^L combinations, whereas phenotype L could be obtained only by the combination G^L/G^L. If the parent plants had both genotypes G^D/G^L, then about one-quarter of the offspring would be light peas.

This statistical analysis of the inheritance of genotypes fits very well with the theory of neo-Darwinism. Assuming that the phenotypes have a different fitness, the connection between genetics and "natural selection" can be understood as follows. In the example above, the phenotypes D and L are associated with different "fitness"—the probability of transmitting a genotype to the offspring—and are prevalent in a specific population and a stable environment of f^L and f^D, respectively. Let us assume that D is more fit than L. For example, pea plants may grow in the shade of some trees and require a darker color to receive more sunlight. Light peas would be at a disadvantage and, consequently, f^L would be low. Let us assume that an environmental change occurred, such as the removal of the trees that provided shade, leaving the pea plants wholly exposed to the sun. Dark peas will now perish because they are exposed to too much light, whereas light peas will have an advantage because they will reflect some of the light.

Consequently, peas with light colors will flourish, increasing f^L. The takeaway message is that, in a Darwinist context, the genotype underlying the phenotype "light" is already present; it simply blooms when the appropriate conditions are met. Instead, in the Lamarckian framework, the condition—the environment—promotes the emergence of the phenotype "light," which is then forged into a genotype and passed down to the offspring.

These fundamental biological discoveries were genuinely life-changing. We would not have genetic tests, recombinant proteins, or *in vitro* fertilization today if it weren't for the work of geneticists in the early 20th century. Lysenko, instead, refused all these scientific advancements merely for political opportunity. He accused many of his opponents of anti-Stalinism, including his mentor, Nikolai Ivanovich Vavilov, who criticized Lysenko's theories. More than 3000 geneticists were imprisoned or executed under Lysenkoism, and crops treated with the Lamarckian approach declined.

2.6 IRON CURTAIN

The Stalinization of the Soviet civilization meant that the key positions in society became occupied by yes-men, driven by political rapacity but limited capabilities. Lysenko was the prototype of the first expurgation by Lenin and the Stalinist Great Purges. It came out that Lysenko had doctored his crop yield results in the 1920s, and in effect, crop yields dropped drastically during Lysenko's appointment. However, his connection with Stalin shielded him from any repercussions. On the contrary, he got rid of all his opponents. Kol'tsov, for instance, was poisoned by the NKVD in 1940, and Lysenko's mentor, Nikolai Vavilov, was arrested in the same year (starving to death in prison in 1943). In 1948, Vladimir Pavlovich Efroimson, Kol'tsov's student and one of the most prominent geneticists who had remained in the USSR then, wrote a detailed report of Lysenko's results to the Ministry of Science. Needless to say, it was Efroimson who was imprisoned. Lysenko's rule over Soviet genetics will last until the mid-1960s.

Not only was agriculture affected by the Great Purges, but so was virtually every other aspect of academic research, including mathematics, economics, and medicine. Remarkably, by executing Eliava, the Soviets significantly weakened their ability to control bacterial infections. Not only did the NKVD eliminate the world's second most prominent expert in phagotherapy but it also prevented the first expert from joining the USSR. D'Hérelle, who was attending the Laboratoire du Bactériphage at the time of Eliava's arrest, was devastated by the news of his friend's death and wisely decided never to return to the USSR.

Within a few years, the world was split into two major political spheres of influence: the capitalist West and the communist East. The schism was more than just ideological. Winston Churchill, who led Britain to victory in the Second World War, observed in 1946 that there was physical separation between the two blocks. "From Stettin in the Baltic, to Trieste in the Adriatic, an iron curtain has descended across the continent," he said to emphasize how very little information regarding what was going on within the USSR and its allies reached the West. Western and Eastern science took utterly different tracks with the outright ban of genetics in the USSR in 1948.

These trajectories started to diverge rather quickly. In 1935, the German company Bayer commercialized a chemical bactericidal: *Prontosil*. Thanks to the work of the German chemist Gerhard Domagk, a chemical compound, sulfanilamide, was isolated that could kill a wide range of bacteria. Even more than that, slight chemical variations of this molecule also displayed bactericidal properties; thus, sulfanilamide became the prototype of a whole class of antibacterials known as "sulfa drugs." Domagk won the Nobel prize in 1939.

The commercialization of antibiotics was another watershed moment in medicine. In 1929, about a decade after the Twort-d'Hérelle phenomenon was unearthed, the Scottish doctor Alexander Fleming noticed a peculiar fact similar to that of his predecessors in the virology field. Fleming also worked in a minuscule laboratory, which he shared with a colleague and one day, he found that his Petri dish had a peculiar appearance. A "white fluffy mass" had grown on a culture of *Staphylococcus*, which he had left on the bench. The contaminated culture was now useless, but he noticed that around the mold, there was a halo where the bacteria had lysed. Later, Fleming identified the mold as *Penicillium notatum* and discovered that it produced a substance ("penicillin") that inhibited or even killed bacteria. Fleming immediately recognized penicillin's enormous medical potential, but he could not continue with his work with penicillin. In 1939, however, Howard Florey and Ernst Chain, working at the University of Oxford, England, managed to purify penicillin and produce it in large quantities. This new antibacterial was mainly employed during the Second World War—essentially testing it on the field—and was marketed after the conflict.

Sulfa drugs and penicillin proved to be even more effective antibacterials than phages. They could kill a wide variety of bacteria, whereas phages were essentially limited to a single species. As a result, instead of needing a different phage mixture for each disease, a practitioner could prescribe the same drug to treat dysentery, otitis, abscesses, pneumonia, and so forth. Furthermore, because sulphonamide and penicillin do not replicate or mutate inside bacteria, their behavior after administration ("pharmacokinetics") is much easier to predict than that of phages. Because of this stability, practitioners could confidently prescribe these drugs that would work as expected in virtually all cases. On the contrary, phagotherapy requires—even today—constant monitoring to ensure that the phages remain active against the targeted bacteria.

The American practitioners began to question the efficacy of this French invention, which was increasingly becoming a Russian specialty. In 1934, a review published in the *Journal of the American Association* (JAMA) by Monroe Eaton and Stanhope Bayne-Jones summarized the clinical evidence available at the time. They rightly pointed out that the high success rate of phage therapy in suppurative infections seemed suspicious since the application of phages in the laboratory was less than satisfactory. However, they were a bit dismissive of the phagial potential: they noted that "the accounts of bacteriophage therapy in urinary infections are not as enthusiastic as those of the treatment of some suppurative conditions," "has, apparently, no striking effect on the course of the [typhoid] disease," and "the therapeutic value of bacteriophage in dysentery are conflicting and for the most part inconclusive."

A second JAMA review by Albert Krüger and Elaine Scribner, published in 1941, was equally dismissive. Despite the publication of six d'Hérelle's monographs on the argument and the visualization of phages by electron microscopy—carried out, however, by Ernst Ruska in Germany—the authors commented that "phage is a protein of high molecular weight" produced "by the cellular synthesis of an inactive precursor which is later changed into the active form." Regarding the "mechanisms of phage formation," they reported that "d'Hérelle and his followers have held to the theory that phage is a living ultra-microscopic parasitic" whereas "Bordet and his school ... advanced the concept that phage is simply a unique by-product of bacterial metabolism possessing the curious property of destroying the cell which formed it." They further commented that "the accumulation of data with the passage of time has strengthened Bordet's stand." Such judgments may have been overly harsh. Statistical methods have improved in the past decades, and it has become clear that data analysis improves by aggregating studies rather than evaluating them individually. Thus, a recent re-evaluation of most of the studies included in the review by Krüger and Scribner confirmed that phagotherapy was more effective than standard methods in reducing infections.

Because the USSR was shielded from the American Medical Association's criticisms, phagotherapy retained in the Soviet Union its halo of infallibility. Furthermore, antibiotics took a long time to be adopted by Soviet society due to their lower quality and higher price than their counterparts in the West. Phages remained the most popular choice of antibacterials. Even today, phagotherapy is still used as a standard antibacterial treatment in Georgia, Eliavas' country. Phages faded from medical practice in the West, but they became the tool of choice for uncovering genetic secrets. The Phage Group was crucial for such a transition.

3 Berlin, Göttingen, and Copenhagen

3.1 STARGAZING

The alarm went off at two in the morning sharp. In the stillness of the night, every rustle is a roar, let alone the blare of an alarm clock. Max Delbrück was the last one in the house to wake at the sound of the bell, pretending not to hear his parents' and siblings' grunting of disapproval at his new hobby. He would dress warmly for the chill of the night, then bring his heavy telescope to the balcony next to his parents' sleeping room and set up to observe the sky.

For young Delbrück, astronomy was a way to make a name for himself and escape his father's considerable shadow. In the 1960s, rebellious teenagers might have become yippies, with Spider-Man as their heroic figure. In the 1910s, though, there was little choice for self-expression in a conservative society such as the Prussian bourgeoisie. Young Max Delbrück appointed Johannes Kepler as his model figure and astronomy as his way out. Born near Stuttgart from a quarrelsome father and a mother accused of witchcraft, Kepler had studied at the University of Tübingen and, about three centuries earlier, improved the heliocentric model of Copernicus by applying ellipses to describe the planetary orbits.

Max Delbrück was born the last of seven children on the 4th of September 1906. The Delbrücks were a significant German family, akin to the Rothschild or Kennedy dynasties, albeit less wealthy. The Delbrücks had roots dating back to the 18th century, and among its members was Rudolf von Delbrück, a Prussian bureaucrat involved in the formation of the German Confederation. Max's uncle was a prominent chemical industrialist, another was the president of the Supreme Court of Justice, and another was a well-known professor of linguistics studies. Then there was Max's father, Hans, a history professor at the University of Berlin, editor of the prestigious Prussian Yearbook, tutor to Crown Prince Frederick (later Emperor), and member of the Prussian Parliament. Lina, Max's mother, also had a compelling ancestry because her grandfather was Justus von Liebig, one of the founders of modern chemistry and synthetic fertilizers, who Karl Marx even quoted in his influential book *The Capital*.

Even Delbrück's neighborhood was imposing. Grunewald was and still is a prosperous residential area of Berlin, home to the German upper class. The young Delbrück was surrounded by ambassadors, politicians, industrialists, lawyers, and doctors. In particular, the Delbrücks were befriended by the Bonhöfflers and the von Harnacks. When the families gathered for Sunday dinners, 22 children played noisily while the elders debated politics. Karl Bonhöffler was a professor of psychology at the University of Berlin. Of his sons, Klaus, married Max's daughter, Karl Jr. became Max's mentor and close friend, and Detrich grew into a renowned theologian and

 DOI: 10.1201/9781003657712-3

fervent anti-Nazi. Dietrich blamed the Church for supporting the Nazi Party, a protest that resulted in his imprisonment in a concentration camp, where he died in 1945. Adolf von Harnack was instead a professor of theology at the University of Berlin and Lina Delbrück's brother-in-law. Emperor William II appointed him as a member of the Prussian Academy of Science, and he was also director of the Prussian State Library and founder of the "Kaiser Wilhelm Gesellschaft" ("Emperor William Society")—which then evolved into the "Max Planck Society," a prestigious research German institute active to this day. Little Max Delbrück knew that Max Planck would profoundly influence his mature years. Back in the day, he was content to steal cherries from a tree that grew in Planck's garden, just down the road from Delbrück's mansion.

Max Delbrück grew up in a nationalist society. The Holy Roman Empire was the forerunner of Germany in the Middle Ages, a transnational kingdom stretching from Cambrai (modern-day France) to Olomouc (Czech Republic) and from Hedeby (Denmark) to Terracina (Italy). Despite its might, the Holy Roman Empire was essentially made up of hundreds of mini-states in a feud with each other. The most powerful of these states were Brandenburg-Prussia, ruled by the Hohenzollern princes, and Austria, headed by the Habsburg princes. The Emperor Francis II of Habsburg, overthrown by Napoleon Bonaparte, dissolved the Holy Roman Empire in 1806. After the defeat of Napoleon at Waterloo, 34 princedoms formed the German Confederation ("Deutscher Bund") in 1815 to replace the Holy Empire. Prussia was the most industrialized and advanced of these states but was also an aggressive and autocratic country with a large standing army and an expansionist policy. By the end of the 19th century, Prussia seemed unstoppable. Under the leadership of Chancellor Otto von Bismarck and King William I, the Prussian army defeated Denmark in 1864, annexing the territory of Holstein—to the north of Lübeck—as well as a coalition of German states led by Austria in 1866, incorporating the region of Schleswig—further north of Holstein.

With the following Peace of Prague, the German Confederation was dissolved, Prussia annexed many of the former princedoms, and Austria—nominally, the Austro-Hungarian Empire following the Hungarian upheaval of 1867—was no longer considered part of the German nation. In September 1870, Prussia avenged Francis II by defeating the French imperial army and capturing Napoleon's nephew, Emperor Napoleon III, at Sedan. In November, the German Empire was established and headed by Prussia. King William I of the House of Hohenzollern was crowned Emperor William I ("Kaiser Wilhelm I," also known as "William the Great") on the 18th of January 1871 in the French imperial residence at Versailles. In 1879, the Prussian Empire further strengthened its power by signing a military covenant with the Austrian Empire ("Dual Alliance"). Like the other European big shots (the United Kingdom, France, Spain, the Netherlands, and Portugal), Germany created overseas colonies in Africa, New Guinea, and the Pacific Ocean as late as 1885. Such a colonial empire, though, was not profitable.

Emperor William I died in 1888, succeeded by Frederick III, a liberal and admirer of the British constitutional monarchy. Alas, he died of cancer 90 days after his coronation. The crown went to his son, William II, a depraved and authoritarian monarch with a crippled arm who compensated for this deformity with an aggressive personality and who deeply despised Britain. The first thing the Kaiser did was dismiss Bismarck, the champion of German unification but also a cumbersome figure, the opposite of a yes-man. The second thing William II did was to end Prussia and Russia's long-standing friendship, hoping to steal territories from the former allies. Russia responded by aligning itself with France, signing a mutual military assistance treaty in 1912.

Count Helmuth Johannes Ludwig von Moltke, General Staff—commander in chief of the Prussian army—backed Emperor Frederick's visions of grandeur by promoting a swift and preventive war against France followed by an attack on Russia (the "Schlieffen plan"). In this way, Prussia would establish itself as the only European superpower. The Kaiser declined the proposal only because Britain clearly stated that it would not have permitted the defeat of the French Empire and the creation of a continental Prussian empire. The only force staying in the way of Prussia was, therefore, the United Kingdom. Another figure that poured gasoline on the fire of war was the commander in chief of the Navy, Admiral Alfred von Tirpitz, who wished to defeat the British Navy—the largest in the world at the time—and make Prussia the leading force in the world. German military expenditures—already massive—skyrocketed to build a gigantic anti-British fleet, an endeavor that rightly alarmed the United Kingdom.

A direct conflict between Prussia and Britain was only avoided because another war broke out first. On the 28th of June 1914, the Austrian archduke Franz Ferdinand was assassinated by a Bosnian extremist, 19-year-old Gavrilo Princip, to avenge the Austrian annexation of Bosnia-Erzegovina in 1908. Princip was posthumously linked to a separatist Serbian terrorist group, the "Black Hand"; thus, the Austrian Empire had the excuse to invade Serbia—a country that was becoming far too strong for the Austrian taste—and gain sea access to increase international trade. Prussia promptly honored the Dual Alliance by joining Austria in this war of aggression, gaining the long-awaited *casus belli* to invade France and fight Britain. The First World War had begun. The Ottoman (or Turkish) Empire—evolved from the leftover of the ancient Roman Empire—and Bulgaria sided with them the following year, forming what became known as the "Quadruple Alliance" (or "Central Powers"). However, these empires were soon surrounded by a cabal of Britain, France, and Russia (the "Triple Entente"). Within a year, Italy joined the allies of the Entente, as later did the US and Japan. Eventually, 30 nations around the globe became involved in the conflict.

Young Delbrück, who was eight years old when the belligerence began, joined the war effort by attending compulsory school meetings of the "Pan-German youth movement" meant to raise a nationalistic spirit in the pupils—a forerunner of the "Hitler Youth" that sprung a couple of decades later. As a shy boy, Max Delbrück did not enjoy these goliardic meetings, and his mother was able to obtain a dispensation for him. Max's father, being a liberal, also did not enjoy the nationalistic bursts of Wilhelmine society. He wrote a petition to the Kaiser, signed among others, by Max Planck and Albert Einstein, in which he pleaded with William II to make the German nation great through intellectual and technological features rather than via military means. Needless to say, the petition remained a dead letter.

The Prussian problem was that the German army was not suited for a long war of attrition, precisely what the conflict had become within a few months. The economic strain on the Empire was striking. The Central Powers had spent nearly 1.8 trillion current euros for the war (about three-quarters given by Prussia), but this sum was dwarfed by the 4.1 trillion euros invested by the Allies. Unlike the Allies, Prussia could not count on subsidies from overseas colonies or bank credit; thus, Prussia and all the German nations bled to death. Moreover, besides the Prussian troops, the Central Powers' armies were obsolete and could not compete with the Allies. These disparities explain why the mighty Prussian Empire eventually lost the war.

On the 10th of November 1918, Emperor William II flew abroad, leaving Prussia without an heir. The day after, the armistice with France was signed in a train carriage near Compiègne. Thanks to the Prussian propaganda machine, the German people believed to be winning the war until the day they capitulated. Such a delusion gave birth to the myth of the "stab-in-the-back" ("Dolchstoßlegende"), for which the German nation had lost the war due to the betrayal of socialists, communists, Jews, and parliamentarian democracy in general. Nonetheless, Kaiser William failed to establish Germany as the world's leading country but managed to liquefy the Prussian Empire, depriving the Hohenzollerns of the crown held since the Middle Ages. The Habsburgs in Austria and Sultan Mehmed VI in Turkey suffered the same fate. Instead, Britain consolidated its imperial power: its economy expanded by 7% during the war years compared with the shrinkage of 27% experienced by Prussia and 24% by France.

After a failed coup led by the German Communist Party of Rosa Luxemburg and Karl Liebknecht and the fall of a Soviet Republic in Bavaria—crushed by the "Freikorps" mercenaries guided by General Franz Ritter von Epp—free elections were held on the 19th of January 1919, resulting in a center-left government guided by Friedrich Ebert. Ebert formed the government not in unsafe Berlin, torn apart by riots, but in the quiet village of Weimar, Thuringia, where the poets Goethe and Schiller lived. A town theater became the headquarters of the constitutional assembly, and the new State became known as the "Weimar Republic." The aristocrat war hero Paul von Hindenburg, also a proponent of the Dolchstoßlegende, was appointed president.

The peace treaty signed in Versailles in May certified the cession of substantial territories in the West to France and Belgium and to the East to Poland. The German nation also had to pay a hefty amount in reparations, but despite these setbacks and the bitterness of the peace treaty—deeply felt by Hans Delbrück—that played a pivotal role in the years to come, life could start again. It won't be an easy life, though. The Kaiser did not just spend every coin in his coffers but also borrowed heavily from the Germans without any capability of paying back the debt. As a result, German families were bankrupt at the end of the conflict, and the young Republic suffered from hyperinflation and widespread famine.

And so, 12-year-old Delbrück was standing alone in the chill of the night, looking into his eyepiece, lucky to have survived the most devastating war humanity had hitherto experienced. His older brother Waldemar had not been so fortunate. Young Delbrück jumped into the field of astronomy with his usual enthusiasm. Astronomy was its only way to remove his father's considerable shadow, and he believed that if one had to do a job, one had to do it properly or not at all.

When the Delbrücks, Bonhöffler, and von Harnacks met weekly, there were lively discussions about many things, mainly politics but not science. As Lord Snow famously stated in his Rede Lecture at the end of the 1950s, culture is divided into humanities and science. Hans Delbrück excelled in the former; Max could make his name only by excelling in the latter. He was influenced in his decision by Karl Bonhöffler Jr., who later became director of the Max Planck Institute for Physical Chemistry in Tübingen. Karl Jr. instilled in young Delbrück a love of pure knowledge associated with science, a value that would guide Max's path in the future.

When he finished the Gymnasium in 1924, Max Delbrück was chosen to give the graduating students a speech ("valedictorian"), and he decided to talk about his hero, Kepler. He was allowed to consult the original manuscripts from the 1600s. This benefit thrilled him and taught him to study only the primary data sources. Alas, he also used large portions of one of the former directors of the Berlin Astronomical Observatory without proper attribution, leading to accusations of plagiarism later on.

The late 1920s were years of normalization, even if not of recovery. In March 1920, there was another attempted right-wing coup led by Wolfgang Kapp and supported by von Epp. The "putsch" was followed by a wave of political assassinations and mob violence under the moniker of "white terror," culminating in the assassination of the Jewish industrialist Walter Rathenau. Another failed coup occurred in 1923 in Munich—the "bier hall putsch." In January 1923, the French army occupied the wealthy region of the Ruhr because Germany refused to pay the reparations of war. Hyperinflation had reached levels where a loaf of bread passed from 63 cents in 1914 to 201 billion in 1923.

Nonetheless, the Weimar Republic managed to survive. Ebert appointed Gustav Stresemann as chancellor, who started paying the Allies and managed to free the Ruhr from the French occupation. Stresemann introduced the "Rentenmark" to stabilize the economy and fight street violence. Stresemann got a hefty loan from the US (the "Dawes plan") that helped reconstruct the German State. Industrial output had doubled since the end of the war. Unemployment was reduced, and strikes also declined. There was also a cultural renaissance—the "golden years" or "Weimar culture"—that could be encountered in the literature of Thomas Mann, the theater of Bertolt Brecht, the paintings of Otto Dix, the films of Friedrich Murnau, and the architecture of the Bauhaus movement.

At the end of the summer of 1924, Delbrück became a student of Hans Rosenberg at the University of Tübingen. Rosenberg had introduced the field of astrophysics in Germany, but despite this modern refinement, astronomy was no longer a hot topic in continental Europe. The state-of-the-art research was done in the US—for instance, by Edwin Hubble at Mount Wilson, California. Max Delbrück further expanded his practical knowledge of astronomy by volunteering as a research assistant at the Postdam Observatory in the winter of 1925. The observatory was built by Erwin Freundlich, who wanted to demonstrate the "red-shift" prediction made by Albert Einstein. Freundlich had already attempted to establish another of Einstein's inferences—that the Sun's gravity would bend starlight—by traveling to Russia in August 1914 to witness a solar eclipse. However, Freundlich underestimated the seriousness of war: being a German behind enemy lines, he was promptly arrested by the czarist troops. Fortunately, he was released a few days later following a prisoner exchange. After the end of the training session, Delbrück moved to the University of Göttingen in 1926 to write his doctoral thesis, which aimed to physically explain supernovae in terms of temperature, pressure, and ionization.

Göttingen was a better place for astronomy than Tübingen since it hosted the nationally renowned astronomer Hans Kienle and the influential mathematician David Hilbert. However, for his thesis, Delbrück needed to translate the work of the British astronomers Arthur Eddington—who investigated the interior of stars, developing a model to explain why the Cepheid stars varied their luminosity—and Edward Milne—who described the stellar atmospheres. Neither Delbrück nor his mentors knew English, so the thesis entered an idle phase. Meanwhile, Delbrück was hooked by the true German scientific novelty of the time: quantum physics.

3.2 QUANTUM LEAP

The publication of Mathematical Principles of Natural Philosophy in 1687 by Isaac Newton represented a significant step change over Aristotelian physics. In the empiricist world of antiquity, where things were done hands-on, the physics of Aristotle was more than sufficient to swing a sword, sail a ship, or erect a cathedral. However, in the more dynamic world of the late 17th century, there was a need for more precise equations capable of delivering cannon balls to their targets or improving transoceanic navigation. With its steam-producing furnaces pushing trains and vessels across the globe, the Industrial Revolution required a field of physics that Newton did not even conceive: thermodynamics. This branch of physics deals with heat transfer and can be traced back to Otto von Güricke, who invented the vacuum pump in 1650. However, the first textbook on the subject was published only in 1859 by William Rankine. By the beginning of the 20th century, it became understood that there were some applications where thermodynamic laws broke down. In particular, thermodynamics failed to explain how radiation was emitted from a black body.

Scientists could explain the common properties of all gases, such as how vapor would behave inside a steam engine, by developing models—based on so-called "ideal gases"—independent of the molecules that comprised the gases used in the experiments. Similarly, there was a need to understand how heat behaved in an object regardless of the material used to build the thing. This ideal object was called the "black body" and was essentially a cavity inside a block of material (iron, porcelain, and so forth). A small hole was drilled in the block to connect the cavity with the exterior so scientists could assess the thermal properties of the black body, chiefly its radiation emission.

When an object is heated, it emits radiation, that is, electromagnetic waves. Some of the radiation is visible: for instance, a poker left in the ember will become first red, then yellow, and finally white. Some of the radiation is not visible: for example, before becoming red, the poker will emit infrared radiation, which we feel as warmth. The thermodynamic model for the emission of a black body is based on the function $S(\lambda|T) = (2\pi ckT)/\lambda^4$; that is, the intensity of the emission of a blackbody emitted at a temperature T is proportional to twice π (Greek pi, about 3.14), c (speed of light, about 300,000 kilometers per second, km/s), k (the so-called Boltzmann constant, about 1.38×10^{-23} joules per degree of temperature, J/°C), and inversely proportional to the fourth power of the wavelength λ. Since there cannot be negative wavelengths, $S(\lambda|T)$ is valid only for positive λ values. At a chosen temperature, $2\pi ckT$ is a term made all by constants; thus, $S(\lambda|T)$ can be simplified to the function $y(x) = 1/x^4 \equiv x^{-4}$.

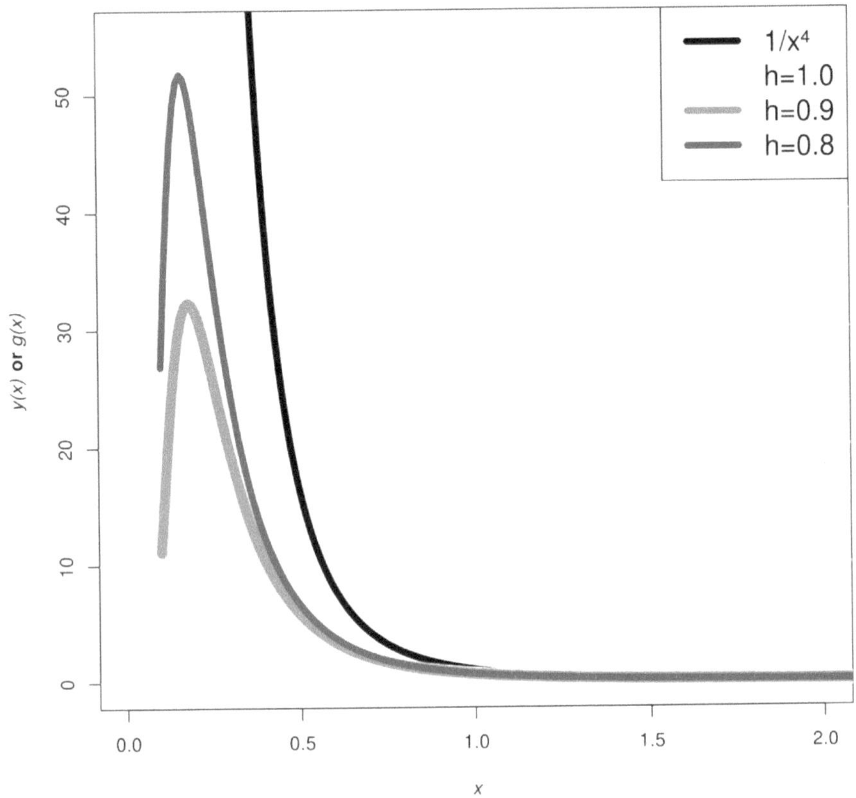

FIGURE 3.1 Model for the black-body emission. Plot of the functions $f(x) = 1/x^4 \equiv x^{-4}$ (in black) and $g(x) = hx^{-5}(e^{h/x} - 1)^{-1}$ for different values of h (in shades of gray). The function $f(x)$ is a hyperbole that reaches zero for increasing values of x, and asymptotically gets to infinity for x values closer and closer to zero. Conversely, $g(x)$ displays a peak whose shape depends on h.

The analysis of the function $y(x)$ shows that the resulting curve of the y values is a hyperbola and that $y(x)$ is defined for every value of x greater than zero (Figure 3.1). In addition, $y(x)$ does not have a real maximum: y goes toward infinity for x approaching zero; conversely, y decreases toward zero for increasingly larger values of x. Using the black body model, scientists predicted that an object heated at about 2400°C, such as a tungsten filament in a light bulb, would remain invisible because the emitted electromagnetic waves would have a wavelength beyond the visible range (380–750 nm). The prediction was clearly wrong and undermined the whole thermodynamics field.

Max Karl Ernst Ludwig von Planck, the guy who lived a few blocks from Delbrück's house, managed to save the day. Plank was born in 1858 in Kiel—in the region of Schleswig-Holsteins, but at the time, it was still part of Denmark. Max Delbrück and Max Planck had many features in common. Both came from families of the upper bourgeoisie, for Planck's father—Johann Julius Wilhelm—was a distinguished professor of law at the University of Kiel who obtained the right to use the aristocratic "von" in the family name—his grandfather and great grandfather were renowned theologians, and his uncle was one of the authors of the Prussian Code of Law. The Plancks moved to Munich when Johann Julius obtained a professorship in law at the Ludwig Maximilian University. Max Delbrück and Max Planck liked hiking and were at ease in all subjects of human knowledge, from philosophy to science. Planck was an excellent pianist, organist, and violinist, sang as a church chorister, and wrote an operetta performed at the Academy Choral Society. Planck did not distinguish himself at the Gymnasium—although he got decent marks—and needed to figure out what to do after high school. He asked the council of a music professor but was dissuaded from entering the conservatory. He asked a professor of physics, Philip von Jolly, who told him that science was dead: there was nothing left to discover; physics was all about making more and more precise measurements. Essentially, von Jolly echoed a speech given by Lord Kelvin at a Royal Society meeting in 1900: "There is nothing new to be discovered in physics now. All that remains is more and more precise measurement." Unconvinced by such a death sentence, the young Planck enrolled in physics at the University of Munich under the supervision of von Jolly. Planck's thesis would be to precisely measure the weight of ammonia, a task just as tedious as von Jolly had predicted.

After three years of boredom, Planck decided to switch to the University of Berlin, which was expanding under the patronage of the Kaiser to rival the French and British universities. In particular, Hermann von Helmholtz was appointed director of the Department of Physics, which boosted the University's research activities. Helmholtz was another giant of science who had made fundamental discoveries in mechanics, physiology, acoustics, and electromagnetism. Planck started to study thermodynamics, having von Helmholtz as a mentor and received his Ph.D. and habilitation. He became a professor first at the University of Munich and then at the University of Kiel. In 1889, he was appointed professor at the University of Berlin after Ludwig Boltzmann and Heinrich Hertz had declined the position. Planck needed to prove himself worthy of the role and decided to crack the problem that his predecessor, Gustav Kirchhoff, had left unsolved: the black body radiation.

In 1900, Planck provided an alternative to the classical model for black body radiation. His definition for the intensity of the emission was $S(\lambda|T) = \lambda^{-5}(2\pi c^2 h)(e^{hc/\lambda kT} - 1)^{-1}$. This formulation has an additional parameter, h, that Planck empirically extrapolated to fit the data, obtaining a value of 6.6×10^{-34} joules per second (Js). Even in this case, it is possible to simplify the equation for a given temperature to $g(x) = hx^{-5}(e^{h/x} - 1)^{-1}$. Like $y(x)$, $g(x)$ is defined for x greater than zero; unlike $y(x)$, $g(x)$ produces a bell-shaped curve with a well-defined peak. Planck's equation not only fitted the data better than classical formulation, improving the thermodynamic theory, but changed physics forever by introducing the field of quantum theory.

Planck considered h as a simple mathematical trick to fit the equation with the data, an "act of desperation," that would be dispensed after some more proper revisions of the equation. Instead, with its unimaginably tiny value, h became essential in many fields of science, becoming a fundamental physical constant and granting Max Planck a Nobel Prize in 1918. In particular, h became crucial to understanding the properties of photons—the particles associated with electromagnetic radiation, including visible light, whose term was introduced by Gilbert Newton Lewis in 1926—and the description of the structure of the atom. In 1917, Albert Einstein reformulated Planck's model for the black body, assuming that energy is not a continuum—as stated by classical thermodynamics—but formed by defined packages (or "quanta") of energy. A similarity might be this: 100 € are made up of 10,000 ¢, so the cent is the quantum of the euro and there is no coinage of, say, ¾ ¢. Such a quantization of physical energy came in handy when physics turned to understanding the intimate structure of the atoms.

In 1897, the English physicist Joseph John Thompson discovered the atom's first building block: the electron. Thompson was the director of the Cavendish Laboratory, founded by Maxwell. In 1904, he proposed an atomic model where negative charges are dispersed into a positively charged mash—the "plum pudding model." Current measurements assign the electron a charge of -1.6×10^{-19} coulomb (C), a mass of 9.1×10^{-28} g, and no real extension—that is, the electron can be considered a mathematical point. A couple of years later, the New Zealander Ernest Rutherford, son of a Scottish farmer emigrant, discovered that what was called α radiation was formed by helium nuclei. Rutherford, who had studied for Thompson, determined that these nuclei were made by two protons, obtaining a Nobel Prize in 1908 for this discovery. It is now ascertained that protons have a charge of 1.6×10^{-19} C, a mass of 1.7×10^{-30} g, and a radius of 8.0×10^{-16} m. The following year, Rutherford and Hans Geiger—the inventor of the famous radiation counter—bombarded a thin gold foil with α particles, and the results astounded both him and the rest of the world: atoms are essentially empty. The α particles passed right through the foil, being only sporadically reflected.

Rutherford devised a model for the hydrogen atom that superseded Thompson's. The model for the atom of hydrogen was similar to a miniature solar system: there would be a sun-nucleus at the center with a planet-electron orbiting around it. Hydrogen is the simplest of all elements and consists of a single proton and one electron, compared with the two protons and two electrons of helium. The Sun contains most of the mass in the solar system, as does the nucleus in the hydrogen atom. In the hydrogen atom, it would be possible to place 65,000 protons between the nucleus and the electron's orbit. To understand how this atomic solar system would look, consider it possible to fit 56,000 planets Uranus between the Sun and its orbit. However, Uranus is 40 million times less massive than the Sun, whereas the electron is only 1836 times less massive than the proton. One can visualize the hydrogen atom as the Sun orbited by a giant planet at a distance between Uranus and Neptune. Since Uranus is barely visible to the naked eye, it is understandable how an atom is essentially an empty object.

Despite such a momentous discovery—humans have tried to understand the structure of atoms since Thales—classical physics had serious problems explaining how an electron could orbit a proton. According to the planetary analogy, an electron would have needed a force to change its rectilinear motion into a curved one around the nucleus. The fundamental law devised by Newton stated that such force F would be equal to the mass m and the acceleration a (that is, the change in velocity over time) of the object (an electron or a planet): $F = ma$. Kinematics stated that for a circularly moving thing, the acceleration would be equal to the mass times the squared velocity over the radius of the orbit (mv^2r^{-1}); electromagnetism demonstrated that for charged particles, such a force would be equal to $e^2(4\pi er^2)^{-1}$, with e being the charge of the electron. Aside from the amount of mathematics, there are no conceptual issues so far.

The problem arises because electromagnetism also dictates that a moving charge emits radiation and, therefore, loses energy. Because the radius of the orbit is linked to the energy of the orbital object, loss of energy means orbital decay. The energy lost at each revolution is related to the radiated power P and the period of the revolution T: $\Delta E = -PT$. The fraction of energy lost over the total amount of energy E possessed by the electron is associated with the fraction of orbital decay: $\Delta r/r = -\Delta E/E$, where r is the radius of the orbit (the distance between the electron and the proton). As the electron radiates energy, it becomes more negative and closer to the proton, meaning that the electrostatic force acting upon the electron will become stronger and stronger, forcing the orbit to decay faster and faster. It has been estimated that, at the first orbit, the electron would decay by about three millionth of its initial level. That does not seem like a tremendous value, considering that even the International Space Station (ISS) is slowly falling toward Earth. However, the ISS has a complex apparatus of rockets that stabilize its orbit; the electron does not have such a luxury. It has been estimated that the electron would spiral down into the nucleus in about 1.6×10^{-13} s. Since electrons have been orbiting nuclei since the origin of the universe some 14 billion years ago, it is clear that classical physics has serious problems describing the structure of the atom.

Meet Niels Henrik David Bohr, savior of physics. In 1913, Bohr proposed a new atomic model in which electrons could orbit the nucleus without emitting energy only when the orbit radius r was set to specific values. Born in Copenhagen, Denmark, on the 7th of October 1885 from Christian Bohr, a professor of physiology at the University of Copenhagen and Nobel Prize candidate, and Ellen, the daughter of a wealthy banker, he obtained a gold medal from the Royal Danish Academy of Sciences for his description of the surface tension of liquids in 1905. Bohr received his Ph.D. in 1911, describing electrons as gases in a groundbreaking thesis that did not have much diffusion since it was written in Danish. Afterward, he visited J. J. Thomson at the Cavendish Laboratory, where he met Rutherford. Bohr impressed Thomson but incorrectly: at a dinner, Bohr innocently pointed out some mathematical mistakes in a manual of Thompson's. Needless to say, Bohr's career at the Cavendish Lab was over.

Still in Cambridge, Bohr met Rutherford during another dinner, and the two clicked immediately. Rutherford invited Bohr to join him at the University of Manchester; Bohr arrived in Manchester in March 1912 and worked on the experimental characterization of the structure of the atom. In 1913, building upon Rutherford's planetary model of the atom, Bohr presented a new atomic model, borrowing from the concept of quanta, where he assumed that the electrons could orbit the nucleus without radiating energy only when the orbit radius r took specific values. The radius of these "stationary orbits" were multiples of a fraction of Planck's constant h defined as $\hbar = h/2\pi$ ($\hbar$ is formally the "reduced Planck constant," but it is commonly spelled h-bar). There was no real scientific reason to assume the stationary orbits: just like Planck's h, these might have been a mere mathematical trick to solve the equations. Besides, even Kepler used ellipses to describe the planets' orbits only because they fit the experimental data better than circles. Arnold Johannes Wilhelm Sommerfeld, a renowned professor of mathematics, then generalized Bohr's model. Bohr obtained a Nobel Prize in physics in 1922. Later on, he also conceived the concept of the "complementarity principle." Formally introduced in 1927 at the International Physics Congress, it is the idea that certain natural phenomena could be explained in seemingly contradictory ways—such as particles or waves—and remain coherent. Bohr then tried to apply such a concept to other fields of human knowledge, ranging from physiology to economics.

Another step forward in the description of the atom was made by Werner Karl Heisenberg. Born in 1901 in Würzburg, Germany, Heisenberg moved to Munich in 1010 after his father, August, obtained a professorship in literature. Werner Heisenberg studied in the prestigious Maximiliangymnasium, which his grandfather had directed for several years. Werner's brother Erwin outperforms him in all subjects except mathematics, and his father secured a physics scholarship for Werner in Sommerfeld's group, obtaining his doctorate in 1923. Heisenberg completed his "habilitation" thesis in 1924 at Göttingen under the guidance of Max Born and formally remained in Born's group until 1927, albeit obtained a fellowship to visit the laboratory of Niels Bohr at the University of Copenhagen, Denmark, between 1924 and 1925. Afterward, he became a professor of physics at the University of Leipzig. The story goes that, in 1925, Heisenberg spent two weeks on the tree-less island of Helgoland—"holy land" in the old German language—in the North Sea to recover from hay fever. During this short recuperation period, he wrote a seminal paper describing how electrons could jump from one stationary orbit to another only when adsorbing a specific number of quanta of energy, again related to $\hbar$. Later on, Max Born and his group further expanded on this work by formally embedding it into the branch of matrix algebra. In 1927, Heisenberg proposed another fundamental aspect of quantum theory: the "uncertainty principle." Given the position of a particle x and its momentum p, the product of their measurement uncertainty is proportional to $\hbar$: $\Delta x \Delta p \sim \hbar/2$. In other words, measurement uncertainties are constant; thus, if one wanted to know the exact position of an electron, one would have to reduce Δx to zero and increment Δp accordingly to obtain $\hbar/2$. Simply put, it is impossible to simultaneously determine an elementary particle's position and speed. Heisenberg received a Nobel Prize in 1932.

Max Born was born on the 11th of December 1882 in Wrocław—currently in Poland, but at the time part of Prussia—from Gustav Born, professor of embryology at the University of Wrocław, and Margarethe Kauffmann, daughter of a wealthy industrialist. He studied mathematics at the University of Heidelberg and at the Swiss Federal Institute of Technology (ETH) in Zurich, joining the University of Göttingen in 1904 for his doctoral thesis under the mentoring of David Hilbert, who developed several processes used in modern mathematics. After receiving his Ph.D. in mathematics in 1906, Born returned to Wrocław to perform his military service and later was invited by Hermann Minkowski—a family acquaintance—as his assistant in Göttingen. Like Hilbert, Minkowski had developed several currently used concepts but was more oriented in the geometrical field than in pure mathematics. Afterward, Born became an associate professor at the University in Berlin, where he met Max Planck and befriended Albert Einstein. In family meetings, Einstein played his violin, and Born joined in playing the piano. Born gained a full professorship at the University of Frankfurt in 1919, and when Sommerfeld joined the University of Munich, Born started his career in Göttingen in April 1921. Within one year, he founded the Institute for Theoretical Physics (ITP)—today converted to a block of private apartments—with Wolfgang Pauli and Friedrich Hund as the first research assistants.

Pauli was born in Vienna in 1900. He was the son of a chemist, the grandson of a physicist, and the great-grandson of a publisher. Soon after graduation, in 1918, Pauli wrote a paper on Einstein's relativity; later, he also prepared a synthesis of relativity for the Encyclopedia of Mathematical Sciences, which Einstein himself praised. He joined Sommerfeld's group for his doctoral thesis, gaining a Ph.D. in 1921. He became a professor at the University of Hamburg in 1923 and at the ETH in 1928. Heisenberg asked Pauli's advice when drafting the seminal 1925 paper. Pauli described the stationary orbits of various atoms, indicating that two electrons cannot have precisely the same properties; even if they have the same energy (because they are on the same orbit), they have other properties that differentiate them. He called these properties "quantum numbers." Pauli also predicted the existence of the neutrino, a particle with no electric charge but with only 1% of the proton's mass. The neutrino must not be confused with the neutron, another electrically neuter particle with a mass slightly higher than the proton. Discovered in 1932 by James Chadwick, neutrons are located in the nuclei of atoms with protons; neutrinos are emitted as radiation during specific nuclear reactions, akin to the α particles. Friedrich Hermann Hund was born in 1896 in Karlsruhe, Germany, and died 101 years later. After his studies in Born's group, he became a professor at the universities of Rostock, Leipzig, Jena, and Frankfurt. He was a lecturer at Harvard University and visited Bohr's group in Copenhagen. Hund then became a full professor in Göttingen in 1957. He specialized in describing groups of atoms, that is, in the quantum description of molecules.

Born's ITP was, therefore, the spearhead of quantum theory that brought together several "Bonzen" ('bigwigs') in mathematics and physics. Apart from Heisenberg, Pauli, and Hund, other distinguished academicians joining Born's group at some time or another included Pascual Jordan, Paul Dirac, James Franck, Walter Heitler, Enrico Fermi, Lise Meitner, and Robert Oppenheimer—who later became the leader of the "Manhattan project" that developed the first atomic bomb. Born introduced the term "quantum mechanics" and, together with Niels Bohr, founded the so-called "Copenhagen Group," which described the atom in terms of energetic levels, essentially considering electrons, protons, and so forth as particles. The Copenhagen Group distinguished itself from Erwin Schrödinger's school of thought, which regarded elementary particles as waves and described electron movement around nuclei in terms of probability. Schrödinger and Dirac shared the Nobel Prize in 1933. Directing the ITP and directing so many "Bonzen" was not easy, and Born suffered from a nervous breakdown in 1928, receiving a Nobel Prize in 1954.

Delbrück became first acquainted with quantum physics at a seminar at the University of Berlin in the spring of 1926. Heisenberg was asked to give a talk at the "Physics Colloquium," a symposium inaugurated by Helmholtz. At the time, the University of Berlin was the world center for quantum physics and relativity, hosting Max Planck and Albert Einstein. Delbrück did not quite understand the mathematics behind Heidelberg's talk, but he had the opportunity to meet all of these scientifically famous people and became fascinated by the subject. Therefore, when Max Delbrück arrived at the University of Göttingen, he encountered a hive of intellectual activity. It would have been like attending Cambridge University in the second half of the 17th century when Newton made revolutionary discoveries while teaching mathematics. Delbrück didn't have to think long when choosing a moribund thesis in the dying field of astronomy or participation in the day's hot topic. He dumped his thesis and leaped toward the new horizon.

One day, at the University's swimming pool, Delbrück met Eugene Wigner. Wigner was born in Budapest in 1902 and was collaborating with David Hilbert. Wigner introduced group theory in quantum physics, describing molecules in quantum mechanics. In 1930, Wigner joined Princeton University with von Neumann, one of the fathers of computer science. During the Second World War, Wigner would later be one of the subscribers of the so-called "Einstein-Szilard letter" that persuaded President Franklin D. Roosevelt to inaugurate the Manhattan Project. Back in that pool, though, Delbrück pointed out that one of Wigner's papers missed some mathematical concepts. Far from being outraged, Wigner proposed that Delbrück collaborate with him on a supplement paper on the subject. Consequently, Delbrück was officially introduced to quantum physics, and that article gained him admission into Born's group.

The subject of his new doctoral thesis was given to him by another of the "Bonzen" in Born's group, Walter Heitler. Just a couple of years older than Delbrück, Heitler had obtained his Ph.D. at the University of Munich under the supervision of Sommerfeld. He had studied with Niels Bohr at the University of Copenhagen and Erwin Schrödinger at the University of Zurich. In 1927, together with Fritz London, Heitler described in quantum mechanic terms the molecule of hydrogen (H_2), discovering the so-called "valence bond." When different atoms combine to form a molecule, some of the electrons in the higher energy stationary orbits—those that are more external—mesh into another type of stationary orbit, the bond orbit. When different atoms combine to form a molecule, some of the electrons in the higher energy stationary orbits—those that are more external—mesh into another type of stationary orbit, the bond orbit. The electrons no longer belong to either of the atoms but are shared with a loss of energy that stabilizes the molecule, explaining why the elements cluster together in molecules rather than remain alone as atoms. Quantum mechanics had entered the realm of chemistry.

Delbrück's task was to describe the bond between two lithium atoms to form molecular lithium (Li_2). Born and Heitler considered this thesis a piece of cake, but although sounding interesting, Delbrück disliked the project because it involved no novelty. He went along anyway, but soon, he realized that it was not even a "quick job" because it involved solving "six-dimensional integrals in polar coordinates" without the help of a computer—which did not exist then.

Despite the project's dullness, Delbrück submitted his thesis in 1929. He still had to pass the oral examination, considered a pure formality. Born managed to obtain a postdoctoral position for Delbrück in the research group of Professor John Edward Lennard-Jones at the University of Bristol. At the time, Bristol was competing with Cambridge as the British hot spot for scientific research, and Lennard-Jones was involved in the quantum mechanics description of molecules.

Delbrück went to Bristol in the fall of that year. As if he had not encountered enough Nobel laureates in Germany, his roommate was Cecil Frank Powell, who discovered another fundamental particle (the "pion"), winning the Nobel Prize in 1950.

Delbrück did not know any English but managed to speak it without formally learning the grammar. Initially, he disliked England but became a devoted anglophile within a few weeks, although his relationship with Lennard-Jones remained cold. Delbrück returned to Göttingen in December for his oral examination. Ill-advised by some fellow students, he did not study a thing for it, and, to everybody's astonishment, he failed the exam.

Nonetheless, he managed to pass the oral examination in 1930. Afterward, with the support of Born and Karl Bonhöffer, Delbrück obtained a fellowship in 1931 to visit Niels Bohr in Copenhagen and Wolfgang Pauli in Zurich. Niels Bohr founded the ITP in 1921 and lived with his family on the top floors of the institute's building. The institute became the Mecca for quantum mechanics research. Delbrück arrived in Copenhagen in February 1931 and immediately befriended George Gamow, an acquaintance from Göttingen.

Gamow was born Georgiy Antonovich Gamov in 1904 in Odessa, Ukraine, to a wealthy family. His grandfather from his father's side was a colonel of the Tsarist empire; the grandfather from the mother's side was a high priest of the Orthodox Church. Unlike his three other brothers, George's father, Anton, did not enlist in the military as had been the family custom; instead, he had become a professor of Russian literature at the University of Odessa, which at the time was a cosmopolitan metropolis of the likes of St Petersburg, Berlin, or Paris. Gamow showed an early interest in observations and science. In 1910, he could observe the Halley comet's passage thanks to a telescope given to him by his father. Gamow also investigated whether the holy wafer transformed into flesh after transubstantiation. Thus, he used a microscope to compare a piece of holy bread taken during a mass and a bit of bread from the kitchen, concluding that transubstantiation was only a symbolic rite.

In 1922, Gamow joined the University of Odessa to study physics, but the lectures were canceled due to the ongoing civil war. Thus, his father enrolled him at the University of St Petersburg. To sustain himself, Gamow worked as a meteorologist at the Forestry Institute first and, later, as a physics teacher at the Artillery School of the Red Army. He started a Ph.D. project under Alexander Alexandrovich Friedman, who introduced the concept of an expanding universe, which was later experimentally verified by Hubble. Unfortunately, Friedman died in 1925, so Gamow had to change his subject of study, choosing the nascent field of quantum physics. However, nobody in the University knew about it, so he organized a self-teaching group with Lev Dadivovich Landau and Dmitri Dmitriyevich Ivanenko, who became outstanding nuclear physicists. Landau won the Nobel Prize in 1962. These "Three Musketeers," as they were known, developed an approach to quantum physics that is still used today. Delbrück had done something similar when he arrived in Göttingen: he teamed up with Victor Weisskopf, Maria Mayer, and the Hungarian émigré Edward Teller to learn quantum physics with the formal guidance of a physics professor. The group organized meetings known as "a little night physics." All these self-taught students became world-renowned physicists. Teller, in particular, will be instrumental in helping Leo Szilard convince Einstein to kick-start the Manhattan Project. Einstein and Szilard had already collaborated in Europe, even patenting together a prototype of a leakage-proof refrigerator in 1930. Later, Szilard became known as the "father of the atom bomb" when he developed the implosion-type of atomic devices.

In 1928, Gamow obtained a scholarship to go to Göttingen and deepen his knowledge in physics. During this period in Germany, Gamow developed a model that explained the production of α particles, first described by Rutherford in quantum mechanics terms. Specifically, he introduced the concept of "quantum tunneling," the property of molecules to pass through seemingly insurmountable obstacles, such as electrons slipping outside atoms. Nowadays, quantum tunneling is employed, for instance, in ultra-high resolution microscopy and quantum computing.

At the end of the scholarship, Gamow changed his ticket back to St Petersburg for one to Copenhagen. By pleading with Bohr's secretary, he obtained a face-to-face snap meeting with the "Bonze" of quantum physics. Bohr, impressed by this flamboyant Russian student, awarded him a scholarship from the Royal Danish Academy of Science. Afterward, Gamow secured another fellowship from the Rockefeller Foundation (RF) to visit the Cavendish Lab and work with Rutherford. There, in collaboration with Fritz Houtermans, Gamow applied quantum tunneling to explain the bombardment of nuclei, the process Rutherford was mastering. John Cockcroft and Ernst Walton translated the theory into practice, winning the Nobel Prize in 1951 for such an exploit. Gamow and Houtermans then applied quantum tunneling to describe the phenomenon of nuclear fusion of hydrogen into helium, extending quantum mechanics into cosmology. Gamow obtained another fellowship from Bohr for the academic year 1930–1931; thus, when Delbrück arrived in Copenhagen in February 1931, Gamow was part of Born's ITP. The two young physicists became instant friends. Not only did they spend time together, but they also started to write joint articles on nuclear physics. Nonetheless, Gamow had to return to Leningrad—as St Petersburg had been renamed in the meantime—near the end of 1931 to extend his visa. What should have been only a short trip to complete a formality became a journey of no return.

The problem was that in the new Stalinist society, the USSR had turned into during Gamow's journey abroad, ordinary people were not allowed to leave the country for fear they would not return. People of value were kept even closer, constantly watched by the NKVD. Gamow obtained several professorships in physics at several Soviet universities during 1932; thus, the USSR wanted to retain a valuable piece of its national science apparatus. Like many other millions of Soviet citizens, Gamow was a prisoner in his land and started to be afraid for his own life and with reason. Despite being a devout communist, his friend Lev Landau had been imprisoned in a Gulag, and Gamow knew it was just a matter of time before suffering the same fate. The NKVD had already spat its sentence, writing in a report that "in his behavior, he is relatively undisciplined and is a typical representative of the literary-artistic Bohemia." For Gamow, the clock was ticking.

At the end of 1931, Gamow married Lyubov Vokhminzeva, a former physics student and dancer. Together, they tried to escape the USSR by paddling with a kayak from Odessa to Turkey (some 540 km) through the Black Sea. The attempt nearly killed them, and when they were rescued, they were not charged with treason because even the NKVD could not conceive that the couple had gone through such a desperate plan. A year later, though, things went differently. Niels Bohr, with the complicity of Marie Skłodowska Curie—Nobel Prize winner due to her work in the characterization of radioactivity—managed to convince the organizer of the Solvay

Conference in Brussels, whose topic was nuclear physics, to invite Gamow as the keynote speaker for the USSR. Allured by the possibility of showcasing the superiority of "proletarian science" over the decadent "capitalistic science," the Soviet authorities promptly issued a visa for the Gamows. Once in Brussels, the couple was secretly dispatched to the US, where Gamow obtained a professorship at George Washington University in 1934.

Back in Copenhagen in 1931, Delbrück felt a "great vacuum" after the unexpected departure of Gamow, missing the "tremendous vital force" of his friend. Nonetheless, he also got along well with Bohr, considering him as somebody "one would wish to have as father"—albeit he was still addressing him as "dear Prof. Bohr"—a figure that profoundly influenced his approach toward science. Nevertheless, he had to move to Zurich in the autumn of 1931 to complete the second part of his fellowship with Wolfgang Pauli. Needless to say, Delbrück befriended Pauli, who was usually harsh with others but not with Delbrück. Delbrück, on the other hand, liked Pauli's directness. Nonetheless, Delbrück realized that, despite his brilliance, he was no match for the "Bonzen" of quantum mechanics. He decided to move into a field that involved less mathematics but more thinking. He turned down a fellowship from Pauli, and in the fall of 1932, he made another leap by joining Lise Meitner's group at the Kaiser Wilhelm Gesellschaft ("Emperor William Institute," KSI) in Berlin. Pauli later commented that this new appointment "would be good" for Delbrück because "he would be forced into greater independence in his thoughts."

3.3 BACK HOME

Lise Meitner was one of the pioneers in radiation biology. She was born in 1878 in Vienna, Austria, and gained her Ph.D. in 1906 with Ludwig Boltzmann as her mentor. However, these steps were not as straightforward as they may appear today. Women started to be accepted at the University of Vienna only in 1897; before that time, higher education was a male prerogative. Meitner became the second woman to obtain a doctorate in 1907 at the University of Vienna—the first was Gabriele Possanner von Ehrenthal. Driven by her thirst for knowledge, Meitner wanted to become a professional physicist; thus, she moved to Berlin and convinced Max Planck to accept her at the KWI as a research assistant, the first woman with such a grade in Prussia. At the Institute, Meitner teamed up with Otto Hahn (Nobel Prize winner in 1944), who obtained a degree in chemistry at the University of Marburg had received a Ph.D. from the University College in London, and worked with Rutherford at the McGill University in Canada. Later on, Fritz Straßmann joined the group. Straßmann graduated in chemistry in 1929 and remained with Hahn even after his stipend had expired in 1932, but Meitner convinced Hahn to open an additional position for the young chemist. Meitner and Hahn discovered the element protactinium (Pa) in 1918, characterized the β radiation, and generated barium from uranium through bombardment with neutrons—a feature their colleagues deemed impossible, akin to alchemic wishful thinking. Later on, she will be the first to realize how the process of nuclear fission—the split of an atom with a large number of protons and neutrons (collectively known as nucleons) such as uranium, into two smaller atoms (barium and krypton)—worked, imagining the separation of a large water drop into two smaller droplets. Nonetheless, despite her achievements, the position she held at the KWI was not paid, and she had lived just above the threshold of poverty.

When Delbrück joined the KWI, Meitner and her research assistant, Heinz Köster, studied the diffraction of high-energy photons from uranium nuclei. Delbrück applied Dirac's formulation of Schrödinger equations to describe the phenomenon—a process now known as "Delbrück scattering." Meitner and Delbrück wrote a nuclear physics textbook and a series of papers (together with Kurt Wohl) on the quantum physics of photosynthesis. Delbrück focused on defining thermodynamics using the concepts of quantum physics. In particular, he was interested in explaining why thermodynamics includes a temporal direction while kinematics—the study of the motion of bodies—does not. According to the classical Newtonian concept, a ball could move from *A* to *B* in a time proportional to its velocity. In formal terms, the speed v of the ball is given by the space x covered in the amount of time t: $v = \Delta x/\Delta t$, where Δ represents the difference between values, so that Δx is the length *A–B* and Δt is the time span of the journey. Nothing prevents the ball from moving from *B* to *A* in a negative time: a negative Δx would simply indicate that the ball is moving in the opposite direction as before; since the ball is coming back, the time Δt will also be negative. However, the speed would be the same as in the previous case: $v = -\Delta x/-\Delta t = \Delta x/\Delta t$.

Thermodynamics does not account for such temporal reversibility. A mug of hot coffee left on a table will cool to room temperature after a certain time. That a cup of hot coffee has more energy than a cold one is intuitive. What is less intuitive is that, in thermodynamics, one has to consider the mug, the table where the cup is placed, and the room that contains both. The set mug-table-room defines a "system." At the start (system *A*), the bulk of the energy in the room is concentrated in a single point (the mug), and the system is defined as highly ordered. At the end (system *B*), the energy is homogeneously distributed in the whole room, and the system is said to be disordered. Formally, system *A* has a lower entropy *S* than system *B*.

Entropy is a concept introduced by the French mathematician Lazare Carnot in 1803, and the second law of thermodynamics dictates that *S* can only increase. It is possible to visualize this process by imagining the molecules of water that make the hot coffee as small balls haphazardly moving at high speed (and, consequently, with high kinetic energy and, in reflex, high temperature) inside the mug. These tiny balls hit the mug's porcelain molecules, which then transfer the energy to the molecules of air, akin to a cue ball setting billiard balls in motion. The energized air molecules will then bounce around the room at a higher speed than before, increasing the room's temperature. It is implausible—a euphemism—that the air molecules will hit the mug simultaneously, transferring the energy back to the water molecules. Therefore, the second law of thermodynamics states that the initial entropy S_i must be lower than the entropy at the end of a process, S_e: $S_i < S_e$. S_e must come before S_i when the process is reversed, but the relation $S_e < S_i$ is impossible, and $S_e > S_i$ is the same process. Hence, thermodynamically, it is inconceivable to go back in time. Delbrück tried translating these thermodynamics concepts into quantum mechanics, but his research did not find anything interesting. Nowadays, though, a whole branch of quantum physics deals with such notions: "quantum thermodynamics."

Later on, Delbrück concentrated more and more on the biological effects of radiation. After the first description of radiation given by Henri Becquerel in 1896 and the stunning pictures provided by Wilhelm Conrad Röntgen of the human bones, including a world-famous one of his wife's hands, radioactivity became quite a hit. People found the fluorescence charming, given off by radioactive material, and some daredevils, such as the golf star Eben Byers, ingested radium for alleged invigorating effects. Today, radioactive elements are seen as exotic substances that should be avoided by all means. Possessing a small amount of uranium at home would almost certainly result in a small-hour raid by the police's special squad. Back in the day, there was no such taboo on radioactive substances. Becquerel originally discovered the properties of uranium after accidentally leaving a piece of this metal on a cabinet containing photographic plates, resulting in an unexpected exposure. Uranium was sold well into the 1950s to children in toy chemistry sets. Even Delbrück was given a sample of uranium by the Phillips Laboratory in the Netherlands, which he brought back to Otto Hahn, who was carrying it in his chest pocket.

Nonetheless, around the mid-1930s, the interest in radiation research shifted from understanding how radioactivity arose to what effect it had, namely on the human body. It soon became evident, in fact, that radioactive materials were harmful. The pioneers in radiation research, Marie and Pierre Curie, showed signs of radiation sickness; Byers died of esophagus cancer; many other people exposed to radiation developed nasty symptoms. For instance, in 1896, merely two weeks after the first Röntgen's picture had been published, Friedrich Otto Walkhoff took the first dental radiography with an exposition of about half an hour. He established, together with Fritz Giesel, the first dental practice with X-ray equipment. Walkhoff noticed that many of his patients started to lose their hair; Giesel died of metastatic carcinoma in 1927. Yet in 1896, a certain Dr. Haws in New York gave a public demonstration of radioactivity with a powerful X-ray emitter, developing blisters on the part of his body that were more exposed to the apparatus. Clarence Delly, who crafted X-ray emitters for the Edison Laboratory, died of metastatic carcinoma in 1904. A pattern started to emerge. It became even possible to measure the absorbed radiation dose based on the amount of skin irritation caused by radiation ("minimal erythema dose"). Today, instead, the radiation dose absorbed by a body is measured by dosimeters containing phosphor crystals that emit light proportional to the amount of radiation reaching them. The amount of radiation itself is detected by Geiger detectors based on the fact that high-energy particles can expel electrons from suitable gases, generating electricity.

Some scientists took advantage of radiation's harmful effects to better understand genetics. They discovered that radiation could be teratogenic, which means that high-energy waves such as X-rays could disrupt the delicate process of fetus formation. Most radiation-derived effects on the fetuses are abortion (because the embryos cannot complete their maturation) or the development of a variety of tumors with early onset. Nonetheless, radiation has been linked to the development of morphological abnormalities such as Marfan syndrome, a condition in which children are abnormally tall and thin, with very long limbs and appendages. Coming back to the example of peas given above, it is like having parental plants with the phenotypes dark (*D*) and light (*L*) and then getting a progeny with the new phenotype iridescent (*I*). The character *I* is encoded neither by the genotype G^D nor G^L, and geneticists can track the transmission of the novel genotype G^I with accuracy. Even more than that, geneticists can work on answering the question: what are genes made of? If radiation can alter a gene, what biological substance can be affected by radiation?

To proceed with this kind of investigation, geneticists have used a range of tricks to produce mutations on demand. It was well understood that mutations could develop naturally—after all, that is the mechanism underlying Darwinism—but only very rarely. High temperatures or chemicals could increase the rate of mutation. Pioneers in genetics, such as Thomas Morgan, mastered such techniques: he had a vast collection of mutants of his preferred testing animal, the fruit fly *Drosophila melanogaster*. This tiny fly typically has red eyes, but Morgan obtained a strain with white eyes and used it to demonstrate that some genes—at least the fruit fly's ones that encode the eyes' color—are associated with sex chromosomes. His collection contained many other types of mutants, including flies with orange eyes or without eyes altogether, curved wings, and black or yellow bodies instead of the typical brown. Morgan set the basis of experimental genetics through his "Fly Room" at Columbia University. However, he always considered the gene to be a theoretical construct. One of his students, Müller, went beyond his master by evaluating the gene as a physical entity.

Hermann Joseph Müller employed radiation to speed up the generation of Drosophila mutants. He was born in New York City in 1890—although his grandfather was an emigrant from Koblenz, Germany—and obtained a Ph.D. in 1915 with Thomas Morgan as his mentor; the two co-authored (together with Alfred Sturtevant and Calvin Bridges) an influential manual on genetics (*The Mechanics of Mendelian Heredity*). Müller was at odds with the academic establishment because he advocated for civil rights (such as wage equality for women and African Americans, tuition reduction, and unemployment insurance). At the time, such tendencies were thought to be dangerously communist. He also publicly condemned eugenics, which was extremely popular in the United States at the time.

As a result, he fled to Berlin, hoping that the apparent freedom of the Weimar Republic would suit him better. However, he soon clashed with the Nazi party; thus, he moved to Moscow in 1932 and became one of the leading Soviet scientists. However, the rise of Lysenko meant the end of his career in the USSR. In a public gathering, he called Lysenko "a fraud," making him not only the target of public uproar but also a walking target for the NKVD. Therefore, he fled the USSR in 1937, joined the Communist International Brigade during the Spanish Civil War, and eventually moved back to the US.

Müller rceived the Nobel Prize in 1946 for his contribution to radiation biology. In particular, he showed that the mutations generated by X-rays were qualitatively identical to those occurring naturally or induced by chemical or thermal means. Consequently, the driving force behind mutation had to be the same in all cases. He also demonstrated that the number of mutations was quantitatively related to the radiation dose used in the experiments and that egg and sperm cells were susceptible to mutagenesis. Furthermore, Müller's work demonstrated once and for all the danger associated with radiation exposure.

Müller conducted the abovementioned experiments regarding the correlation between radiation dose and the number of mutations in Berlin with Nikolaj Timofeev-Ressovskij, whom he met during a trip to Moscow in the 1920s. They discovered that when they used chemicals, mutations appeared after a certain threshold in drug concentration was reached; conversely, radiation dose was linearly related to the number of mutations. Since interpreting the experimental results required a good understanding of physics, Müller went to Copenhagen in 1933 to talk with Bohr, but the meeting was fruitless.

Nikolaj Vladimirovich Timofeev-Ressovskij (or Timofeev-Ressovsky), nicknamed the Wild Boar, was born in Moscow in 1900. He studied population genetics under Sergei Chetverikov in Moscow. During the Bolshevik revolution, he enrolled in the "Green Army," which followed the ideas of the Russian anarchist Peter Kropotkin and did not align with either the Tsarist White Army or the Soviet Red Army. In 1919, however, the Green Army joined the Red Army, and Timofeev-Ressovskij fought the Russian Civil War on the Polish front. He got involved in science teaching, although formally, he did not obtain a degree due to the onset of the war. In 1921, he joined the State Institute of Public Health, directed by Kol'tsov, and became his assistant in 1924. Timofeev-Ressovskij joined the KWI in Berlin in 1925, thanks to a science exchange between the Soviet Union and the Weimar Republic and with the mediation of Oskar Vogt. Timofeev-Ressovskij had collaborated with Vogt when the latter went to Moscow to dissect Lenin's brain, who died in 1924. Timofeev-Ressovskij and Vogt studied the genetics of the inheritance of mental diseases. Later on, in 1937, Timofeev-Ressovskij refused to go back to the USSR due to the widespread Lysenkoism and continued his research in genetics even during the Nazi regime. He was forced to come back home in 1947 when he was coerced to join the NKVD equivalent of the Manhattan Project, which was carried out in Sungul, a Gulag specifically designed for atomic research.

Delbrück and Timofeev-Ressovskij met in Berlin at one of the informal scientific gatherings the former organized in his family's mansion. One of the driving forces that led Delbrück to invite fellow physicists and other scientists to his place was a speech given by his mentor, Bohr. In August 1932, Bohr gave a seminar entitled Light and Life in Copenhagen, attended by Delbrück, where Bohr asserted that living beings were not merely the sum of their parts. Since physicists could only describe atoms, the synergy obtained by gathering atoms in a living body was beyond modern physics. "The existence of life must be considered an elementary fact that cannot be explained," he claimed. Bohr based his reasoning on what the Danish philosopher David Favrholdt defined in 1999 as the "thanatological principle:" In-depth research into the structure of living beings kills them, making it impossible to understand living entities.

Delbrück was galvanized by such statements and sought to determine whether it was a faithful proclamation. In particular, he was enthralled by the prospect of revolutionizing biology with concepts derived from quantum physics, just as quantum principles reinvented classical mechanics. Nonetheless, there was a fundamental difference between the master Bohr and the student Delbrück: the former focused his speculation on philosophical grounds, and the latter sought empirical evidence to "reach an understanding of life based on physical experience."

Furthermore, Delbrück had to act quickly because esteemed competitors followed the same path. One of his former colleagues in Born's department in Göttingen, Ernst Pascual Jordan, had also been inspired by Bohr's talk and had funded "quantum biology," spoiling the originality of Delbrück's plan. Jordan was one of the "Bonzen" of quantum mechanics. Born in Hannover in 1902, he formalized, together with Born, Heisenberg's original model of quantum mechanics and later merged it with Schrödinger's formulation. He then focused on quantum electrodynamics and cosmology. In 1928, he became a professor of physics at the University of Rostock, Germany, and in 1933 joined the Nazi party. During the Second World War, he conducted applied research for the Air Force ("Luftwaffe"), and, in 1944, he became a professor at the University of Berlin. After the war, he was elected parliamentary member of the Christian Democrats conservative party.

Jordan took Bohr's approach a step further. He conducted his biological research entirely on theoretical speculation, treating living beings like abstract entities and describing them as he described quantum mechanics objects. He also expanded the thanatological principle by introducing the theory of the "amplifier": specific effects do not have a cause, just like nuclear decay occurs "acausally." When this principle was applied to biology, it was impossible to comprehend the inner workings of life because some biological effects occurred without explanation. Jordan's dogmatism did not appeal to either Bohr or Delbrück.

To disprove Jordan and verify Bohr's assumptions, Delbrück ought to learn as much as possible about biology as soon as possible. In essence, he needed to perform a topic review regarding the intersection between physics and biology at a time when scientific literature was not indexed in public repositories. In pure Socratic spirit, he chose the best next thing: ask the experts. Thus, at the end of 1934, Delbrück "arranged a private seminar with biologists, biochemists, and physicists in which rather able people participate, so that all of us learned a lot." These home seminars became regular meetings, a common feature in the physics community. Bohr, for instance, organized yearly meetings in Copenhagen (the "Bohr festival") attended by the most prominent physicists in Europe. At the festival, physicists could discuss informally hot topics in quantum mechanics while hiking or drinking. It was customary for young students to mock their professors to avoid the generation of a chaste of untouchable barons in quantum physics.

One of the attendees to Delbrück's meetings was Karl Günther Zimmer, born in Wrocław in 1911 and who obtained a Ph.D. in biophysics in 1934 at the University of Berlin. Zimmer became assistant to Timofeev-Ressovskij in 1933, and the two collaborated until 1945 when the Russians deported Zimmer to Songul to work on the development of the Soviet atom bomb. In 1947, Zimmer and Timofeev-Ressovskij worked again together in the same team in Songul until the former was released in 1955. Zimmer moved to the University of Copenhagen and then to Karlsruhe, Germany.

At the "Grünewald seminars," the three were in sync—Zimmer, the experimentalist; Timofeev-Ressovskij, the geneticist; Delbrück, the theoretician. The trio had the numbers to crack the secrets of life and verify Bohr's theory.

3.4 THE GREEN PAMPHLET

Delbrück, Timofeev-Ressovskij, and Zimmer collaborated to investigate the structure of the genes within the field of radiation biology. The project culminated in a long and articulated scientific article published in 1935 that is widely regarded as the funding stone of "molecular biology," a term officially introduced in 1938 Warren Weaver in an internal report of the RF.

The article was entitled "On the nature of gene mutation and gene structure" and was published on the organ of the Göttingen Academy of Sciences: *Nachrichtigen von der Gesellschaft der Wissenschaften zu Göttingen, mathematisch-physikalische Klasse, Fachgruppe VI: Biologie* ("News from the Göttingen Society of Sciences, Mathematical-Physical Class, Division VI: Biology"). Since three authors wrote it, it was commonly referred to as the "Dreimännerwerk" ("Three men work" or "Three-man paper"). Although today, it is almost expected to have multidisciplinary scientific endeavors, back then, the norm was to have a single author per paper. There were exceptions: for instance, the seminal article on the formalization of quantum mechanics using matrix algebra was carried out by Bohr, Jordan, and Heisenberg in 1925 and was nicknamed the "Dreimännerarbeit" (the "Three men job"). In biology, though, such multidisciplinary collaborations were hitherto unheard of.

The authors stated their goal at the beginning of the paper:

> we attempt to construct a general picture of the nature of the gene and of mutation, on the basis of experimental investigations ... Compared to previous hypotheses ... we believe we have taken a step forward, in that our ideas are constructed only from experimental results ... this has enabled us to avoid methodologically questionable conclusions ... Of course, we are far from regarding our ideas as conclusive.

It was well understood that there were various types of mutations, ranging from organelle aberrations to chromosomal rearrangements. Still, the trio stated that "only gene mutations shall interest us," focusing the paper on what is now known as point mutations. Moreover, the authors restricted their investigation only to mutations mapped to the X chromosome to make the analysis more straightforward. It was also common knowledge that multiple alleles formed some genes, and there were "back mutations" that neutralized a given mutation. X-rays were chosen as the investigative tool because they were "reliably reproducible."

Nonetheless, the paper did not use X-rays to investigate the structure of the genes, as will be done a decade later by Rosalind Franklin, James Watson, and Francis Crick. Instead, it just enumerated the number of mutations generated for given amounts of radiation, quantified in units of measures no longer used today (r, röntgens). The paper reported, for instance, that the percentage of X-linked mutations was, on average, 0.19% in the absence of exposition to 3000 r of X-rays and 8.84% upon irradiation. Similarly, the frequency of direct mutations was 4.6×10^{-4} upon irradiation with 5000 r, whereas that of back mutation was 3.2×10^{-5} in the same conditions. Using such a simple approach, the trio could make a series of assumptions regarding the nature of mutations. Unlike natural mutations, those caused by radiation are not affected by temperature or time of exposure but rather by the final dose of radiation.

Furthermore, the rate of mutation is specific to each gene. From these premises, the authors inferred that genes are ensembles of atoms in a particular conformation, where "the atoms in the assemblage have determinate central locations and that the electrons are in determinate state" and "a particular mutation is a particular rearrangement of a particular molecule." Therefore, "a mutation is produced by an external infusion of energy or a fluctuation of thermal energy … and it consists of a rearrangement of atoms into a new equilibrium configuration." In modern terms, the gene can be considered a metastable structure. The transition from one stable configuration (the wild-type allele) to another (the mutated allele) necessitates surmounting an energy barrier. Radiation provides the energy to overcome such a barrier (activation energy). The equation describes the probability W of this transition: $W = Ze^{-U/kT}$, where U is the activation energy, kT is the thermal energy of the gene, and Z is a constant (10^{14} s^{-1}).

We now know that genes are made of nucleic acids. Specifically, genes are embedded in a polymer of the molecules composed of a phosphate group (H_3PO_4), a five-carbon carbohydrate (ribose), and a nitrogenous base that can be either an adenine (A), a guanidine (G), a thymine (T), a cytosine (C), or a methylcytosine (C^{me}) (Figure 3.2). The phosphate group, ribose, and nitrogenous base make what is known as nucleotide, further classified according to the nitrogenous base. The nucleotides adenine and guanidine are collectively called purines; thymine and cytosine are grouped as pyrimidines. Mammalian and bacterial genes are made of deoxyribonucleotides (thus, they are referred to as deoxyribonucleic acids, DNA) because the nucleotides miss a hydroxyl group (OH) in the ribose; specific viruses have genomes instead made of ribonucleotides (thus, they are referred to as ribonucleic acids, RNA), where the hydroxyl group is present.

However, genes are unstable because of their inherent resilience: contrary to the author's statement that the "gene is not readily disposed to reacting chemically," nucleic acids react chemically in many ways, even with water. Each day, the human genome within a single cell undergoes 18,000 events of depurination, 600 events of depyrimidination, 100 events of cytosine deamination, and ten events of methylcytosine deaminations. In addition, the intrinsic reactivity of DNA generates the following aberrant nucleotides per day in each cell: 2000 8-hydroxy-guanine (8-oxoG), 2000 ring-saturated pyrimidines, 1000 lipid peroxidation, 6000 7-methyl guanidine, 1200 3-methyl guanidine, and 100 O^6-methyl guanidine. DNA can even react with ultraviolet light, leading to the fusion of adjacent pyrimidines.

The trio was correct in considering the gene "as an assemblage of atoms," but they missed the scale of such aggregation. The nucleotides have the following molecular formulae: A: $C_{10}H_{12}O_5N_5P$ (33 atoms); G: $C_{10}H_{12}O_6N_5P$ (34); T: $C_{10}H_{13}O_7N_2P$ (33); C: $C_9H_{12}O_6N_2P$ (31); C^{me}: $C_{10}H_{14}O_7N_2P$ (34). Thus, on average, the nucleotides contain 33 atoms. The 24 human chromosomes (22 autosomes plus the sexual chromosomes X and Y) comprise 3,088,269,832 nucleotides; since the release of a hydrogen atom follows their polymerization, but one is left at the end of each chromosome, the human genome is formed by 98,824,634,648 (9.9×10^{10}) atoms. With so many particles, quantum physics is no longer a practical tool for describing this ensemble, just like a round bristle brush is not the right instrument to whitewash a ceiling. Unless someone is explicitly studying nucleic acid arrangements, it is common practice to ignore the structure of the nucleic acids entirely and focus on the sequence of nitrogenous bases. As a result, when referring to a gene's sequence, one refers to the combination of A, T, G, and C (C^{me} is required for regulatory purposes and counts as C) present in that gene.

X-rays are an example of ionizing radiation, which can expel an electron from a molecule. Since the expulsion of the electron implies a quantum leap, quantum

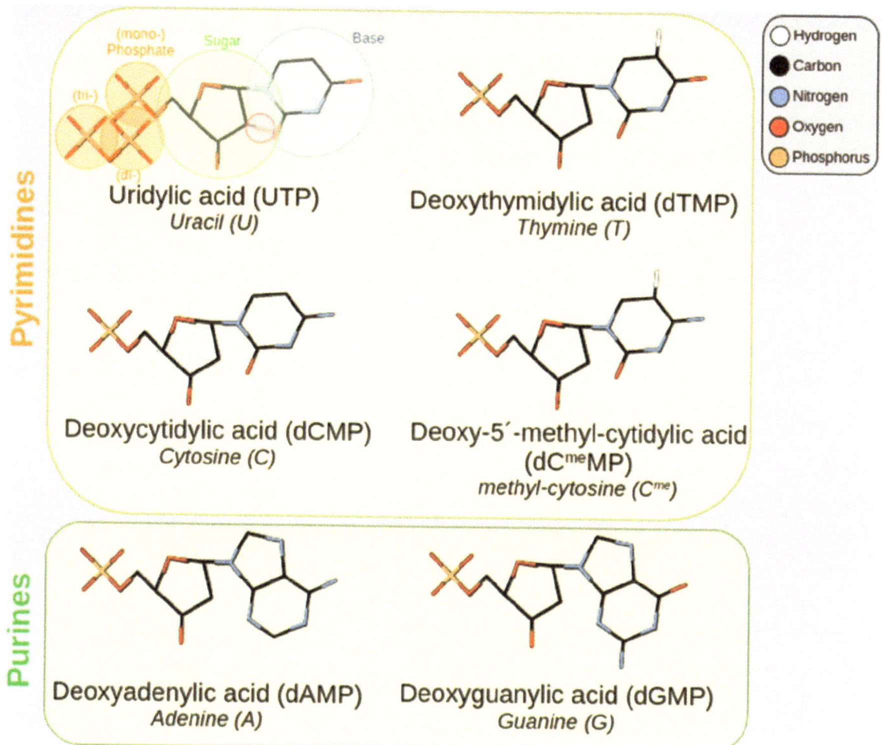

FIGURE 3.2 Structure of nucleic acids. Like all biological molecules, nucleic acids are formed by combinations of hydrogen, carbon, nitrogen, oxygen, and phosphorous atoms. The nucleic acids are characterized by an entity containing nitrogen (base) that changes among the different molecules, a ribose molecule (sugar), and phosphorous-based moiety (phosphate). The base variation allows differentiating among different nucleic acids, the phosphate moiety can contain from one to three phosphorous atoms, and the sugar can include or not include oxygen (red circle). Ribose not containing the additional oxygen atom is known as deoxyribose and takes the prefix "d"; nucleic acids can have either ribose or deoxyribose, except for thymine (T), which is replaced by uracil in biological entities. Methyl-cytosine (C^{me}) is a modification of cytosine (C) used to regulate gene expression. Nucleic acids are further subdivided according to the structure of the bases: those containing a two-ring base are known as purines, whereas the others are called pyrimidines. The figure identified the different nucleic acids with full and short names, including the one-letter code. Because uracil contains three phosphorous moieties, it is also known as uracil tri-phosphate (UTP). All other nucleic acids, on the other hand, have a single phosphorous moiety and are thus mono-phosphates.

mechanics describes the process appropriately. However, given the number of molecules involved in the process, ionization also lies well within the realm of chemistry. X-rays can disrupt the bonds within and between nucleotides, causing a chemical anomaly that leads to a mutation. However, since a layer of coordinated water surrounds the chromosomes, the first radiation target is not the gene but a water molecule. Thus, $H_2O + \nu \rightarrow H_2O\cdot^+ + e^-$, where ν is the quantum of radiation, e^- is an electron, and $H_2O\cdot^+$ is a water radical, which is unstable and degenerates into hydroxyl radical (HO·) and a proton (H^+). The electron can either attack the nucleotides or react with oxygen to generate super-oxide radical ($e^- + O_2 \rightarrow O_2^-$), which can also react and damage the nucleotides. Alternatively, the radiation can dissociate water directly into hydroxyl and hydrogen radicals: $H_2O + \nu \rightarrow HO\cdot + H\cdot$. One does not need quantum mechanics to describe these processes; chemistry is a handier tool.

All in all, the paper was a seminal work, albeit it was soon outdated by more focused biological research. Nevertheless, the paper had to be read to be influential. Back then, scientific articles were not available online as "pdf" files but only embedded within the volumes of journals, in this case, the first volume. *Nachrichten* was not renowned and had a minimal audience, so much so that Delbrück commented that they gave a "funeral first class" to their seminal work. Nonetheless, the publishers provided the authors with several reprints of the articles. The reprints were enclosed in green covers, and the authors sent them to selected colleagues. Given the color of the cover, the piece was dubbed the "green pamphlet."

One of the recipients of the reprints was Erwin Schrödinger, who used it as part of a series of seminars at Trinity College in Dublin during 1943, then collected it into a book (*What is life?*) that became very influential within the scientific community; in particular, it inspired Watson and Crick in their quest for the structure of DNA. According to Schrödinger, life is just the juxtaposition of atoms; thus, quantum physics has all the credentials to explain the secret of life. This view was diametrically opposite to Bohr's.

Schrödinger argued that it was possible to describe the properties of some objects because they display an inherent order. For instance, we can explain the expansion of gases or the conductivity of metals according to their temperature. This is because the particles inside these aggregates show an intermediate demeanor. Some molecules might be slower than others, and some atoms have fewer electrons than others, but on average, we can predict the behavior of these aggregates. Genes also follow rules: the rules of inheritance defined by Mendelism and statistical genetics.

Nevertheless, there was a critical distinction. Cytogenetics and microbiology have shown that each cell has only one genome: one chromosome in the case of bacteria and a set of chromosomes in the case of eukaryotes. This starkly contrasts with the billions of molecules in a cloud or a wire. Where did the laws governing gene behavior originate? Schrödinger used the three-man paper as an experimental basis to demonstrate the intrinsic stability of the genes, assuming that the laws of genetics came from their structure. Schrödinger suspected that the inner order of genes is "the real clue to the understanding of life."

Based on the experimental findings provided by the three-man paper, Schrödinger assumed that the genes were located on "aperiodic crystals" of limited extension. Schrödinger took a reductionist stance and considered that all that was essential to the cells were the genes, whereas the trio clearly pointed out that genes were parts of a complex enzymatic system. In the following years, the experimentalists disproved many of the features in Schrödinger's book, primarily the structural nature of genes. In 1987, Nobel laureate Max Perutz commented that "in retrospect, the chief merit of What is life? is its popularization of the Timofeev-Ressovskij, Zimmer and Delbrück paper." Regardless, the three-man paper was surpassed by later experiments, but it opened the road for applying physics in biological research.

Salvador Luria was another recipient of the green reprints. He was given a copy of the paper by his colleague at the Fermi's department in Rome, Franco Rasetti, in 1938. Rasetti was another of Meitner's researchers in Berlin simultaneously with Delbrück. Like many other physicists, Luria was inspired by the three-man paper to pursue a career in biology. If one might think studying human genes using something as tiny as a fruit fly is far-fetched, Luria and Delbrück will further investigate gene structure using something even tinier: viruses. But to understand such a step, it is necessary to travel to Italy and thence to America.

4 Rome and Los Angeles

4.1 A VESSEL TO THE PROMISED LAND

"Tardy with age were I and my companions, when we came to the strait pass, where Hercules ordain'd the bound'ries not to be o'erstepp'ed by man … You were not form'd to live the life of brutes, but virtue to pursue and knowledge high." Dante's description of Ulysses' journey beyond Hercules' pillars came gloomily back to Luria's mind as he watched the jolly lights of Lisbon zoom out of sight. Lisbon: an oasis of vitality guarding the straight of Gibraltar—the "Pillar of Hercules"—just outside the bleak shadow of the Nazi empire. "Old indeed," he thought, considering the steamer's passengers around him. There was no happiness aboard. There was no excitement about going to the land of sheriffs and gangsters, Indians and tycoons, to the country quickly overtaking Britain as the most advanced in the world. There was only remission and despair. Even the gleaming uniforms of the Argentine officers around him on the observation deck were not perky but pretentious. "Gentile souls surrounded by brutes," he reflected.

The various writers and freedom fighters he had encountered in his journey—an authentic odyssey—from Rome to Lisbon came to his mind. He also remembered the composer Paul Hindermith, who was downstairs playing a melancholy sonata at the piano allocated in the first-class saloon just for him so he could keep up with his compositions during the Atlantic crossing. He had met a bunch of cultured and compelling people forced to flee their homes and occupations due to a horde of racist zealots who had pushed once more the world to the brink of annihilation. Like himself, all these people were renouncing their established lives in Europe to start afresh in another land, just like the ancient Greeks left their motherland in Greece to reach the "Magna Graecia" in southern Italy. Would they manage to begin a new life? Would they be successful? Or would they end up as enslaved people or beggars? Luria sighed. How can one not be lost in classical memories when one embarks on a steamer called *Nea Hellas*?

Nea Hellas ("New Greece") was a veteran who carried hope across the Atlantic. She had been built just after the end of the First World War in Glasgow, Scotland. She was put at sea in 1921 and Christened Turbine Steam Ship *Tuscania*. With her 17,000 tons and 175 m in length, she was the little sister of the gargantuan RMS *Queen Mary*, which weighed 80,000 tons (310 m), also built in Glasgow. The *Tuscania* was also smaller than the notorious *Titanic*, built in Belfast, which weighed 46,000 tons (269 m). She was the little sister among these ocean liners even in terms of speed (16 knots against the 21 knots of the *Titanic* and the 28 knots of the *Queen Mary*) and capacity (1400 passengers against 2140 of the *Queen Mary* and 2435 of the *Titanic*).

Nonetheless, the *Tuscania* managed to serve for over 40 years. In contrast, the *Titanic* did not even finish her maiden journey—but she could not beat the *Queen Mary*, which is still anchored in Los Angeles, serving as a museum. Despite being a lightweight in the class of ocean liners, the *Tuscania* mounted the state-of-the-art

DOI: 10.1201/9781003657712-4

naval technology of the time. And despite never reaching the glamour of her bigger sister, *Queen Mary*, she was luxuriously dressed, decorated with taste, and spacious, even in the third class.

After the economic crisis of 1929, the *Tuscania* was sold to the Goulandris brothers from Greece, who owned the *General Steam Navigation Company*. The ship was renamed *Nea Hellas*, and a new maiden journey was made in May 1939. The Second World War began in September, and the *Nea Hellas* joined the war effort as a troop transporter. She became part of convoys and was chased by U-boats—the nefarious German submarines that did not falter in opening fire on passenger ships. By the time the war was over in 1945, she had transported some 60,000 soldiers across the Atlantic, who dubbed her *Nelly Wallace*. The Allied Powers, who commandeered her in 1941, officially returned her to her owners in 1947, and she remained in service until 1955. Afterward, she was renamed New York and placed into service in a Germany-USA tract, making her last trip in 1959. She ended up in a Japanese scrap yard in 1961, but during her lifetime, *Tuscania* carried about 800,000 people from deprived Europe to the American Dream.

Heading back to his first-class cabin, Luria wondered how the world could have fallen for another global conflict so soon after the first, which was supposed to be the war to end all wars. Like so many scientists around the world, Luria took little interest in politics. Thus, the swift expansion of Nazi Germany came as a sheer surprise to him. Everyone knew Germany had always been economically strong and was at the forefront of science, arts, and philosophy. Consequently, Germany was inhabited by plenty of bright people. How could a bunch of vulgar thugs take control of such a remarkable country? More or less in the same period, the playwright Bertold Brecht posed the same question and wrote a satirical piece about the sudden spread of Nazism that was published after the war (*The resistible rise of Arturo Ui*). In it, he referred to Adolf Hitler, the leader of the Nazi movement, as:

> The bigger gangster of all times
> Whom heaven sent us for our crimes
> For weakness and stupidity!

But then again, his own nation, Italy, has gone along the same path about a decade before Germany. As the Spanish painter Francisco Goya had said, "the sleep of reason breeds monsters."

Salvatore Luria was born in Turin, Piedmont, in 1912. Turin was Italy's industrial capital, home to the once-famous FIAT automotive and the royal House of Savoy, the oldest in Europe. The Savoys had annexed the whole of Italy in the late 1800s—much like the Prussian Hohenzollern had done with Germany—and now Victor Emmanuel III was the country's ruler. Salvador's father was a small-firm accountant, and his mother was a high school teacher. The Luria family had solid albeit overlooked Jewish roots and accounted for several rabbis within the Piedmontese community. When Salvador enrolled in high school in 1925, the Fascist Party had already started negatively influencing the Italian intellectual community. For instance, his professor of Latin at the grammar school Massimo d'Azeglio, Augusto Monti, who was also a writer, was jailed for antifascistic activities. Soon, many other intellectuals of Jewish origin were forced to leave Italy altogether, for instance, Emilio Segrè (Nobel Prize winner in 1959), Rita Levi Montalcini (Nobel Prize winner in 1986), and Enrico Fermi (Nobel Prize winner in 1938). It was not only Jews that were purged by the Fascist state. In 1926, the Italian government led by the only allowed political organization, the Fascist Party, introduced the so-called "confino di polizia" (confinement), a sort of exile within the Italian borders.

Such confinement was a measure of preventive detention, that is, a sentence in the absence of a crime other than non-adherence to fascist conformism. It was simply a measure of repression that was tied to ubiquitous propaganda. The fascist legislation (the "leggi fascistisime") passed in 1926 included both the Border Police and a special tribunal for political crimes. But while such a tribunal required a crime actually carried out and proven evidence of it, confinement was at the complete discretion of the executive bodies—chiefly prefects, police commissioners, or fascist cadres. There were examples of people just released from a sentence imposed by the special tribunal and sent immediately to the border. In other cases, there was not even a need for antifascist activity: it was enough to be disliked by a fascist hierarch or tell a joke about Benito Mussolini, the founding father of fascism, to end up in confinement. Between 1926 and 1943—the year of the fall of the fascist regime—nearly 12,000 people had been trapped in confinement, primarily communists but also Jews and homosexuals.

Benito Mussolini made grim irony by calling these Italian Gulags a "seaside vacation." Even Silvio Berlusconi, who served as prime minister of the Italian Republic in several right-wing cabinets about half a century later, cynically commented in 2003 that the internees enjoyed "a vacation paid for by the State." The antifascist politician Carlo Rosselli, who was imprisoned on the island of Lipari and later murdered by the fascists in 1937, defined confinement as "a cell without walls." The internees were deported to isolated villages or directly to small islands, closely guarded by the army—the Corps of Royal "Carabinieri"—and were deprived of their passports, effectively seeing their rights as citizens abolished, a vilification imposed not even on convicts. The internees were crammed into dilapidated dormitories with flea-filled beds and were not allowed to interact with the inhabitants of their imposed villages; they could only move along predetermined routes at certain times of the day. Contact with the outside world was severed: the internees could send only one letter a week—censorship was applied to outgoing and incoming mail—they could not go to cinemas or restaurants or speak a foreign language. It was common for the internees to be beaten to a pulp by the guards for no reason whatsoever.

The fascist state would establish, over two decades, more than 200 places of confinement. Among the illustrious names of those confined there could be counted Sandro Pertini and Pietro Nenni (who would become presidents of the Italian Republic), Altiero Spinelli (who would become one of the founding fathers of the European Community), Curzio Malaparte (journalist and writer), and Antonio Gramsci (secretary of the Italian Communist Party).

If the "confino" could be compared with the Soviet Gulags—which were also prisons with no walls—in 1931, the Fascist government enforced a sort of Lysenkoism on academia by imposing an oath to the fascist government. Of 1,225 professors employed in the University and high schools by the Italian government at the time, only 12 refused to take such an oath, even because leader antifascists such as the philosopher Benedetto Croce and the secretary of the Italian Communist Party (PCI) after Gramsci, Palmiro Togliatti, suggested not to abandon the academia to the fascists and take the oath regardless of personal taste. These dozen brave hearts who did not bend to the wills of the fascists lost their jobs and were forced to relocate abroad. The oath sparked an outcry from academics in democratic countries—as a prime example from Albert Einstein. Such loyalty requirements toward the government might appear strange today since nobody in Western countries requires an official statement when signing up for a job. However, it is not limited to Fascist Italy or Nazi Germany. In 1955, a Senator of Illinois, Paul Broyles, put forward a similar law in the US. Thousand of academicians, including Luria, signed the oath without complaints.

Back on the deck of the *Nea Hellas*, Salvador Luria had a sigh of relief. Yes, he was crossing the pillars of Hercules as Ulysses had, but while the latter was going to his death, Luria was going toward a new life. America was the land of opportunities, after all. Millions of Italians had moved there to find jobs that the motherland could not provide. Even his former teacher, Enrico Fermi, had left for New York three years earlier and had become a professor at Columbia University. Instead, throwing Luria to the "confino" would have been like sending him to a rotten purgatory. That would have been the end of his story. There was hope, after all. Then, Luria slipped into a bad mood when he remembered the mastermind behind fascism, Mussolini.

Benito Amilcare Andrea Mussolini was born in Predappio, Emilia-Romagna (northern Italy), to a blacksmith, Alessandro, and a primary school teacher, Rosa. Alessandro was a staunch socialist and named his first-born son after Benito Juárez (a revolutionary leader who ordered the execution of the Mexican emperor Massimiliano in 1867), Amilcare Cipriani, and Andrea Costa (socialist leaders). Alessandro named his second-born son after Arnaldo da Brescia, a heretical monk who harshly criticized the Church in the 12th century. It is said that the apples do not fall far from the tree, but they made a very large parabola in this case since both sons became far right leaders, the opposite of socialists.

Benito Mussolini presented himself as tenacious anticlerical, albeit his mother Rosa had him baptized and educated by the Salesian friars. Eventually, Benito was expelled from the Salesian's school because he had stabbed one of his schoolmates. Later, he obtained habilitation as a primary school teacher; he was proud of being called "professor" although the actual title in Italy for such a role was simply "maestro" ("teacher"). He then joined the Italian Socialist Party (PSI), not so much to make his father proud but because this was a vital and rising organization that provided many opportunities for social climbing.

To avoid military service, he took refuge in Switzerland in 1902 but ended up jailed for vagrancy in Lausanne. Mussolini has been depicted as a "dirty tramp," a "crook," and a rough and boisterous opportunist and megalomaniac. Angelica Balanoff, a Russian dissident activist who took refuge in Switzerland and became one of Mussolini's many lovers, portrayed him as "eager for notoriety and filled with a spirit of revenge against society." The historian Alan Taylor described him as "a vain, blundering boaster without either ideas or aims." Mussolini despised manual work—albeit he later portrayed himself as a hard worker by posing as a farmhand in front of the press—and postured himself as an intellectual, above all the others, as an Italian Lenin. Although a member of the Socialist party—a direct emanation of Karl Marx's teachings—Mussolini was more inclined to the anarchism of Babeuf and Blanqui, both harshly criticized by Marx. More than toward the dictatorship of the proletariat in Leninist terms, Mussolini was inclined toward the revolution as a beautiful, utterly destructive force, a view that at the time was circulated by the art movement of the Futurists. Thus, in 1904, he returned to Italy with a pro-war idea and enrolled in the Italian army.

Between 1910 and 1912, he directed the socialist review *Lotta di Classe*; from 1912 to 1915, he was the director of the socialist newspaper *Avanti!*. He wrote a book against the Pope—albeit later, he became a fervent defender of the papal throne. In those years, Prime Minister Giovanni Giolitti had declared an occupation war on Libia, and Mussolini's support of the war resulted in his expulsion from the Socialist Party in 1915. He obtained funds from Ansaldo, an Italian armament firm, to start a newspaper (*Popolo d'Italia*)—of Blanquist inspiration first, outright right-wing afterward—that sustained that war and later praised the Italian involvement in the First World War. During the conflict, he served as a journalist, but later, when he gained power, he appointed himself corporal and created a myth of his wartime heroism.

After the war, on the 23rd of March 1919, about a hundred jobless veterans met in Saint Sepulchre Square in Milan to protest against the government that had left them down. The protesters gathered as a movement called "Fascio Rivoluzionario d'Azione Interventista" ("Revolutionary League for Interventionist Action"). The "fascio" ("bundle") was a symbol first introduced by the Etruscans and later adopted by the ancient Romans. It symbolized joint strength: one straw can be easily bent, but a bundle won't. Initially, the bundle was adopted as a symbol also by socialist organizations. The connection to the mighty Roman empire enabled Mussolini to construct a nationalist narrative based on re-establishing a powerful Italy that would matter on the international stage. The Saint Sepulchre rally had essentially a socialist program, and Mussolini rode the tiger and became the movement's leader, again changing his political direction. He renamed the movement "Fascio di Combattimento" ("Fraternity of Combat") to have a more appealing name for the local elections of November 1919. The Fascist party obtained less than 5000 votes against the over 170,000 of the Socialist party. Mussolini realized he could not compete with the socialists in their backyard: he needed to differentiate. Thus, the party became overtly conservative, and what the party lacked in intellectual rigor was propped up by physical violence.

The rally in Saint Sepulchre was just one of many socialist-style protests between 1919 and 1921, a period known as the "Biennio Rosso" ("Two Red Years"). Despite socialist propaganda claiming an impending Bolshevik revolution, the PSI could not channel the discontent toward revolutionary means, an incapacity that baffled even Lenin and Trotsky. The PSI was further weakened by the scission that gave birth to the PCI, but even the PCI could not spark a Soviet-style revolution in Italy. All in all, the "Biennio Rosso" was just general unrest due to the shock the First World War had blown on society but without a breath of development. Such a characteristic was well understood by Giolitti, who did not authorize the intervention of the Police or the Army—as was custom back then—to suffocate the strikes. Giolitti was proven correct since the strikes eventually faded spontaneously, but the inactivity gave Mussolini the political opportunity to present himself as the country's savior.

In 1920, the Confederation of the Italian Industrialists ("Confindustria") recruited veterans, particularly those belonging to the special forces known as "Arditi," as private security guards for their factories. These private guards ensured that the plants would not be damaged during strikes and started to take active measures by beating trade unionists. Mussolini—who obtained secret funding not only from these industrialists but also from wealthy landowners—embedded this irregular corp within the fascist organization and used it systematically against the socialists. The targets of this paramilitary organization—known as "Blackshirts" or "Squadracce"—were striking workers, syndicalists, rioting peasants, journalists, and politicians who were non-sympathetic to the fascist cause.

Thus, the "Biennio Rosso" was followed by the "Biennio Nero" ("Two Black Years"), characterized by diffuse street violence. There were about a dozen assaults per day, and by October 1922, the police had recorded 3000 dead socialists against 300 dead fascists. Aside from beating their targets to a pulp with sticks, the Blackshirts enjoyed forcing victims to drink entire bottles of castor oil, a strong laxative. In August 1922, the fascists assaulted and devastated the *Avanti!* Headquarters. Giacomo Matteotti, a socialist member of Parliament, accused the violence implemented by Mussolini in Parliament's plenary sitting by saying that "the government, and above all its authorities, watch impassively as the law is being massacred."

Moreover, Matteotti was gathering evidence of the widespread corruption among the fascist echelons; thus, he was a rather cumbersome presence. Matteotti was later kidnapped and killed by a fascist group led by Amerigo Dumini. Mussolini claimed the moral paternity of the assassination by giving a famous speech in the Parliament:

> I assume, I alone, the political, moral, historical responsibility for everything that happened … If fascism was a criminal association, I am the leader of this criminal association

Nonetheless, King Victor Emmanuel III refused to prosecute either the fascist organization or Mussolini for this crime. When he was updated regarding the investigation into this high-level murder case, the king covered his face with his hands and cried that he was both blind and deaf. The king's acquiescence can be explained not only by his fearful disposition but also by the fact that his dashing and vigorous cousin, the Duke of Aosta, was a genuine fascist. If Victor Emmanuel III had not supported the fascists, he could easily have been dethroned and the Duke installed as the new ruler.

Mussolini's strategy was straightforward, at least in retrospect: to create mayhem while blaming the socialists for causing it and the government for failing to deliver peace. Only the fascists could provide the cure to such social chaos. Yet, the simplicity of this political ordeal was successful: in May 1921, Giolitti formed a coalition government that included the fascist organization—meanwhile renamed the National Fascist Party, NFP. The Blackshirts committed dozens of homicides during the day of the election alone, but Giolitti ordered the police not to intervene. The NFP gained only 35 seats but evolved from a street gang to a legal nationwide entity.

Despite presenting himself as a natural-born leader, Mussolini simply and systematically snatched somebody else ideas. For instance, his favorite posture—where he stood with fists on his hips and the chin up—was inspired by the *Maciste* film franchise (top-rated back in the day), whereas his haranguing the crowd from a balcony, the Roman salute, and even his famous moniker "Duce"—from the Latin for "leader"—were all copied wholesale from D'Annunzio, who was widely known as the "Commander."

Gabriele D'Annunzio, born in 1863, was an army general, a Freemason, a member of the Parliament, and a poet. He led the Italian "Decadentism," an artistic movement that praised aesthetic excesses and bravado that today can be defined simply as "machismo." D'Annunzio was a larger-than-life character akin to the film stars of our age: a womanizer who lived above his means in luxurious villas and threw lavish parties. He exalted war to purify humankind from weak people. Hence, he was an interventionist and volunteered as a fighter pilot in the First World War, becoming a war hero. He survived a plane crash, losing his right eye—a feature that gave him an even greater charismatic halo.

In September 1919, D'Annunzio led a gang of "desperados"—unemployed veterans and regular soldiers of the royal army that rebelled over their superiors—to seize the city of Fiume in Croatia. Fiume—Rijeka in Croatian—was founded by the ancient Romans, although settlements had been present since the Paleolithic and had been part of the Austro-Hungarian empire for centuries. After the First World War, Fiume became part of the Kingdom of Slovenia, Croatia, and Serbia—the forerunner of Yugoslavia—but the Italians felt it was part of Italy, albeit the Savoys did not claim it. D'Annunzio and his bunch effortlessly conquered the city and proclaimed its annexation to the Kingdom of Italy. However, the Savoys could not take the gift to avoid international confrontation, so the situation stalled. Officially, Italy was against this sleight of hand; behind the curtains, though, the Italian army supplied D'Annunzio with money, provisions, and weapons. In August 1920, D'Annunzio declared Fiume an independent city-state under the name of "Reggenza del Carnaro" ("Carnaro's Regency") with a socialist inspiration. The Regency was the first state to recognize the Russian Soviets as a State. In the same month, the independency of the Reggenza del Carnaro was acknowledged by the Italians and Yugoslavians with the Rapallo Treaty, but that was not recognized by D'Annunzio, who refused to dismantle his turf. Thus, the Italian government sent the regular army to free the city of Fiume from D'Annunzio's personal brigade, and there were several days of battles. On the 31st of December 1920, the Regency was evicted, and D'Annunzio fled to Lombardy with the cashier of the expedition. Fiume was then annexed to Italy in 1924 by Mussolini but would go back to be part of Yugoslavia in 1947 and of Croatia in 1991.

The expedition in Fiume ("impresa di Fiume") taught Mussolini some suitable lessons. The first was what fascism should look like. Mussolini, an opportunist by nature, had swayed the NFP between the political left and right. He had proudly stated that "we [fascists] are ultimate relativists." After 1920, Mussolini realized that an amalgam of chauvinism and militarism, with a dash of romanticism given by an ideal connection to the grandeur of a mythicized ancient Roman empire, would provide him plenty of political maneuverability. The second lesson was that the Savoys would not move a finger if they could gain something. Land annexation qualified as a gain, but also containing the socialists—with their requests for shorter and better-paid working hours as well as civil rights that undermined the power of the crown—would also be a significant earning. Moreover, fighting the socialists provided a stream of hidden money provided by industrialists—such as Giovanni Agnelli, owner of FIAT—and wealthy landowners, who had all the interest in keeping workers and peasants at bay. The bottom line was that the fascists, emboldened by the passivism of both King Victor Emanuel III and Giolitti, stepped up their game rather quickly.

4.2 ALL THE WAY TO ROME

During the Two Black Years, the fascists began to raid the municipal councils of several cities in northern Italy (chiefly Ferrara, Cremona, Parma, Ravenna, Livorno, Milan, and Bologna) to avoid the rightful settlement of socialist town councils. The Blackshirts also stormed in Trentino and South Tyrol against German minorities. The chronicles of the time were reporting occurrences such as the "Estense's Castle Massacre," "Parma's barricades," and the "d'Accursio's Palace Massacre." The new government led by Luigi Facta did not move a finger against the NFP. On the 26th of October 1926, taking advantage of a cabinet reshuffle, four high-ranking cadres of the NFP (Michele Bianchi, Italo Balbo, Cesare Maria De Vecchi, and Army general Emilio De Bono) established a shadow government—known as the "quadrumvirate"—in the city of Perugia, with the intent to launch a nation-wide fascist revolution and force king Victor Emanuel III to give the power to the NFP.

While the revolution took place, in the form of ever more violent actions, Mussolini remained safely hidden in his newspaper's headquarters in Milan; meanwhile, the king was on holiday. The general staff Pietro Badoglio assured the king that the Army could easily take down the rioters, but Victor Emanuel refused to act. On the 27th of October, Facta produced an emergency law declaring the state of siege, which would have given Badoglio the required authority to stop the revolution. Still, the king again refused to sign it. The prominent Italian newspapers of liberal inspiration, such as *Corriere della Sera*, *Stampa*, and *Messaggero*, supported the fascist uprise. Eventually, King Victor Emanuel III sent a telegram to Mussolini, begging him to come to Rome and establish a new government. On the 30th of October, Mussolini arrived in Rome by night coach train, while the Blackshirts sent by the triumvirate camped in miserable conditions in the countryside around the city. On the 31st of October 1923, the new government was established with the king's blessing: Mussolini became not only Prime Minister but also Minister of the Interior and Minister of Foreign Affairs ad interim; there were four NFP ministers and ten ministers from the conventional liberal and conservative parties. In his first speech to the Chambers, Mussolini declared that the Parliament was useless and he would recast it into a bivouac. Matteotti was the only one who stood up to him, shouting, "Up with the parliament!" Little did he know that his days were numbered.

Afterward, the NFP strengthened its grip on Italian society. Some of the most important supporters of the NFP were rewarded with a place in the Senate, chiefly Agnelli and Facta. D'Annunzio was remunerated—and taken out of public life to not antagonize the Duce—with the title of Prince of Montenevoso and a luxurious villa on Lake Garda. Afterward, NFP elements were installed in crucial positions in the judiciary and executive branches. The royal guard and the national militia—the so-called "Blueshirts"—were dismantled. The National Party was integrated into the NFP. Freemasonry was declared illegal while Pope Pio XI talked against his catholic reference party—the Italian Popular Party founded by Don Luigi Sturzo in 1919—in favor of Mussolini, presenting him as the "Man of Providence." The government started a series of privatizations welcomed by Confindustria, and Mussolini claimed that he was "for the man and against the State"—although he would establish a State behemoth in 1925.

In August 1923, the "Acerbo Law" introduced a first-past-the-post system, which assigned two-thirds of the seats to the list that had exceeded 25% of the votes. The Parliament accepted the law with a majority of 235–139, while the Blackshirts played with pistols and cleaned their fingernails with daggers in the lobby of the House of Parliament. The law was applied in the election of April 1924, characterized by the usual rounds of beatings and homicides by the Blackshirts. The NFP gained the favor of 65% of the votes and obtained a majority of 403–107. In 1925, strikes were forbidden by law, and, unsurprisingly, the salaries of the Italians became the lowest among the whole of Western Europe except for Spain. In 1930, workers were forced to ask permission from the prefecture to travel, a measure that de facto reintroduced serfdom in Italy.

After the 1924 elections, the Blackshirts started to openly harass the members of Parliament from the opposition, culminating in the assassination of Matteotti in June. The homicide was a bad public relations coup for Mussolini, and the Party was about to collapse. Giovanni Amendola, from the PSI, convinced the opposition members to leave the Parliament permanently as a protest (the "Aventine Secession") to force the king to intervene. Not only did King Victor Emanuel III not do much of anything against Mussolini, but the government did not fall because it still had the support of the Liberal Party led by Giolitti. Yet, even Giolitti had to regret his support of Mussolini. In November 1924, the NFP restricted the freedom of the press. Giolitti was furious, but Mussolini replied: "If his majesty were to call me today and order me to resign, I would stand at attention." Such an order was never given. On the contrary, a vote of confidence gave Mussolini a majority of 208–54. Giolitti's Liberal Party was declared illegal in January 1925, the secessionists led by Amendola were deprived of their status as members of the Parliament, and the PCI parliamentarians were arrested.

By 1934, the NFP had 95% of the preferences because it was the only Party running for the elections. Membership in the NFP was necessary to get a job. In 1936, by conquering Ethiopia through a bloodbath, the kingdom of Italy became officially an empire. The Rosselli brothers, the last socialist leaders who still openly opposed the regime, were killed in 1937. In early 1939, the Parliament was supplanted by the "Chamber of Corporations," a directorate that controlled the emanation of laws and directly regulated the job market. The Parliament was never officially abolished; it just ceased to be employed. Since the country's economic life was decided into the Chamber of Corporations, which was nothing less than a private organization owned by the NFP, such an institution quickly became a cradle of cronyism and corruption.

In all this, Mussolini presented himself as a practical man who had brought social peace and sped up the work of a useless parliament. Nonetheless, he knew that was just appearance and that what he delivered, instead, was a plain and straightforward regime. He conned the Italians with an inefficient and corrupted plutocracy where the reckless few lived on the shoulders of the suffering many, passing it off for a dynamic and daring empire. Mussolini knew that and proudly commented to his biographer Emil Ludwig that:

> people today don't have as much time to think as they used to. Modern man's capacity for belief is unlimited. The crowd loves strong men. The crowd is a woman.

An epitaph of democracy that was as meaningful yesterday as it is today. Therefore, when Luria moved to Rome to study at the University of La Sapienza under Enrico Fermi, the fascist dictatorship was in full swing.

After graduating from the lyceum, Luria enrolled in the faculty of medicine of the University of Turin in the laboratory of Giuseppe Levi, a fervent antifascist. Later on, Rita Levi Montalcini and Renato Dulbecco will join the same research group. Meanwhile, one of his high school buddies, Ugo Fano, joined the physics department; Fano later became a professor of Physics at the University of Chicago. Luria was not excited about medicine and later found his specialization field, histology, unappealing. On the contrary, he was fascinated by Fano's vivid accounts of quantum physics, portrayed as a "romantic landscape." Thus, after completing medical school, Luria signed up for a radiology course in Turin to get closer to the area of quantum physics.

Luria had to pause his studies due to compulsory military service, which he carried out as an army doctor, albeit "army life was not inspiring" either. At the end of the service, in 1937, Fano convinced Luria to join Fermi's laboratory in Rome together. Toward the end of the year, the two friends moved to the capital of the fascist empire. Yet, despite the universally recognized beauty of the "eternal city," Luria failed to enjoy his time in Rome. Timid by nature and raised in a restrained environment such as the Piedmontese, he was uneasy with the extroverted way of life of the Romans. He could not reconcile the "hazy grayness" of Turin with the "crude openness" of Rome.

Moreover, he realized, "professionally speaking the Rome experience was a torment." Working in a world-renowned research group was a demanding job. Just like Delbrück understood not to be at the levels of Göttingen's "Bonzen," Luria admitted that "it soon became clear that my ability to deal with physics was weak."

Enrico Fermi was in the league of the "Bonzen" of physics. Born in Rome in 1901, he began studying physics to cope with the death of his brother, who died after a minor surgery. In 1922, Fermi obtained a degree in physics from the Scuola Normale Superiore, a prestigious institute for the gifted. Afterward, Fermi received a scholarship from the Italian government to visit Max Born in Göttingen. He then became a professor of Physics at the University of Florence in 1924 and at La Sapienza in 1927. In 1928, Fermi married Laura Capone, who had Jewish roots. In 1929, Fermi enrolled in the NFP, and the day after, he was nominated by Mussolini to be a member of the Academy of Italy, an institute created by the regime in 1926 to diffuse fascist science in direct opposition to the Academy of the Lincei, which had instead an antifascist orientation. The Academy of Italy gathered renowned personalities of Italian science, such as Guglielmo Marconi, who had patented a development of

Nikola Tesla's radio—in 1943, the patent was void by the US Supreme Court. Other big shots in the Academy of Italy were Luigi Pirandello, who received the Nobel Prize in literature in 1934, and Giovanni Gentile, an eminent liberal philosopher who became the ideologist of fascism.

Fermi, like Born, gathered a group of outstanding young men around him. Indeed, while Göttingen was the reference center, in collaboration with Copenhagen, for theoretical physics, Fermi's laboratory at 90 Panisperna Street, Rome, became the world reference center for applied physics. While Born's people were known as "Bonzen," Fermi's collaborators were called the "Boys from Panisperna Street." The group was characterized by an exuberant attitude toward life and science, in sharp contrast to the conformism of the fascist way of life. The leading members of these "boys," each with a moniker, were Franco Rasetti ("Cardinal Vicar"), Emilio Segrè ("Abbot"), Ettore Majorana ("Grand Inquisitor"), and Bruno Pontecorvo ("Puppy")—Fermi went under the name of "Pope," Luria as "Signor Garzone" ("Mr. Busboy").

Fermi studied the characteristics of the particles that undergo Pauli's exclusion principle (now known as "fermion" particles) but then specialized in the properties of neutrons. The idea was to convert common (non-radioactive) elements to radioactive ones by bombardment with neutrons. Fermi discovered that it was possible to slow down the speed of neutrons, for instance, by passing them through a layer of paraffin. While neutrons move at about the speed of light (300,000 km/s), neutrons that have undergone a series of collisions with paraffin molecules slow down to about 2 km/s. It turned out that slow neutrons were better at radio-activating atoms and, later on, splitting atoms.

Luria's visit to Fermi's laboratory did not give him deep insight into quantum physics, but it was formative nonetheless. For starters, he learned to use physics's rigorous approach to his research. In addition, he was introduced to Delbrück's quest to identify the intimate structure of genes. Luria became acquainted with Delbrück's work on radiation biology thanks to Rasetti, his spectroscopy teacher at La Sapienza, who passed Luria a copy of the green pamphlet. "It seemed to me that these writings opened the way to the Holy Grail of biophysics, and it was then," Luria recalled "that I swore to myself that I would become a knight of that Grail."

Moreover, attending Panisperna Street gave Luria a chance encounter that changed his life. Luria was acquainted with Geo Rita at a stop of the streetcar's line "circolare rossa." Rita was born in Vignanello, north of Rome, in 1911 and was deeply antifascist. He obtained a degree in medicine in 1935. In 1937, Rita started his research in virology. When Luria and Rita met, the latter was isolating anti-*Shigella* phages from the Tiber River to fight dysentery bacteria in classic d'Hérelle style.

"Between bacteriophages and myself, it was love at first sight," Luria remembered. "Bacteriophages are actually beautiful: each has a head and a tail, and whiskers." Luria started working with bacteriophages immediately, trying to grow and count them. "I had been looking for some biological objects on which to test Delbrück's ideas on the gene, something small enough to be like a gene and yet easy enough to work with." Phages were perfect for the job because one could make billions of copies of them and quantify them precisely. Thus, Luria and Rita started a prolific collaboration that turned into a sincere friendship. In June of 1938, "they wanted to prove that one single phage particle could trigger lysis of a bacterial culture … we used thousands of test tubes … in the sleepy Institute of Microbiology nobody had never seen anyone working so frantically."

The demonstration was utterly statistical, just like statistics was used in physics to describe the characteristics of gases and atoms. Statistics was introduced to improve gambling since it provides the probability of obtaining a given outcome. For instance, the probability of drawing an ace of spades from a standard deck of 52 cards is 1/52 = 0.019 (or about 2%), whereas the probability of getting a three from a toss of a die is 1/6 = 0.167 (or about 17%). These examples show uniform probabilities because each event has the same chance of occurring: the probability of drawing a queen of hearts is still 0.019, just like that of tossing a six is 0.167. Let us suppose, though, of tossing two dice: in this case, the probabilities are no longer uniform because some values are more likely to occur than others. For instance, the chance of getting a two (one pip on each die) is 1/36 = 0.028, whereas that of tossing a three is 1/36 (one pip on one die, three pips on the other) plus 1/36 (three pips on one die, one pip on the other); thus, 1/36 + 1/36 = 2/36 = 1/18 = 0.056. Seven is the number most likely to be drawn when tossing two dice, with a probability of 1/6 = 0.167. Plotting the possible occurrences by the associated probabilities gives a bell-shaped curve, with a peak at the most likely occurrence. Such a type of distribution of probabilities is known as "normal." Calculating the probability of getting values below a certain level is also possible. For instance, the likelihood of getting a four or less when tossing two dice is 1/36 (probability of getting a two), plus 2/36 (probability of getting a three), plus 3/36 (probability of getting a four); thus, 6/36 = 1/6 = 0.167. Such a sum of probabilities goes under the name of "cumulative distribution." In general, the cumulative probability is given by the ratio between the number of wanted outcomes over the total possible combinations.

However, Rita and Luria's research differed from tossing two dice. Borrowing from radiation biology, where physicists count the number of hits (mutation generated by ionizing rays) during a given time frame, they were counting the number of bacteria infected by phages after a specific period. The French mathematician Siméon Denis Poisson described such cases in the 19th century. Specifically, Rita and Luria were interested in determining the exact number of phages x in a test tube containing 10^8 Shiga bacteria, given that the phages had been diluted to an expected concentration of n. The probability P of having exactly x phages instead of the nominal n is:

$$P = \frac{n^x}{x!} e^{-n}$$

where e is the Euler's number (also known as the base of natural logarithms and has a value of about 2.718) and $!$ is the factorial function—the multiplication of terms equal or below x, for instance, $3! = 1 \times 2 \times 3 = 6$. The probability of having less than m particles is given by the cumulative distribution Q, defined as:

$$Q_m(n) = \sum_{x=0}^{m-1} P$$

where Σ represents a sum of terms where x takes the values from x to $m-1$ sequentially. In particular, for $m=1$, the sum is limited to $x=0$:

$$Q_1(n) = \frac{n^0}{0!}e^{-n} = e^{-n}$$

By setting m to 1, Rita and Luria wanted to know how likely it was that a single phage infected a bacterial culture, causing the suspension to lyse. They set up 100 tubes, counted the number of tubes with complete lysis, and calculated the probabilities. Rita and Luria demonstrated that a single phage could lyse an entire bacterial culture because the model based on the Poisson distribution fits the data (Figure 4.1). In contrast, the normal distribution will not fit the data. Since Rita and Luria used 100 test tubes, comparing their experiment with a bingo drawing is possible. Rita and Luria

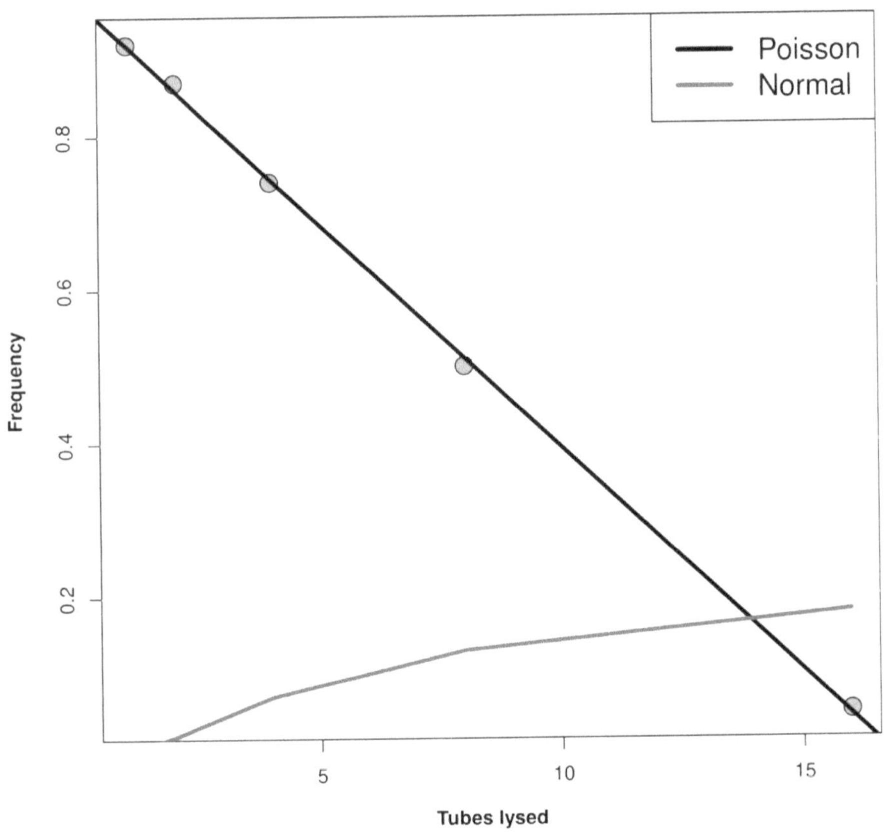

FIGURE 4.1 *Demonstration of lysis due to a single phage.* Plot of the number of tubes initially containing 10^8 *Shigella* bacteria showing complete lysis (dots). A model based on the Poisson cumulative distribution Q for values less than one fits the data (black line), but one based on a normal distribution is completely off (gray line). The Q function was adjusted by a factor $k=0.25$; thus, the model is $Q=e^{-kN}$, where $N=(2, 4, 8, 16)$.

determined, for instance, the frequency of obtaining eight completely lyzed suspensions out of 100; the probability of drawing at least eight in a bingo round would be 8/100 = 0.080.

Afterward, Rita worked not only on the characterization of growth and infectivity but also in other fields of virology and immunology. He became a microbiology professor at the University of Siena in 1953 and, in 1969, established the first Italian chair of virology at La Sapienza. He passed away in 1991.

4.3 NUCLEAR FISSION

Luria obtained a scholarship from the Italian government on the 17th of July 1938. He planned to use that money to go to Berkeley University in the United States. Alas, the next day, Mussolini published a "Racial Manifesto" that preached against Jews and collocated, with an intellectual acrobatic twist, the Italian ethnicity among the Nordic races of Aryan lineage. This manifesto was no slight intellectual stretch since the Aryanism preached by the influential occultist Jörg Lanz added the "Mediterraneananoids" to "Negroes" and Jews as satanic races that opposed the Aryans. One might think that Mussolini introduced racial laws merely to appease his stronger partner, Hitler, while Italy remained an essentially tolerant nation of "brava gente" (good people). In reality, the Italians had demonstrated a deep-seated racism against the African people of their empire and the apartheid laws implemented in the colonies were then enforced internally with the decree of 1938. One of the drivers in the implementation of the racial laws was Mussolini's intention to shatter the myth of the Italians as good people.

Discrimination against Jews has been commonplace in Europe since the Middle Ages. In 1215, Pope Innocent III introduced distinguishable badges for Jews, a practice that was notoriously re-implemented by the Nazi regime. In addition, Jews were segregated in ghettos and did not have access to several types of jobs. This practice was concluded only with the instauration of the Italian Kingdom in the late 19th century. With the rise of fascism, anti-Semitism took a new stance. The historian Ruth Ben-Ghiat remarked that "il Duce was an anti-Semite of long-standing;" historian Gene Bernardini commented that "while racial anti-Semitism was largely unforeseen and probably unwanted by most Italians, it evolved from Fascist principles and policies which they generally supported."

Thus, on the 18th of July, Luria and many other Italians—including Margherita Sarfatti, one of Mussolini's many lovers who was also force to leave Italy—were fiercely reminded of their ancestry: he lost his scholarship ("like an ephemeral flower, [it] had lasted l'espace d'une nuit"), and hundreds of thousands lost their jobs, including his father and brother. The Panisperna boys, many of them Jews, were also struck by this wave of hatred.

On the 17th of November, the NFP ratified a racial law that implemented Mussolini's manifesto; Fermi's group melted away. Without means—already in June, the government had refused Fermi's proposal to create a national institute of radioactivity—and co-workers (who had been expelled from La Sapienza) as well as fearing for his wife's safety, Fermi realized that his career in Italy was over. On thc 6th of December 1938, he traveled with his wife to Stockholm to receive the Nobel Prize. Fermi did not wear the Fascist uniform—required by the etiquette because he was a member of the Academy of Italy—as a protest against Mussolini's regime. But the objection went well beyond the form. Instead of returning home, the Fermis secretly moved to Britain, and, on the 24th, they embarked on the steamer Franconia for the US. A few months later, he became a professor of physics at Columbia University. In December 1942, in collaboration with Leo Szilard, he carried out the first controlled chain reaction of atomic fission, building a uranium-based reactor nine meters tall underneath the university's stadium. In 1944, he became a naturalized American citizen and enrolled in the Manhattan Project.

In 1934, the Curies generated radioactive phosphorus from ordinary aluminum through neutron bombardment under the reaction $Al + He \rightarrow P + n$, where Al is the chemical symbol for aluminum, He for helium, P for phosphorus, and n represents a neutron. Atoms are made up of both protons and neutrons. Atoms with the same number of protons (isotopes) have the same chemical properties but grow heavier when the number of neutrons increases. Because protons and neutrons comprise the nucleus of atoms, they are grouped as nucleons. It is customary to define the number of protons of an element with a subscript; thus, $_{13}Al$ indicates that aluminum contains 13 protons, helium comprises two protons and it is represented by the symbol $_{2}He$, whereas phosphorus contains 15 protons and the symbol $_{15}P$ indicates it. Note that there are 15 protons at the beginning of the Curies' reaction and 15 at the end. A superscript instead represents the total number of nucleons; thus, ^{27}Al and ^{4}He means that aluminum has $27-13=14$ neutrons and helium has $4-2=2$ neutrons. An atom of helium with two neutrons is known as α particle ($^{4}_{2}He$). Since phosphorous has 30 nucleons (^{30}P), and one is released in the reaction, the reaction described by the Curies is also balanced at the neutron's level. The Curies also noted that the artificially produced phosphorous decayed within three minutes to the more stable atom of silicon with the emission of a positron (e^{+}), a particle with the same mass as the electron but an opposite charge: $^{30}P \rightarrow {}^{30}Si + e^{+}$. Even in this case, there is a balance in the number of nucleons.

Ever since the Curies presented these results, there had been a race to convert natural elements into their radioactive versions. Fermi was the champion of this race since his group systematically bombarded all types of elements to produce artificial radioactive elements. Among the elements he used, there was protactinium; thus, Hahn and Straßmann—who felt being called into action since their element had been used—aimed to replicate Fermi's experiments to determine "whether Fermi had found isotopes of transuranium elements, or isotopes, of the next-lower element, protoactinium." Later, they extended their research, and in early December 1938, Hahn and Straßmann obtained a strange result after neutron-bombardment of uranium.

They believed to be dealing with the reaction $^{238}U + n \rightarrow {}^{235}Th + {}^{4}He \rightarrow {}^{235}Ra + {}^{4}He$, where Th is the chemical symbol for thorium, and Ra stands for radium. In all steps of the Hahn-Straßmann reaction, there are 239 nucleons, but the results that the two researchers obtained did not match the expectations: the radium did not have the chemical properties of radium. Delbrück was involved even in this phase: he was still in the group at the time and thought the problem was due to different isotopes. Hahn was unconvinced of this explanation and wrote a letter to Meitner stating that "our radium isotopes do not act like radium but like barium … perhaps you can come up with some fantastic explanation?" During a walk in a forest in Sweden in the company of his nephew, also a physicist, Otto Frisch, Meitner did just that. Meitner imagined the atom of radium as a large drop of oil and Hahn's process splitting it into two droplets. The droplet model was introduced by Gamow in 1930 and then refined by Heisenberg and Bohr a few years later; it is still in use today in the physical community. In essence, the large atom of uranium (U) had been broken down into the smaller atoms of barium (Ba) and krypton (Kr). Such a simple visualization revolutionized the field of physics and shook humanity to the core. The corollary was more important than the actual explanation of the phenomenon: the energy release.

The uranium atom contains 92 protons ($_{92}U$), but it also includes 146 neutrons with a total of 238 nucleons (^{238}U). Barium and krypton are instead symbolized by $_{56}Ba$ (^{144}Ba) and $_{36}Kr$ (^{89}Kr), respectively. Meitner realized that the reaction carried out by Hahn and Straßmann was $_{92}U \rightarrow {}_{56}Ba + {}_{36}Kr$. There are 92 protons at the beginning and end of the reaction, so there is nothing remarkable here. Things are different when including the neutrons: $^{235}U + n \rightarrow {}^{236}U \rightarrow {}^{138}Ba + {}^{84}Kr + 3n$. In essence, a neutron—especially a slow one—is absorbed by uranium to form a radioactive element that quickly decays into barium and krypton by forming two additional neutrons. The released neutrons could then induce the fission of surrounding ^{235}U atoms, and since each step produced more neutrons than the preceding one, a chain reaction is ignited. However, while there is a total of 236 nucleons at the beginning, there are only $138 + 84 + 3 = 225$ nucleons at the end; thus, there has been a loss of 11 neutrons. The answer to what happened to these missing particles is implicit in Einstein's equation $E = mc^2$: the missing mass m of the neutrons has been transformed into energy E in quadratic proportion to the speed of light c.

The mass of a neutron might look tiny (1.7×10^{-27} kg), but a single gram of uranium contains 2.6×10^{20} atoms—compared with 10^{12} stars present in the Milky Way galaxy—each converting eleven neutrons into pure energy. It has been estimated that the energy released from one gram of uranium corresponds to 8.2×10^{10} J, equivalent to 20.1 tons of trinitrotoluene (TNT), the yardstick of comparison of explosives. Let us consider the following example to know how much damage such an amount of TNT can do.

In 2015, a chemical factory in the harbor of Tianjin, China, exploded, releasing energy equivalent to 21.9 tons of TNT. The blast obliterated the factory itself, leaving a crater of 97 m in diameter, and caused damages to all surrounding buildings in a radius of 2 km—and a toll of 173 victims. The bottom line is that a bit of uranium can cause more damage than a lot of explosives.

Such information was critical since the world was fast approaching another global conflict. Meitner and Hahn agreed to keep secret this scientific breakthrough, which they christened "nuclear fission." It soon became an open secret. Otto Frisch told Bohr, who told the physicist Lèon Rosenberg, who literally told everybody: he not only made a public seminar about it in the US but also replicated and published Hahn's experiment. The now publicly available results fitted with predictions made in 1933 by Leo Szilard, who then started the chain of events that gave birth to the Manhattan Project. Meitner became known as the "German Marie Curie."

Following the racial laws and Fermi's hasty departure, the Panisperna's group vanished. Majorana suddenly disappeared, and, to this day, nobody knows what happened to him. Segrè followed Fermi to the US, also joining the Manhattan Project. Even Ugo Fano went to the US. Pontecorvo instead flew first to Paris (where he worked at the Curie's Laboratory), then to the US and Canada. In 1948, he became a British citizen. He worked at the Atom Energy Authority in Berkshire, but Pontecorvo and his family suddenly went to Helsinki during a holiday in Italy and disappeared. Pontecorvo surfaced five years later in Moscow as a member of the Soviet Joint Institute for Nuclear Research by giving interviews about the peaceful use of atomic energy. Pontecorvo always denied having given away any of the many secrets in atom research he gained during his work in crucial sectors in the West, but few believed him. Rasetti went even further than Pontecorvo by refusing to contribute to developing an atomic bomb. He was convinced that "nothing good could come from new and more monstrous means of destruction … no matter how perverse the Axis powers were, it was clear that the other front was sinking to a similar moral (or immoral) level." Thus, he declared himself "disgusted with the latest applications of physics" and went to Canada to study paleontology, becoming a world leader in the field.

Similarly, Lise Meitner refused to participate in the Manhattan Project: "I will have nothing to do with a bomb," she declared, but ironically, the Americans dubbed her the "German mother of the bomb." Years later, Hahn became the promoter of one of the most compelling peace appeals, the Mainau Declaration (1955), signed by 52 world-renowned scientists, recognizing that "a mentally ill or power-obsessed dictator could … doom the civilized world … in a bomb war there are no longer victors and vanquished." Some speculated that Ettore Majorana outperformed all of these scientists in his opposition to nuclear warfare. His disappearance, rather than being the outcome of suicide—the most accredited hypothesis—could have been a form of self-exile from nuclear research because he understood, before anyone else, that such research would have led to mass extermination.

Following the introduction of the racial laws, the Luria family did not have the means to expatriate and was hidden from fascist violence through some Gentile—Italians of no Jewish ancestry—friends until the end of the Second World War. Luria's parents agreed that it was safer for their son to go abroad. Thus, in November 1938, Luria was hosted by a family friend, the physicist Sergio de Benedetti, in Paris. There, Luria met Fernand Holweck, an expert in radiation biology at Curie's Laboratory. Holweck managed to obtain a scholarship from the French National Research Fund. At Curie's Laboratory, Luria was reunited with his former colleague Bruno Pontecorvo and could continue his phage work. "The work was exciting and demanding," he recalled. "I felt pleased with myself, discovering in my own personality an element of adaptability and efficiency." In Paris, Luria was also able to publish his research with Rita. Due to the racial laws, Rita was forbidden to co-author a publication with a Jew, but such a law did not exist in France at the time.

Nonetheless, even "Paris was a dull experience." Although living just a few blocks away from the scandalous writer Henry Miller, Luria was "not even capable of taking full advantage of the sexual freedom so much greater in France than in Italy." Regardless, there was little time left for such frivolities. At the beginning of September 1939, Germany, after having swallowed Austria and Czechoslovakia in 1938, went a step too far by invading Poland, an act that ignited the Second World

War. Luria enlisted for military service, but he was never drafted. Nonetheless, he joined the war effort by collaborating on research projects in the Holweck group, such as developing high-quality periscopes for tanks and naval defense systems to avoid magnetic mines.

Luria also acted as a consultant in the process of patenting a novel additive for airplane fuel. The patent did not go forward, but the revenue he gained in the administrative process was enough to pay for his passage to America. This patent was a strike of luck because German troops, after Poland, had also conquered the Netherlands and Belgium and were penetrating fast into France. The Germans entered Paris on the 14th of May 1940. Then began what has been called the "grande fuite" ("the great escape"), the exodus of the French population. "By then the Parisians had been leaving for days … the scene of Paris being emptied of its people was heartbreaking." Luria had to flee that evening on a bicycle trip to Marseilles. It was "a month through a hallucinating summer landscape … in my peregrinations, I met all kinds of people, from gruff, hostile peasants to tender landladies." Once in Marseilles, he was involved in the

> battle for visas … the only protracted ground battle to be fought in France that year. One had to be ready with a French exit visa, Spanish transit visa, Portuguese transit visa, and these required an entry visa from one overseas country … For me, alone and young, the battle of the visas was one more experience … some could not take it, and committed suicide in front of the often closed doors of the consulates.

After crossing Francoist Spain, he arrived at the New York harbor on the 19th of September 1940, where the *Nea Hellas* was rammed by a Norwegian freighter. "We had left our first mark in the promised land," he commented.

4.4 THE EMPEROR'S NEW CLOTHES

"Did you see them?" Delbrück asked Lillian Vaughan Sampson, who had married Thomas Morgan in 1904. A few days before, she had participated in a seminar on phages that Delbrück and his colleague Emory Ellis had organized. Like all other attendees, she claimed to have seen the halos in the bacterial lawn that unveiled the presence of phages. "You know," she admitted, "the light was very poor." Delbrück laughed and told her, amused, that the situation resembled Andersen's tale about the emperor's clothes. "I am sorry, what tale is that?" she queried. Delbrück replied that, once upon a time, an emperor went along naked, convinced by his cheeky tailors to wear a beautiful dress that appeared invisible only to stupid people. Everybody claimed to be seeing the clothes to avoid being taken for dummies. Mrs. Morgan blushed and smiled back, acknowledging that it was a situation like that, indeed. In any case, he couldn't pretend Americans were aware of this old European tale. This story was probably unknown to many Europeans. He was now a visitor to another country and had to adjust to new customs. Delbrück pledged to buy her a copy of Andersen's omnibus as a memento of such a sketch.

Delbrück had joined the hoard of Europeans who sought safety in the United States. There were millions of them. During its expansion toward the Wild West, the US needed plenty of workforce, but after the spread ended, immigrants became a nuisance. In 1921, Warren Harding's administration introduced the Emergency Quota Act, which imposed a quota of migrants for each country. Similar legislation is enforced in the European Community today. Back then, as now, skilled people—namely scientists—encountered much less red tape than ordinary citizens. The likes of Einstein and Fermi were very much welcome in the US in the 1930s, and prestigious positions in top-ranking universities awaited them. Such a policy of open doors to migrating top European scientists was the tipping point that propelled the United States to the position of technological leader that it has today.

While European migration had been a cornerstone in building the American States, the cause for leaving the homeland in the late 1930s was no longer the absence of jobs but the fascistization of Europe. The social cataclysm and economic disruptions caused by the First World War—which a catastrophic financial crisis would follow in 1929—jointly with inept liberal parliaments and socialist parties emboldened by the Bolshevik revolution were fertile grounds for the growth of fascist movements. The "red peril" was, back then, a clear and present danger: apart from the revolution in Russia, there had been a Soviet Republic in Bavaria, an attempted coup by the Spartacists in Germany, the "Two Red Years" in Italy, and a Soviet Republic in Hungary in May 1919 (led by Béla Kun). The conservative forces began to support the organizations forming to combat the "red peril."

As a result, fascist movements could be found almost everywhere between the two World Wars. France was a particularly fertile ground for the rise of fascism, probably owing to the intellectual freedom that this country enjoyed following the Revolution of 1789. Among the others, Georges Valois founded the "Faisceau," taking inspiration from Mussolini's "Fasci di combattimento"; Charles Maurras established the "Action Française"; Colonel de La Rocque created the "Croix de Feu"; and Henry Dorgères brought upon the "Greenshirts." In Spain, there was the "Falange Española" (founded by José Primo de Riviera); in England, the "British Union of Fascists" (led by Sir Oswald Mosley); in Belgium, "Rex" (established by Léon Degrelle); in Hungary, "Arrow Cross" (created by Ferec Szálasi); in Ireland, the "Blueshirts" of Colonel O'Duffy. The list also includes the "National Syndicalism" in Portugal, the "Ustaša" in Croatia, the "Greyshirts" in Iceland, the "Nationaal Socialistisches Beweging" in Holland, and the "Nasjowal Samlig" in Norway. Fascism was on the rise also outside Europe, for example, the "Union Party" of Rev. Charles E. Coughlin and the "Silver Shirt Legion" of William Dudley Pelley in the US, the "Blackshirts" of Seigo Nakano in Japan, and the "New Guard of New South Wales" in Australia. The most successful and influential of the fascist movement, though, was the "German Worker Party" ("Deutsche Arbeiterpartei," DAP), which managed to bring Europe to the brink of annihilation.

DAP was founded by Anton Drexler, a tool maker from Munich, in January 1919. Like the "Fascio Rivoluzionario d'Azione Interventista" in Italy, DAP was a national socialist movement. "National socialism" was a term introduced by the French author Maurice Barres in 1896 to describe a new breed of socialists. National socialists, like all socialists, desired better working conditions and higher wages. However, unlike classical socialists—who acted following Marx's motto, "proletarians of all countries, unite!"—they wanted these improvements only within their country of origin; thus, they were nationalists. The national socialists were inclined to conquer other countries to enrich themselves by enslaving foreign workers. Unlike the Fasci, though, DAP was characterized by a stark anti-Semitism from the start. Like the Fasci, DAP remained a marginal organization until a charismatic frontman was brought to the national spotlight. In Italy, such "Duce" was Mussolini; in Germany, the "Führer" was Hitler.

Adolf Hitler was born in 1889 in Braunau-am-Inn, Austria. In 1887, his father, Alonis Schicklgrüber, changed his surname to Hitler, and he had no idea who his father was. Alonis' mother claimed that his father was a Jew named Frankenberger from Graz. Alonis married three times; the last wife, Klara Polzi, was Adolf's mother and almost 30 years younger than her husband. At school, the teachers described the young Adolf as a lazy loner—although he later claimed they considered him a natural-born leader—whose only interests were art and history. Alonis died in 1903, and Adolf Hitler went to Linz, living out of his mother's allowance. At 16, Hitler moved to Vienna several times, trying to join the Academy of Arts as a painter, but he was always rejected. His lifestyle was erratic, and, like Mussolini, he always refused to find regular employment. Apart from a life-long fondness for the music of Richard Wagner—a notorious anti-Semite—Hitler became enchanted with the writings of Jörg Lanz von Liebenfels (neé Adolf Lanz) and Georg von Schönerer, who founded an occult movement known as "Ariosophy" that praised the Aryan race—a mythicized race supposedly representing an ideal pure breed of Germans—against "inferior" races such as the Jews.

Hitler's mother died in 1905, and the allowance ran out in 1907. Penniless, he ended up as a tramp, just like Mussolini. In 1912, he inherited money from an aunt and went to Munich, living as a bohemian, to avoid compulsory military service. However, he was arrested for draft evasion and dispatched to Salzburg, where he failed his physical examination. Hitler was sent back to Munich, but when the First World War started, he volunteered in the German, rather than Austrian, army and was enlisted as a courier. Although his commanding officers said he lacked leadership qualities, he was a brave soldier—more of what could be said for Mussolini—who obtained two Iron Cross medals for military valor and survived a poison-gas attack. After the war, he remained in the army and was assigned to the military intelligence service to ensure that Bolshevik's ideals did not take over among the troops.

In September 1919, Hitler attended a DAP meeting at the Sternecker Brau pub in Munich. During the gathering, his polemical oratory impressed everyone, including Drexler, who appointed him head of the party's propaganda department. Hitler's silver tongue was so good that, within a year, he replaced Drexler as party secretary. Hitler changed the party's name to "National Sozialistische Deutsche Arbeiterpartei" ('National-socialist DAP,' NSDAP, or Nazi party for short) and introduced the "swastika" (or crooked cross, "Hakenkreutz")—an ancient symbol of good luck in the Far East culture—as the symbol of the organization. He also established, in 1921, a group of private security guards—a de facto paramilitary organization—known as stormtroopers (Sturmabteilung, SA) or Brownshirts. Stewarding is common among political parties to defend themselves from attacks; using the service order to beat political opponents physically and Jews was instead an NSDAP innovation.

On the 8th of November 1923, the NSDAP organized a coup (or "Putsch") during a gathering at the Bürgerbräukeller in Munich. Hitler, backed by the Brownshirts, forced the governor of Bavaria and the chief of police to declare a fascist revolution and march to Berlin to demand power, as Mussolini had done. Mobs overthrowing a democratic government might appear to be a far-fetched idea, but it can happen even today in Western countries. Unlike the Italian case, though, the Weimar government did not falter and sent the police, which opened fire against the rioters. Although the Putsch was later mythologized to create a halo of bravery around its organizer, the historian Joachim Fest gave another account:

> Hitler fled, leaving behind a few thousand followers and sixteen dead. The legend, obviously put about later by himself, that he carried a helpless child out of the firing line ... has been proven false. Whilst he was hiding at Uffing am Staffelsee ... he declared that he must end it all and shoot himself but the family succeeded in making him change his mind. Soon after he was arrested.

Hitler went on trial in February 1924, using the court of justice as a megaphone to spread his message to the masses. His rhetoric worked even on the judges since he was sentenced to only five years of minimal-security imprisonment and was released on parole on the 24th of December of the same year. He spent a few months in prison, writing an influential book, *My Struggles* (*Mein Kampf*), that presented three main leitmotifs. The first—neglecting to mention his possible Jewish origins—was anti-Semitism, an ever-green for right-wing ideologists. The second was the supremacy of the Aryan race—an empty container that could be filled at will, in this case, pure-blooded Germans. The third was the complaint that the war reparations imposed on Germany were too harsh, omitting that Germany had levied similarly in previous years. For instance, after defeating Napoleon III, Prussia set an equivalent of 12 billion of today's euro for reparations to France (the "French indemnity"). Bismarck commented, back then, that "France being the richest country in Europe, nothing could keep her quiet but effectually to empty her pockets." The indemnity was defined to be equivalent to the one imposed by Napoleon on Prussia in the Treaties of Tilsit in 1807. Thus, war reparations had always been employed as a post-battle weapon, even by Germans. Prussia was indeed forced to pay the equivalent of 31 billion euros, a sum that was agreed to be too much even by the Triple Entente, but the First World War had also been a much more devastating conflict than all the previous ones.

Mein Kampf received poor critical acclaim; for instance, the historian Louis Snyder commented that it was written with the style of a "half-educated man." Yet, the book became a best-seller translated into 11 languages, giving Hitler a wealthy revenue. Remarkably, Hitler did not invent anything: his were concepts wholesale borrowed from the right-wing ideology. The "stab-in-the-back" myth, for instance, was a leitmotif among the conservative and fascist German communities, sprung by the fact that the Prussian war-time propaganda had been so good that most of the German population was unaware of losing the war until the day that they actually lost the conflict. Since the socialists were part of the governmental coalition during the war, the people blamed the Marxists—remarkably, Marx was of Jewish origins—for the empire's collapse under the assumption that the socialists had obtained personal favors to sell their country to the enemy. Even the swastika that Hitler introduced as the NSDAP logo, as the historian Reginald Phelps highlighted, was "a widespread symbol among racists long before Hitler had been heard of." He simply shouted these fascist ideas louder than the others and had the strength, supported by like-minded ruthless people, to put them into action.

After the release, Hitler went into business as usual; the NSDAP remained a marginal movement whose primary political tool was street violence. The main innovation that Hitler introduced was the institution, in 1926, of a personal bodyguard force known as "Schutzstaffel" (SS) under the direction of Julius Schreck, replaced in 1929 by Heinrich Himmler. The SS was an elite corp whose members were hand-picked to meet the high standards of the Nazi ideology: these soldiers were supposed to be pure Aryans—that is, with proven lack of Jewish ancestry, preferably with blond hair and blue eyes—be well mannered—unlike the SA, who behaved just like thugs—and be fervent Nazis. The SS played a central role in spreading and defending Hitler's creed in the years to follow and perpetrated brutal criminal acts during the Second World War.

It was only after the death of the quite capable chancellor Stresemann on the 3rd of October 1929 and the crash of the American financial market 21 days later that the NSDAP took over. Struggling to meet ends after the financial crash, the US voided the loan that Stresemann had obtained to deal with the war repayments without wrecking German society. Consequently, unemployment and social unrest arose again in Germany, and the Nazis were quick to cash the turmoil in their favor. The Weimar Republic had always been plagued by low governability: no single party could reach the majority of the seats in the parliament, and each government was prone to votes of no confidence. Starting from March 1930, it was not even possible to reach any level of agreement in the parliament ("Reichstag"), and President Hindenburg was forced to implement article 48 of the constitution to allow the chancellor to pass laws by decree (without the vote of the parliament). In such a chaotic political landscape, any bundle of votes that could win a majority in the parliament was an immense asset, and the NSDAP became a strategic political resource.

In April 1932, Chancellor Heinrich Brüning of the "Catholic Centrum" party enforced a ban on both SA and SS. Without the support from street violence provided by the Brownshirts, the NSDAP started to shrink, but not fast enough: the Nazis were still a treasured political force. In July of the same year, the aristocrat Franz von Papen deposed the government and forced Hindenburg to appoint him as Prime Minister, aided by a vast network of high-level connections rather than political skills. Kurt von Schleicher, a general of aristocratic ancestry—thus, Hindenburg's peer—was instrumental in the appointment because he persuaded Hindenburg that von Papen could easily be maneuvered as a puppet. The von Papen administration, established in June, was a technical government known as the "Cabinet of Barons"; von Schleicher was appointed Minister of Defense. One of the first deeds taken by the cabinet was to lift the ban on the SA and SS. Flash elections were held the same month to legitimize the government. However, the NSDAP, not von Papen's coalition, received the most votes: 37% of the preferences, corresponding to 230 seats in parliament. The Nazis even surpassed the SPD as the first political force in Germany.

In September, the Cabinet of Barons was dissolved after another vote of no confidence, and in November, new elections were held. The NSDAP plummeted to 33% of the preferences, revealing that the Nazis were losing momentum. Regardless, still being the leader of the largest German party, Hindenburg offered Hitler the chancellorship under the condition that Article 48 should have been suspended: Hitler had to govern by gaining a parliamentary majority, as democracy dictated. Hitler refused: he wanted free hands to rule his way. The chancellorship was given to von Schleicher, who unsuccessfully tried to win NSDAP's support in his cabinet. Unable to reach a parliamentary majority, von Schleicher attempted to convince Hindenburg to establish a military dictatorship. However, the 87-year-old president, even if an army aristocrat, believed in democracy and refused. The new cabinet lasted barely a month (December 1932 to January 1933). Von Papen, who had held secret meetings with Hitler in Cologne to split the power, convinced Hindenburg that Hitler, a semi-literate, was the best choice as a chancellor because they could easily maneuver him, champions of the elite class. This reasoning allured Hindenburg as von Schleicher had convinced him to appoint von Papen as a chancellor just a few months earlier. Thus, Adolf Hitler became Prime Minister on the 30th of January 1933, precisely a decade after Mussolini.

Like Mussolini, Hitler proved to be everything but a puppet. In less than a month, he encountered the first excuse to limit the democratic rule of Germany. On the 28th of February 1933, a Dutch communist, Marius van der Lubre, set on fire the Reichstag. The common knowledge after the Second World War was that it was the Nazis who burned the parliament to blame the communists for trying and making a coup. More recent historical evidence suggested instead that van der Lubre acted alone, but the Nazis, together with a large portion of the conservative class, genuinely believed that a communist coup was underway. While no Bolshevik revolution was ongoing, Hitler convinced Hindenburg to reinstate Article 48 to suspend freedom of speech, assembly, and personal liberty. The communist party was outlawed. The Brownshirts' brutality was unleashed against socialists, communists, and Jews. On the 24th of March, Hitler proposed a law to rule by decree for four years, the so-called "Enabling Act." Only the socialists opposed the proposal, which was praised even by Pope Pius XI. Once the law was passed, Hitler could rule at will, and the parliament's function was effectively abolished. All political parties but the NSDAP were dissolved, and a detention center for political enemies was established in Dachau under the supervision of the SS—it was later transformed into a concentration camp for Jews. The rule-by-decree law was extended in 1937 and then in 1942. The democracy of the Weimar Republic never died: it was transformed into its polar opposite through the legal institutions provided by its constitution. As the journalist Ezra Klein wrote on *The New York Times*: "Democracy does not die in darkness—it degrades through deal-making."

After having gained absolute power, the march of the Nazis to accumulate even more dominion became unstoppable. In April 1933, the secret political police "Gestapo" (Geheime Staatspolizei) was formed under the direction of Hermann Göring; it became an essential means of oppression for the regime. Then, the trade unions, like the political parties, were outlawed. In June 1934, von Papen talked against the Nazi regime in a gathering at the University of Marburg; in retaliation to this insult, the SS later assassinated his speechwriter, and von Papen was sent away as ambassador to Austria, an empty position that neutralized his political influence. The Brownshirts were also dealt with. The SA had remained independent from Hitler's will, which is why the SS force was formed in the first place. With the NSDAP now in charge of the whole nation, the SA was asking for more and more influence.

Furthermore, Hitler, now legitimately in power, did not need the SA's brutal force any longer. Thus, the Brownshirts became a nuisance rather than an asset. On the 30th of June 1934, Hitler unleashed the SS against the SA in what became known as the "Night of the Long Knives": about 200 Brownshirts were killed, including the SA's highest echelons, and many more were arrested.

On the 2nd of August, Hindenburg died. Instead of electing a new president, Hitler held a referendum to appoint himself as Prime Minister and President of the Republic, the Führer of the nation. Hitler gathered 95.7% of the preferences. After years of post-war hardship, Hitler must have seemed to many Germans as the hope of a new life. Even Otto Hahn was bewitched by Hitler's charisma: in an interview with a Canadian journal, Hahn declared, "I am not a Nazi. But Hitler is the hope, the powerful hope, of German youth ... In his daily life, he is almost a saint. No alcohol, not even tobacco, no meat, no women. In a word, Hitler is an unequivocal Christ."

Anti-Semitism, which was the core of Nazi ideology, was implemented very early. On the 7th of April 1933, two weeks after the introduction of the Enabling Act, a bill ("Law for the Restoration of the Professional Civil Service") was approved that initiated the discrimination of Jews. Paragraph 3 of the law stated that "civil servants of non-Aryan descent are to be retired," meaning that everybody—but evidently, not Hitler, who was not sure who his granddad was—who counted parents or grandparents of Jewish ancestry could not hold governmental positions, including academic ones. Later, things became even worse. On the 6th of May 1933, the Nazis organized several bonfires to burn books written by enemies of the State, in particular Jewish authors. On the 15th of September 1935, the so-called "Nuremberg Laws" were implemented through two bills: the "Empire Citizenship Law" and the "Law for the Protection of German Blood and German Honor." Jews were now forced to wear recognition marks: a David's star was to be worn on the outer clothes at all times. The NSDAP arranged an organized assault on Jewish properties on the 9th of November 1938, which became known as the "Night of the Crystals" ("Kristallnacht") after thousands of shop windows were smashed. The same month, fascist Italy approved the racial laws.

The impact of the Professional Civil Service law on the scientific community was massive. Specifically, because most of the scientists involved in quantum physics, beginning with Einstein, were of Jewish origin, the forefront of scientific research was the most affected. About one-quarter of the physicists and 1600 researchers, including twenty Nobel Prize winners, lost their jobs. The loss of high-quality scientists from Germany worried Max Planck, who met the Führer on the 6th of May 1933 to complain about the scientific bleeding caused by the civil servants law. He told Hitler that not all Jews were "worthless;" some, instead, were "valuable to mankind." To our modern taste, such a remark might already sound too discriminatory, unworthy of a man of science. To Hitler, it was altogether too little: he went berserk and shouted at Planck that "a Jew is a Jew!"

The "Bonzen" were also part of the discrimination. Max Born, who "never felt particularly Jewish," rediscovered his ancestry the hard way and left for England on the 15th of May 1933, accepting a position at the University of Cambridge first and later at the University of Edinburgh. Wolfgang Pauli, Paul Dirac, and James Franck followed suit. Even Erwin Schrödinger, who did not have Jewish origins, decided to leave Germany, obtaining a position at the University of Oxford. On the other hand, Werner Heisenberg chose to remain in his homeland since he was a pure-breed German. During a tour of scientific conferences in the summer of 1939, Heisenberg met Enrico Fermi at Columbia University. Fermi proposed Heisenberg to relocate to the US, "You would be on the side of decency," Fermi tried to persuade him. "It would feel like treason," Heisenberg responded. "Germany needs me."

Nonetheless, life in Germany was difficult also for Heisenberg, for he was accused of being a "White Jew." He was forced to ask Himmler, a family acquaintance, for support. After obtaining protection from the SS chief, Heisenberg was appointed director of the KWI and then joined the "Uranverein" ("Uranium Club"), an association of top-rated German scientists researching applications for atomic energy.

In September 1941, Heisenberg went to Copenhagen to meet his former mentor, Niels Bohr, who also has Jewish origins. Heisenberg showed Bohr some drawings for a nuclear reactor. Bohr feared it was a sketch of a process to produce uranium for an atomic bomb. After all, Meitner and Hahn had introduced the concept of nuclear fission three years earlier. Bohr suspected that Heisenberg wanted to enroll him in the project, so he cut the meeting short. After the Nazi empire had swallowed Denmark, Bohr flew to the US, where he alerted Edward Teller that Heisenberg, one of the most excellent experts in quantum physics, was working on a nuclear project, possibly constructing an atomic bomb. Such a warning will put in motion the chain of events that gave birth to the Manhattan Project, which included Bohr as a collaborator. After the war, it became clear that a Nazi atomic bomb was never in sight. Unlike the massive coordinated effort of the Manhattan Project, the German scientists of the "Uranverein" acted in competition with one another and did not even manage to construct a functional nuclear reactor. However, such a misunderstanding of German capabilities ushered in the atomic age.

The former Delbrück's mentor, Lise Meitner, was also overwhelmed by the Nazi anti-Semitism. Although she was Lutheran, she had Jewish origins. She was shielded from the laws of 1933 and 1935 by her Austrian citizenship. However, Hitler annexed Austria ("Anschluß Österreichs") on the 13th of March 1938, and consequently, Meitner became a Nazi subject. She tried to remain in her place at the KWI as long as possible, even because her mentor Max Planck had been appointed director of the KWI, but then she realized that she was genuinely jeopardizing her life by staying in Germany. Eventually, she risked fleeing to Denmark with an expired passport. Given her financial straits—she has been working virtually for free at the KWI—Hahn gave her a diamond ring he inherited from his mother to help her escape. At the border, the Nazi officer sequestered her document for a check-up. "My heart almost stopped beating," she recalled, because "I knew that the Nazis had just declared open season on the Jews." Luckily, after "ten minutes that seemed like so many hours," she was given back her passport, and she could reach Stockholm, where Bohr found her a job at a university in Sweden. The Nazi discrimination did not affect only Meitner. When the Society of German Chemists became a Nazi-controlled organization, Straßmann withdrew from it; he also refused to join the Nazi party, losing the prerogative to obtain the habilitation for teaching and access to jobs in the chemical field.

Max Delbrück, too, although of non-Jewish origins, realized that academic life in Germany was no longer possible. The regime had transformed the exam for the habilitation to teach at the university into an assessment of the level of commitment to the Nazi cause. Delbrück attended a preparatory school, a "Dozentakademië, an indoctrination camp," he recalled, in Kiel, on the Baltic Sea, where "after three weeks of 'free' discussions, they decided whether you were sufficiently politically mature." For him, the process was a sort of field trip: "we lived on a very nice estate with a large park ... and in a way it was a marvelous thing, because it was the first time in my life that I got thrown together closely with people from other disciplines." But the picnic was not a successful one:

> of course there was the business of having these wonderful lectures by reliable party members ... I obviously was too incautious, and I was informed afterward that I wasn't quite mature enough. So I tried again [the following year]. The next time it was in an equally beautiful place, Thüringen ... still I must have shot my mouth off. It must have been transparent that I wasn't in great love with the new regime ... it was pretty clear that a university career was not likely to be open for me.

Luckily for Delbrück, the RF established a special fund supporting German scholars discriminated against by the regime. The scheme, which ran from 1933 to 1945, provided about 300 scholarships not only to Jews but also to other researchers who lost their access to the European academic system. Delbrück was not only eligible, but having already gained a Rockefeller scholarship, he was well positioned to win another. He was contacted by the representative of the Foundation for Europe, Dr. Harry Miller, in 1936 to explore whether Delbrück might be interested in a new scholarship. Delbrück jumped at this opportunity. Initially, he wanted to visit the geneticist John Haldane in England, but no position was available there. Thus, Delbrück proposed to visit Thomas Morgan in Pasadena and Milislav Demerec in New York to improve his understanding of genetics. The RF was quite interested in genetics; thus, the plan was enthusiastically accepted. Lise Meitner also agreed to the trip, and the German authorities promptly provided him with a visa. Leaving Nazi Germany was, for Delbrück, neither a political stance nor the result of active discrimination; instead, it was simply a job opportunity. "I don't think that it was anything to my credit that I left," he recalled. If he had remained in Germany, he probably would have gone along like the other scientists—for instance, Hahn or Heisenberg. "Many nasty things have been said about those who could have left and didn't leave," he said. More than Nazi collaborators, they were simply people doing their jobs.

4.5 A VISIT TO THE BASEMENT

Delbrück arrived in New York in September 1937 and joined Demerec's group at the Cold Spring Harbor Laboratory. Demerec put him to dissect larvae of *Drosophila*, a job that did not excite the young researcher. In addition, autumn was a sort of low season for the laboratory, and "nobody really talked to each other, so, after working hours, there was just nothing to do except for a walk in the woods." Unsurprisingly, Delbrück quickly joined Curt Stern at Johns Hopkins University, then Wendell

Stanley at Princeton University. In 1935, Stanley had crystallized the tobacco mosaic disease virus (TMDV), showing a distinctive characteristic of this virus, and probably all viruses: TMDV could be crystallized as other inanimate molecules, such as graphite, but the crystals of viruses retained—up to a point—their infectivity. Delbrück had read about these crystallization experiments in Germany and even discussed them with Niels Bohr. Both physicists agreed that viruses might be for microbiology what hydrogen was for physicists: a simple model that could open the gates of understanding life's intimate essence. Therefore, the stop in Stanley's laboratory was a crucial step in Delbrück's journey toward becoming a biologist.

Afterward, Delbrück joined the laboratory of Louis Standler and Barbara McClintock at Columbia University. Standler had applied ultraviolet (UV) light to mutate corn, just like Timofeev-Ressovskij and Zimmer had used X-rays to mutate flies. Delbrück went along very well with Standler but clashed with McClintock, who dismissed the young student with a bitter comment: "he didn't understand genetics." McClintock was studying the role of centromeres and telomeres in the replication of chromosomes and, a decade later, demonstrated the existence of transposons, genetic elements that can move between different loci of the genome.

Delbrück arrived at the California Institute of Technology ("Caltech") in Pasadena on the 15th of October 1937, just a month after entering the US. Morgan was very cordial with him and assigned him to the research group of Alfred Henry Sturtevant, who had described the first genetic map—the sequence of genes present in a specific chromosome—of *Drosophila melanogaster* in 1911. Sturtevant assigned Delbrück to re-examine the genetic map of chromosome 4 because the previous experimental data were not wholly satisfying. Delbrück had to read the literature to catch up on the argument but struggled to understand the geneticists' jargon.

Delbrück socialized very readily at Caltech. In particular, he befriended one of Sturtevant's collaborators, Calvin Bridges, who Delbrück described as a "hippie," almost two decades before the hippie movement reached its apex. Nonetheless, there was an abyss between Delbrück's approach, which was based on mathematical abstraction typical of theoretical physics, and that of the Caltech geneticists from the medical field. Morgan invited Delbrück to introduce himself to the department by discussing his previous work. Delbrück started his speech with the phrase, "let us imagine a cell as a homogeneous sphere," causing a burst of hilarity in the audience.

Nonetheless, somebody, another student at Caltech, was ready to accept Delbrück's radical approach to biology. The two young scientists only needed a chance to meet. Back in the laboratory after a field trip to Arizona and New Mexico together with the plant physiologist Fritz Went, Delbrück learned that he had missed a talk by a fellow student whose interest was in virology, specifically in the biology of bacteriophages. Delbrück, intrigued by the crystallizing capability of viruses, decided to meet this speaker in person. The student's name was Emory Leon Ellis, and his laboratory was in Caltech's basement.

Ellis was born in 1906 in Grayville, Illinois, and gained a Ph.D. in biochemistry at Caltech in 1934, obtaining a position afterward at the Food and Drug Administration (FDA). In 1936, the director of Caltech, Robert Millikan, approached Ellis with a research grant provided by Dr. Seeley Greenleaf Mudd, the son of a wealthy miner. Mudd had recently graduated from the Harvard Medical School and was interested in the antitumoral potential of X-rays. Mudd convinced her mother to establish a fund for basic research in oncology. On the other hand, Ellis was deeply interested in oncology and wanted to determine the intimate mechanisms that drove cancer development.

Nonetheless, he understood that working with animals such as monkeys or mice was too cumbersome: "to make even a very approximate assay," he said, "required a large animal colony, with all its attendant problems and expense ... [and] a long period of time ... compared to the few hours required for an assay on phage suspension." Ellis knew tumors could be efficiently induced by infecting chickens with a retrovirus isolated by Payton Rous or rabbits with a papillomavirus isolated by Richard Shope. Both researchers were based at the RF and later won a Nobel Prize.

As a biochemist, Ellis had no background in microbiology and learned about the biology of bacteriophages as an autodidact. He obtained a strain of *Escherichia coli* known as K-12 from Carl Lindegren, a colleague at Caltech, who acquired it from Edward Tatum, who got it from Charles Clifton, both at Stanford University. *Escherichia coli* was first isolated by the German-Austrian pediatrician Theodor Escherich, who isolated it from newborn stools and described it in his dissertation of 1886 with the name of *Bacterium coli commune*, renamed *Bacillus coli* in 1895 and finally, *Escherichia coli* in 1919 in honor of his discoverer on a proposal by the Italian Aldo Castellani and the English Albert John Chalmers, co-authors of the book Manual of Tropical Diseases. This bacterium, which was virtually unknown in the 1930s—but is widely used today—had the advantage of growing readily in standard laboratory conditions. Interestingly, *E. coli* was isolated by Escherich not because this bacterium is particularly abundant in feces but because it grows easily, unlike most other bacteria that die outside the colon. The strain K-12—the K stands for "Kapsel," the German word for "capsule"—was first isolated in 1922 from a diphtheria patient from Palo Alto, California. K-12 had the peculiarities of carrying a prophage—a phage's genome integrated into the bacterial chromosome—together with additional plasmids, and it was capable of conjugation, all features that will later help the understanding of molecular biology. With his wife's help, Ellis isolated phages from samples obtained from the Pasadena sewage treatment authority and applied d'Hérelle's methods to count the viral particles.

When they met, Ellis explained to Delbrück the life cycle of phages, which at the time, despite the exhaustive work of d'Hérelle, were still considered enzymes by the mainstream scientific community. In the 1930s, biology was still undiscovered, and genes were supposed to be made of proteins rather than nucleic acids. "This seemed to me just beyond my wildest dreams of doing simple experiments on something like atoms in biology," Delbrück recalled.

Having quickly clicked, Delbrück and Ellis decided to do some work together. Morgan readily approved of this collaboration, so the two young scientists started analyzing the growth process of phages. It was "a tremendous time because it was all really new," Delbrück said. The article derived from their collaboration is widely considered the birthmark of molecular biology nowadays. In it, the two scientists essentially repeated d'Hérelle's experiments on the growth of phages, which they called "one-step growth."

First, they looked into phage growth in solid media. Growing and counting phages in a Petri dish necessitates forming a homogeneous layer of bacteria all over the dish's surface, similar to having a lawn in English grass. Since phages cause the bacteria to explode, their growth would show as holes in the bacterial layer, as pits in English grass can infer the presence of moles. Bacteria can be mixed with phages and disseminated on the surface of a Petri dish, or phages can be spread on a Petri dish first, followed by bacteria. Ellis and Delbrück found that these two approaches provided essentially the same results, even if the former method increased the number of plaques by 20%. Mixing bacteria and phages before plating increased the chances of interaction between the host and prey but did not alter the interaction mechanism. They also noted that increasing the agar concentration in the Petri dish reduced the size of the plaques but not their count: they counted 392 plaques on average in 0.7% agar plates, 423 in 1.5%, and 438 in 3.0%—to be fair, based on the data provided by Ellis and Delbrück, there is a difference between 0.7% and 3% agar concentration (the p-value obtained by student t-test is 0.008 compared with a significant level of 0.025) due to a trend in increasing plaque numbers with concentration (Figure 4.2). On the other hand, the diameter of the plaques passed from 2 mm at 0.7% agarose to 0.2 mm at 3% agarose. The authors also stated that "changes in the concentration of bacteria spread with the phage on the agar plates had no important influence on the number of plaque obtained."

Secondly, Ellis and Delbrück analyzed the replication of phages on a liquid medium. Phages were adsorbed to bacteria in liquid suspension, and then the number of plaques counted from aliquots withdrawn at specific time points. The trick was to start the experiment with very few phages to avoid a quick complete lyse of the bacterial suspension. The authors noted that "if the suspension with the added phage was allowed to stand 5 minutes at 37°C (or 20 minutes at 25°C) the number of plaques obtained on plating the suspension was found to be 1.6 times higher." However, "this initial rise is not to be confused with the first 'burst' which occurs later and increases the plaque count 70-fold" and was supposed to be due to "an increase in the probability of plaque formation." Like d'Hérelle, Ellis and Delbrück had found that the phages replicated in steps, not continuously like bacteria: each step increased the virus count 70-fold instead of the doubling fashion observed in their hosts. The young researchers calculated the rate of adsorption of phages to their hosts using the equation $\dot{P} = -\delta PB$, where $\dot{P}$ is the number of phages attaching to the bacteria in the unit of time, P is the concentration of phages, B is the concentration of bacteria, and δ is a constant that is characteristic of the phages and depends on the temperature (it was 1.2×10^{-9} mL/min at 15°C and 1.9×10^{-9} mL/min at 25°C in Ellis and Delbrück's case). In addition, the authors reported that

> a single bacterium can adsorb many phage particles. The subsequent growth of phage in these 'multiple infected' bacteria might conceivably lead to (a) an increases in burst size; (b) a burst size at an earlier time, or (c) the same burst size at the same time, as if only one of the adsorbed particles had been effective … No significant difference was found either in the latent period or in the size of the burst. The bacteria which had adsorbed several phage particles behaved as if only one of these particles were effective.

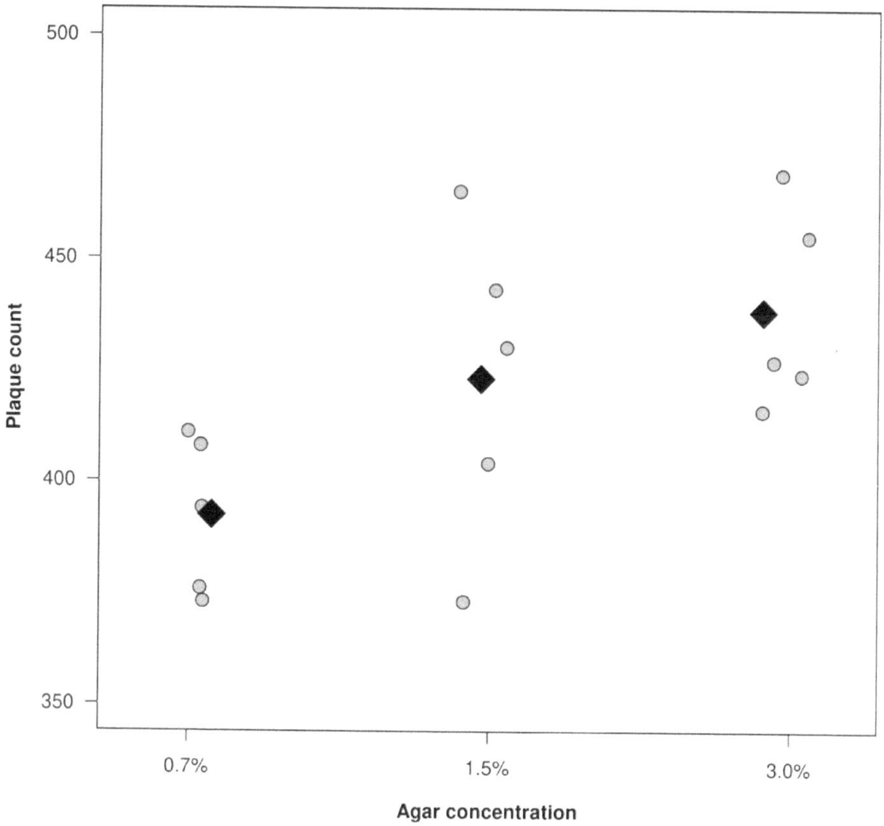

FIGURE 4.2 *Plaque counting due to agar concentration.* Plot of the number of plaques based on the amount of agar mixed in the solid medium. The data are based on the values reported by Ellis and Delbrück in 1939 for 0.7% (394, 408, 376, 411, 373 plaques), 1.5% (373, 430, 443, 465, 404 plaques), and 3.0% (424, 427, 455, 416, 469 plaques) agar over five experiments. The raw data are represented by dots and their average by the diamonds. Using the statistical test known as student *t*-test, the plaques counted at 0.7% and 1.5% did not significantly differ (p-value = 0.137), as did not between 1.5% and 3.0% (p-value = 0.448), but the counts at 0.7% and 3.0% did (p-value = 0.008). To be significant, the p-value of these tests had to be lower than 0.017 (a base value of 0.05 divided by three comparisons).

Ellis and Delbrück calculated the efficiency of plating—today referred to as "plaque-forming units" (PFU)—using Poisson distribution, in parallel to, but independently from, Rita and Luria, by calculating the number of plates infected with only one phage. By diluting a phage suspension and counting the number of dishes with no plaques, Ellis and Delbrück determined the plating efficiency by comparing two attempts. For instance, they reported 23 out of 40 plates without phages; thus, a proportion of absence of phages equals $p_0 = 23/40 = 0.57$. From this, they could determine the initial number of particles with the equation $n = \ln(p_0)$; thus, $\ln(0.57) = 0.56$ phage in a sample of 1 mL.

Although the paper was not scientifically original, it cleared the fog of the enzyme theory introduced by Bordet and still widely believed in the US. The young authors noted that,

> our experiments confirm in the main the picture proposed by d'Hérelle, according to which a phage particle grows in the following way: it becomes attached to a susceptible bacterium, multiplies upon or within it up to a critical time, when the newly formed phage particles are dispersed into the solution.
>
> In the plaque counting method a single phage particle and an infected bacterium containing any number of phage particles will each give only one plaque. This method therefore, does not give the number of phage particles but the number of loci within the solution at which one or more phage particles exist. These loci will hereafter be called 'infective centers' … we shall call the fraction of infective centers which produces plaques the 'efficiency of plating.'

Ellis and Delbrück criticized the experiments based on liquid media, carried out, for instance, by Krüger and colleagues, because it did not provide enough sensitivity to observe the step-growth of phages. For example,

> let us take a sample of a growth mixture in which is suspended one infected bacterium containing fifty phage particles. If this sample is plated, it can show but a single plaque. However, if the sample is assayed by the lysis method [in liquid medium], this single infective center soon sets free its fifty particles … and the time required to attain lysis will approximate that of fifty free particles rather than that for a single particle.

The article by Ellis and Delbrück was even reported by the *Los Angeles Times*. A few weeks after the first phage seminar, Ellis gave another talk on phages and brought some Petri dishes to show what plaques look like. The plates "were passed around and everybody said 'Ah!'," including Morgan's wife. "It turned out that nobody had been able to see them," Delbrück commented. "Everybody had taken it on faith that there were plaques there, which I thought was quite hilarious."

The team-up between the two young researchers did not last long, though. After about a year, Ellis was forced to return to studying oncogenic viruses in mice rather than phages. Later, Ellis and Seeley Mudd disagreed on continuing the research. The latter wanted to establish a large menagerie to overcome the doubts raised by the former in using animals by brute force. Such a disagreement led to the termination of Ellis' grant in 1941. Afterward, Ellis became Executive Director of the Office of Industrial Associates in Caltech between 1963 and 1965, retired in 1969, and passed away in 2003.

4.6 FLYING SOLO

Unlike Ellis, Delbrück was given plenty of time to pursue his research. The context here resembled Twort and d'Hérelle about two decades earlier: the former was denied the opportunity to investigate his discovery, and the latter could pursue his exploration to the full. Regardless, Morgan was enthusiastic about the results of his young students and presented a very positive report to the RF, suggesting the extension of funding for Delbrück. The Institute agreed, and another year's tenure was provided, then renewed in the autumn of 1938.

Fully primed in the laboratory techniques needed for phage research by Ellis, Delbrück could continue to investigate the mechanisms underlying the virus's replication, a research question that fully matched the RF's goals. Delbrück wrote three papers, the only ones he made as sole author, that expanded his characterization of phage growth. These studies were aimed at countering the enzyme theory of the Belgian group. Based on Bordet's model, Northrop pushed forward, well into the late 1930s, the similitude between enzymes derived from inactive precursors ("zymogens") and the formation of bacteriophages. "Pepsin and trypsin are immunologically distinct from their precursors just as phage and virus are immunologically distinct from their host," he stated. Northrop also observed that "the time required to cause lysis is inversely proportional to the log of the phage concentration … as is usually the case with enzyme reactions." He also reported a coliphage's molecular weight of "about 500,000" Dalton (Da)—the unit of measure of bio-molecules weight. Such a weight might appear enormous compared with the 28,000 Da of trypsin and the 4.2 million Da of tritin, the biggest human protein—incidentally also an enzyme. However, it was well below the value found for the "tobacco mosaic virus protein, which is known to have a molecular weight of 17 million," as he admitted, the weight of phage lambda DNA alone (33 million Da) and that of phage T4 head's alone (194 million Da). Northrop went further, describing the properties that he—and the Belgian Group—believed were accountable to phages: "If a solution of this inert protein is inoculated with a trace amount of active trypsin all the inert protein is gradually converted to the active enzyme, trypsin. This reaction is autocatalytic just as is the rate of increase of bacteriophage."

In the first study, Delbrück extended the analysis of the one-step growth curve, including direct visualization of the bacterial lysis: "a map was drawn of about 100 bacteria in the field of vision and these were checked every 5 minutes for changes." He reported that bacteriolysis was qualitatively dependent on the ratio between bacteria *B* and phages *P*:

> *P/B* = 1. No changes in the shape of the bacteria were observed … Up to about 30 minutes … only a few bacteria disappeared, the others showed no changes. Between 30 and 70 minutes 70 per cent of the bacteria disappeared. They disappeared by a process of fading out, *without noticeable change in shape.* The fading out takes about 2 minutes.
>
> *P/B* = 200 or greater. The bacteria kept their normal size and rod shape up to about 20 minutes. Then suddenly within 1 or 2 minutes the large majority was transformed into spherical bodies.
>
> *P/B* between 1 and 200. For these intermediate cases observations on agar plates showed not an intermediate type of lysis but a gradual shift of the *fractions* of bacteria that are lysed by fading or by swelling respectively.
>
> [There are] two entirely different types of lysis, which we designate as 'Lysis from within' and 'Lysis from without.' *Lysis from without* is brought about almost instantly by adsorption of phage at a threshold limit, which is equal to the adsorption capacity of the bacterium. No phages are liberated … *Lysis from within* is brought about by adsorption of *one* (or few) phage particle(s) … When the threshold value is reached, and not before, the phage is liberated by a sudden destruction of the protoplasmic membrane.

In essence, Delbrück described what now goes under the name of "passive" ("lysis from without") and "active" ("lysis from within") phage therapy. In the former case, the bacterium is overwhelmed by a massive phage attack. The bacterium dies almost immediately, but there is no active replication of the phage. In the latter case, one phage particle infects the host, followed by active replication. Then, the bacteria cell lyses when a determined number of virions (equal to the specific burst size of the phage) is reached. The P/B ratio is now called "multiplicity of infection" (MOI). Thus, phages actively replicate themselves at a low MOI. Moreover, Delbrück, using Poisson distribution, noted that "if $P/B = n$ there will be a fraction e^{-n} of the bacteria uninfected." It was the same result reported by Luria and Rita.

The second part of the solo trio was "the study of a detail of the adsorption process that had not previously received attention, namely the dependence of the rate constant of adsorption on the physiological state of the bacterial host." Essentially, different groups reported varying phage adsorption rates. For instance, "Krüger reports no difference between the rates of adsorption of live and killed bacteria, whereas [Martin] Schlesinger finds a 2.5-fold greater rate for the live than for the heat-killed bacteria." Delbrück noticed that the "difference may be due to the fact that Krüger measurements with live bacteria were done at 10° … whereas Schlesinger measured at 37°." Delbrück highlighted that "such variations in rate" might arise "from differences in the physiological state" of the hosts. Thus, he studied the adsorption rate under various conditions, including temperature and bacterial density.

Today, it is well-established that phage replication is interwoven with the growth rate of bacteria. For instance, fast-growing bacteria have a larger surface than starving bacteria, increasing the chance of encountering phages. In addition, fast-growing bacteria have a well-developed protein synthesis apparatus that the phages can hijack. Thus, in culture media rich with nutrients, it has been measured that the eclipse phase (τ)—the time between the infection and the production of viral components—could be as short as 23 minutes in phage T4 infecting *E. coli*, with a burst size of 110 virions per host. In nutrient-poor media, the latency increased to 136 minutes with a burst size of 12. Thus, fewer viruses are produced in a more extended period. The dilution of bacteria is also relevant to assessing the adsorption rate because few bacteria have more nutrients than crowded bacteria. Moreover, it has been discovered that phages infecting starving bacteria can establish a "hibernation" state, where a lytic phage infects a semi-starving host, starts encoding its proteins, but then halts its replication until a more favorable intracellular environment—one where nutrient and building blocks for DNA and proteins are available—is restored.

Delbrück also confirmed that Krüger's method of assessing phage activity—the time needed for a small number of phages to lyse a whole material suspension—was not sensitive enough to describe phage replication and was inferior to the plaque method. Delbrück concluded that "the adsorption rate constant of phage to bacterium is found to change between wide limits, depending on the physiological state of the bacterium." The third part of the triptych was a mathematical essay on the statistical fluctuations underlying the measurement of the adsorption rate.

4.7 QUANTUM BIOLOGY

Delbrück obtained another job extension that lasted until September 1939. He had initially planned to go back to Germany via Japan—which was also involved in the war since it had signed a military treaty with Nazi Germany back in 1936 and had carried out cruel expansionist politics since the early 1930s that ignited a long-lasting war with China—but the dismissal of his line manager Lize Meitner from the KWI convinced him that the situation in his homeland was dire. Already in February 1939, he contacted his supervisor at RF, Warren Weaver, to declare his wish to remain in the US, a desire reiterated in June. Nonetheless, his grant expired just at the onset of the global conflict. Despite his complacency with the young scientist, Morgan had no funds for another position. Delbrück had to live with borrowed money to avoid returning to a Europe torn by another world clash.

Due to a crude splitting of the Prussian empire after the First World War, several German communities ended up outside the German nation, creating a "German question" tacitly recognized by the leading European countries. The merger between Germany and Austria in March 1938 looked like an internal German affair akin to the unification of kingdoms within the British Isles. After all, most Austrians had already voted favorably for Germany's unification in a referendum held in 1918. Annexing the German-speaking part of Czechoslovakia in October of the same year was understood as a way to resolve the German question peacefully. By letting Hitler do his business, the European powers—namely England and France—hoped to have granted stability in the continent. On the contrary, instead of quenching the demand for additional territories, the annexations encouraged the Nazis to launch an aggressive expansionist operation within Europe. Nonetheless, one should not consider the British Prime Minister of the time, Arthur Neville Chamberlain, as a sweetheart subservient to Hitler's wishes but rather as a capable and honest civil servant horrified by the carnage of the First World War who was determined to avoid another bloodbath.

The appropriation of the rest of Czechoslovakia in March 1939 demonstrated that Hitler was not so interested in reunifying the German people but rather that he was driven by the desire to conquer the whole of Europe. Another sword of Damocles that hung over Europe's peace was that a piece of Prussia had been handed to Poland as part of the Versailles Treaty. It was transparent that Hitler would soon complain against Poland to achieve the reunification of East Prussia with the motherland, conceivably swallowing up all of Poland in the process. Consequently, Chamberlain signed a military agreement with Poland on the 28th of August 1939, followed by French Prime Minister Édouard Daladier.

After securing a non-aggression pact with the Soviet Union, an ideologically shocking but strategically logical move to avoid fighting a vast enemy army on an eastern front, Hitler ordered the invasion of Poland—Stalin, on the other hand, invaded what was left of Poland and since he was at it, also Estonia, Latvia, Lithuania, Finland, and part of Romania. Hitler hid himself behind a finger by staging a fake Polish aggression against Germany. Himmler's right hand Reinhard Heydrich selected a covert SS commando team to dress up as Polish soldiers, attack a German border post, and kill some German inmates purposely shipped from the concentration camp of Sachsenhausen to provide a *casus belli* that could justify a Nazi military reaction. The invasion of Poland would have looked like a rightful reaction should somebody fall for such a piece of active measure—what we call today "fake news." Hitler was sure Chamberlain and Daladier would not dare implement the just-established military agreement—Hitler mistook Chamberlain's good faith for weakness. Instead, England and, France, Australia, New Zealand, and India declared war on Germany on the 3rd of September.

The Second World War had begun, and within a few months, 33 states around the globe would be engulfed in the bloodiest conflict humanity had ever witnessed.

Having neutralized the risk of an eastern front against Russia, Germany implemented a revised Schlieffen plan (the "Sichelschnitt" plan devised by General Erich von Manstein) to conquer France. In April 1940, Germany invaded Denmark in only six days and Norway, despite the heavy British support, in a couple of months. The disastrous Norwegian campaign led to the dissolution of the Chamberlain cabinet and the election of Winston Churchill as the new Prime Minister on the 10th of May 1940—although Churchill organized the military defense of Norway. On the same day, Germany attacked France—swallowing the neutral states of Belgium, Luxembourg, and the Netherlands on the way—conquering Paris on the 14th of June and forcing the country's capitulation on the 23rd.

Hitler reenacted the First World War armistice by moving the train wagon where the peace treaty between Prussia and France had been signed to Compiègne and forcing the French command to sign an armistice. With this theatrical act, Hitler could prove to the Germans that he had avenged the "stab-in-the-back" of the First World War. The war he had unleashed was not about conquering the world but simply the rightful conclusion of the previous global conflict. The Nazi Minister Joseph Goebbels remarked that "the disgrace is now extinguished." Ironically, the French population experienced its version of the "stab-in-the-back" narrative. Oblivious to the debacle of the French army due to censorship, the Frenchmen realized they had lost the war at the very last moment. Like the Germans, the French people started to blame traitors in the high echelons of the government and the military establishment. The keyword became "nous sommes trahis" ("we have been betrayed").

The Nazis split France between an occupation area in the north and a puppet state led by Marshal Philippe Pétain—a First World War hero and fervent anticommunist—with the small city of Vichy as the capital. The motto of the French Republic, "Liberty, Equality, Fraternity" was replaced by the more Orwellian "Work, Family, Country." The German empire (the "Deutsches Reich") was now formed by a greater Germany, accounting for 86 million people—compared with the 42 million in France before the occupation and 47 million in the United Kingdom—and a continental empire of occupied countries of an additional 34 million. By early 1940, Nazi Germany appeared to be an invincible war machine, facing just the British empire—because Churchill stubbornly refused to sign an armistice with Hitler—which was resisting with the covert help of the United States.

In the summer of 1940, Delbrück was approached by Linus Pauling, also an RF fellow at Caltech. Pauling confided with his former colleague that he had some reservations—"That's baloney," he commented—on an article written by another of Göttingen's Bonzen, Pascual Jordan. Pauling asked Delbrück whether he would like to join him on a reply on such a paper. Delbrück, who "didn't want to be impolite," accepted and worked on smoothing the edges of the piece written by Pauling. Jordan introduced the concept of quantum biology in the 1930s, and the term was popularized by Schrödinger's *What is Life?* Pauling and Delbrück stated that

> Jordan has advanced the idea that there exists a quantum-mechanical stabilizing interaction … of great importance for biological processes … able to influence the process of biological molecular synthesis in such a way that replicas of the molecules present in the cell are formed.

Bohr's complementarity principle had deeply influenced both Delbrück and Jordan. Bohr remarked that a corpuscle such as an electron could be described either as a particle or a wave according to the experimental method used. Nonetheless, an electron could not be a particle and wave simultaneously. In a similar fashion, one might say that a dog can be considered a pet or a predator, but not both at the same time because a predatory pet is a contradiction in terms. As a result, declared Bohr, the various experimental approaches used in physics complemented each other in portraying corpuscle properties. Another parallel could be that astronomical bodies like the Sun or Jupiter can be investigated using optical or X-ray telescopes: the object is the same, but the information obtained by these instruments is quite different and complements each other, giving us a better understanding of these celestial bodies.

Because the complementarity principle requires some gauging, Bohr, both a physicist and a philosopher, questioned the underlying idea of measurement, expanding his inquiry into the epistemological realm. In particular, Bohr followed the teaching of Immanuel Kant, an influential German philosopher who, in the late 1700s, investigated the basis of human knowledge. According to Kant, matter is made up of constituents that are essentially the atoms of physics. We do not directly understand these constituents, the so-called "thing-on-itself": we can only access the sensory projections provided by our senses or gauging equipment. Hence, a thing-on-itself might not appear the same to different observers. Bohr discussed these aspects with the Danish neo-Kantian philosopher Harald Høffding.

These ideas culminated in Bohr's seminar *What is light?* attended by both Delbrück and Jordan. During the conference, Bohr remarked that "if one looks at even the simplest kind of cell, one knows it consists of the usual elements of organic chemistry and otherwise obeys the laws of physics." However, life is more than the sum of its parts, for, as Bohr added, "one can analyze any number of compounds in it [the cell], but one might never get a living bacterium out of it." This observation is accurate even today: nobody managed to construct a living entity simply by mixing chemical compounds. The most achieved was hijacking a bacterial cell and forcing it to express a synthetic genome. Such a feat was carried out at the John Craig Venter Institute (JCVI), founded by biotechnologist and businessman Craig Venter, who was deeply involved in the Human Genome Project. The researcher at JCVI observed that the simplest bacteria are mycoplasmas and assessed that *Mycoplasma genitalium* has 525 genes, of which only 250 are deemed essential. The team at JCVI synthesized a reduced genome containing only these vital genes, removed the native genome from the bacterial cell, introduced the synthetic genome, and obtained a novel life form christened JCVI-syn1.0 that has been upgraded to JCVI-syn3.0 in later years. Despite such a success, the consensus is that the transition from organic chemistry to living systems (abiogenesis) occurred only once (about 4.3 billion years ago) and in only one place (probably in an abyssal thermal vent on Earth).

In the 1930s, there was a clear distinction between "reductionists" and 'vitalists.' The former arose from the mechanism philosophy (or mechanistic materialism) of the French thinker René Descartes, who viewed living beings, except for humanity, as organic machines. According to reductionists, living beings are the sum of their constituents, the organic molecules that make up their cells. It is feasible to understand precisely how life functions once the chemical behavior of these building parts is determined. On the other hand, vitalists assumed that living beings contained an additional yet unidentified element that did not belong to chemistry. As a result, biological beings were more significant than the sum of their organic parts. Parallel to the rise of quantum mechanics, a third school of thought, led by the Austrian biologist Ludwig von Bertalanffy, spun off from the reductionist movement: "organicism." The organicists claimed that the vitalist element was simply the reflection of a poor understanding of the laws of physics and chemistry underlying the biological processes. Pascual Jordan belonged to the organicism school, but he developed his approach to the argument. For instance, in 1932, Jordan claimed that

> we know that there are in a bacterium ... a very small number of special molecules endowed with dictatorial authority ... [that] can bring the entire organism to death and dissolution—similar to the way a successfully executed assault against a leading statesman can set an entire nature into a profound process of dissolution.

In 1939, these words assumed a much bleaker outlook. Jordan referred to his "amplifier theory," which stated that quantum mechanics processes occurring at specific loci within a cell generated random—that is, not following a cause–effect fashion—reactions that avalanched into macro-processes reverberating throughout the entire cell.

Jordan sought to introduce an unforeseeable element ("indeterminism") in cellular biology to oppose the strict cause-effect chain underlying materialism, the funding ideology of Marxism and liberalism. Ironically, the young Marx was fond of the Greek philosopher Epicurus exactly because he conceived atoms that swirled on their account rather than in a causal manner, as instead taught by Democritus, thus displaying self-determination. Nonetheless, Jordan found a parallelism between quantum mechanics and the Nazi regime that he believed would revolutionize the whole field of biology.

The quantum spots he conceived within the cell could subdue the cell to their will, just like the NSDAP's political centers were leading the German nation. Jordan, however, failed to recognize a fundamental contradiction underpinning the Nazi regime: despite its reliance on technological improvements, post-Weimar German society remained conservative and superstitious. Thus, in defiance of Jordan's eager membership in the NSDAP in 1933, the Nazi upper echelons did not see him favorably. To the NSDAP's top brass, quantum mechanics would always remain a Jewish science, no matter how Jordan tried to spin the story. Unsurprisingly, Jordan's academic career did not take off. Notwithstanding his outstanding scientific credentials, he became a docent at the University of Rostock in northern Germany, which was not exactly a hotbed of quantum physics or biological research. Moreover, after the war, his engagement with the NSDAP was not well-received, and he again struggled to stay afloat. However, he later became a professor at the University of Hamburg.

Nonetheless, Jordan did not give up his political activism and was elected a member of parliament with the conservative Christian Democratic Union in 1957. In the same year, he promoted the introduction of atomic weapons within the German army, which prompted a reply (the Göttingen Manifesto) signed by some of the most renowned German nuclear scientists, including Born and Heisenberg, that condemned the employment of nuclear arsenals.

In their joint paper, Pauling and Delbrück regarded protein–protein interactions as chemical processes rather than quantum mechanisms. Thus, they stated, "we have reached the conclusion that the theory cannot be applied in the ways indicated by him." They summarized the protein–protein aspects of Jordan's theory:

> Let us consider two identical proteins ... A and B, which interact with each other ... if both are in their lowest states the interaction is normal. If, however, one is in its lowest state and the other in an excited state there occurs a resonance phenomenon ... For one of these symmetry types there is a resonance stabilization and attraction between the two molecules, and for the other a resonance repulsion.

The two physicists then counter that

> we find that under the conditions of excitation and perturbation prevailing in aqueous solutions the resonance interaction could not cause a specific attraction between like molecules ... It is our opinion that the processes of synthesis and folding of highly complex molecules in the living cell involve, in addition to covalent-bond formation, only the intermolecular interactions of van der Waals attraction and repulsion, electrostatic interactions, hydrogen-bond formations, etc, which are now rather well understood.

While atoms can stabilize themselves as molecules by forming a bond orbital, the resulting molecules, especially macromolecules like proteins, are too massive for quantum mechanical description. In essence, Pauling and Delbrück restricted quantum physics by consigning biological processes, such as phage production, to the realm of chemistry.

The quest for quantum mechanical explanations of biological processes eluded Jordan and Delbrück. In 1962, the theoretical chemist Christopher Longuet-Higgins gave an epitaph to this branch of research: "Nowadays, we smile at the concept of a 'vital force' ... when a biochemist begins to use quantum-mechanical language in this nebulous way, we may justifiably suspect that he is talking nonsense." Jordan tackled the issue of quantum biology in the same way as Delbrück did during the preparation of the three-man paper: by studying everything there was to know about biology as quickly as possible. Despite being motivated by Bohr's discourse, Jordan and Delbrück took different approaches. Jordan imposed a political stance on his quest, which alienated other researchers and rendered quantum biology obsolete. In contrast, Delbrück declared in the green paper that his research on the nature of genes was based solely on "experimental investigations." The circumstances were similar to those of Plato and Aristotle: both were students of the same instructor, Socrates, but the former asserted that worldly objects were reflections of ideals beyond our reality, while the latter became a superb experimentalist.

However, neither scientist could adequately explain the role of quantum mechanics in biological processes. Ironically, several occurrences of biological processes chiefly attributable to quantum mechanics have been described today. For example, photosynthesis depends on a fundamental property of quantum mechanics: the quantum leap of electrons from one orbital to another due to the absorption of a photon. Furthermore, enzyme complexes use quantum coherence back-scattering—the mechanism by which input scattered light produces an intense coherent reflected ray due to the addition of the wavelengths—to boost photosynthetic efficiency. Furthermore, enzymes such as aromatic amine dehydrogenase use quantum tunneling, which is involved in the transfer of electrons between molecules, to deaminate amines to form aldehydes. Magnetoreception—the use of the Earth's magnetic field for navigation, as employed, for instance, by birds—also involves chemical reactions directly based on quantum mechanical processes. These enzymatic processes directly tied to quantum physics have no exceptional role in cellular biology; they are just reactions that occur concurrently with thousands of others. Instead, the sum effect of these reactions decides a cell's fate. One chemical reaction might cause a deleterious effect on cell viability, leading to cell death if left alone. Nevertheless, chemical reactions—even those directly attributable to quantum mechanics principles—do not occur in a vacuum. On the contrary, cells have evolved several mechanisms, often redundant and overlapping, to counteract deleterious reactions' effects. For instance, the genome of each human cell undergoes about 30,000 reactions daily that cause double-stranded DNA damage, the most serious of the insults a DNA molecule can sustain. If left alone, such damage can lead to cell death, but the cell has developed a series of sensitive chemical reactions that neutralize the genomic insult.

To continue Jordan's analogy, one can say that a democracy rather than a dictatorship regulates cellular biology. Quantum mechanics is the fundamental mechanism governing atoms' existence and interactions as molecules. Because chemistry is the discipline that details the interchange of electrons between molecules, and quantum mechanics laws best explain electrons, one may argue that chemistry is a subfield of quantum physics. On the other hand, the scale of the interactions is critical: as previously said, the more molecules in a system, the less helpful quantum mechanics becomes in describing the system.

After working on the joint paper with Pauling, Delbrück remained in Pasadena until, on the 21st of December, he received a job offer from Francis Slack at Vanderbilt University in Nashville, Tennessee. Delbrück promptly accepted the offer and "arrived in Nashville on New Year's Eve in a driving snowstorm." What happened was that the RF, given its strong interest in the nascent field of molecular biology and the talent demonstrated by the young German, had agreed with Vanderbilt University to subsidize his salary until 1942. Delbrück was to teach physics, but, he recalled, "in return for a gentleman's agreement that I would have half the time free for research." Thus, "I had an office in Physics, and did some teaching there, and I had a lab … kindly put at my disposal by the head of the Biology Department, [Edwin Eustace] Reinke."

Delbrück attended an American Physical Society convention in Philadelphia on the 28th of December 1940, where he reunited with his old mentor Wolfgang Pauli. Pauli, like Meitner, was an Austrian citizen of Jewish origin who was reasonably concerned about becoming a German citizen following the Anschluss in 1938. Pauli was in a far better situation than Meitner because he lived in Switzerland. However, in 1940, the Switzers anticipated a Czecolovacchia-style invasion from Hitler's side; thus, Pauli decided it was better to be safe than sorry and relocated to the United States, promptly obtaining a chair at Princeton University.

There was another reason Delbrück went to Philadelphia. A young Italian, only a few years his junior, had contacted him to request a meeting at the convention. The Italian physician had read the green paper and expressed his admiration for Delbrück's achievements. Like Delbrück, this Italian scholar was thrilled to be working with phages, a field of research that, as Ellis had learned, was still too marginal to gain widespread acceptance. After months of lone work, Delbrück might finally meet some like-minded scientists with whom to collaborate. The Italian doctor's name was Salvador Luria.

5 Nashville and Bloomington

5.1 *HOMO SCIENTIFICUS*

"Yes, I do!" These simple words sealed the marriage between Max Delbrück and Mary Adeline Bruce, colloquially known as Manny, a "radiant, fun-loving bride," as Luria later described her. It was the summer of 1941; Max and Mary Adeline would live happily together for several decades, giving birth to four children: Johnathan, Nicole, Tobias, and Ludina. Mary Adeline was born in Butte, Montana, but had been raised in Cyprus, where her father, James Latimer, worked as a mining engineer for the same Mudd family that would institute Emory's research grant a few years later. Afterward, the Latimers moved to Beirut, Lebanon, before returning to the US.

Max and Mary Adeline met at a party thrown by the German émigrée Arnold Bergstrasser, her mentor at Scripps College, where she studied art. At the time, Delbrück was required to upgrade his visa from tourist to immigration grade, so he needed to go abroad and re-enter the US. Thus, he decided to take a trip to Mexico to visit some friends; along the way, he was invited to that party. It was love at first sight. Delbrück recited Goethe's poetry to impress her, and she, ten years his junior, found this foreign scientist a most interesting fellow. After the wedding, the Delbrücks befriended the wealthy entertainment entrepreneur Alfred Starr. Since Max was still an instructor, Starr proposed to Vanderbilt University's provost that he cover the difference between Max's current salary and that of a full professor. Essentially, the Delbrücks' income doubled overnight just like that.

The Delbrücks were a prototypical scientist's family: they possessed neither television nor radio sets, owned only a utility car rather than the conspicuous American ones, and did not have a dish-washing machine because they employed the time for cleaning the dishes as a chatting opportunity. Mary Adeline not only carried out all the mundane things of family life so that her husband could concentrate on his science, but she was also engaged in the town's politics and contributed as a freelance to the local newspaper. Such a kind of marital dedication was by no means an exception: for instance, even Albert Einstein's first wife, Mileva Marić, helped her husband draft his first papers in physics without being even mentioned in them. Mary Adeline teased her husband and his colleagues by theorizing about a new human breed: *Homo scientificus*. Even during their short honeymoon, Mary Adeline had the impression that his husband "couldn't wait to go back to Cold Spring Harbor," she commented. She was not far from the truth, for her husband had a compelling reason to return to the CSHL: he and Luria had arranged to start their combined work on phages that very summer.

DOI: 10.1201/9781003657712-5

Meanwhile, soon after arriving in New York, Luria visited his former boss, Enrico Fermi, at Columbia University. Fermi immediately interceded to find him a job, contacting the RF's human resources administrators with a terse, "I believe he will make good use of whatever help you may give him." Very soon, Luria found himself employed at Columbia University in a laboratory with "a spectacular view over the Hudson River and the George Washington Bridge," one of the most iconic of New York's sights. Luria might have been impervious to the charm of Rome and Paris, but he fell in love with New York, considering—like a Woody Allen *ante litteram*—Manhattan "a gray-and-rose cathedral." He had finally found his dimension.

Luria also applied for American citizenship, and during the application process, the immigration clerk asked him whether he wanted to change his name. Luria seized the moment to get rid of Salvatore, "which he never liked." With an act of foreign branding, he split his Christian name as "Salvator-E," then associated it with "Salvador, toreador, conquistador." For the outstanding "E," he "asked the person next in line to come up with a name starting with E," and he got the answer, "Edward." Dr. Salvador E. Luria—Salva for the friends—was born.

Afterward, Luria had met Delbrück with Pauli and George Placzek over a dinner a few days before New Year's Eve 1941 in Philadelphia. Luria commented, "Delbrück struck me as a dominant personality … tall and looking even taller because of his extreme thinness, moving and speaking sparingly and softly but in great precision. His seriousness was occasionally broken by sparks of amusement." During the supper, the physicists spoke about quantum mechanics in strict and fast German, so the young Italian did not understand anything. Nevertheless, Luria and Delbrück spent the first day of 1941 together, sharing notes and planning to meet again in the summer at the CSHL to answer the question, as Delbrück put it, of "how does one phage makes hundred or more phages in twenty minutes or so inside a bacterium?"

Delbrück and Luria convened at CSHL in June 1941, determined to publish few articles but of high quality. Both Delbrück and Luria loved Cold Spring Harbor at first sight. "The place was something of a hermitage," Luria remarked. "The living quarters were monastic, some on the verge of collapse … the tennis court, always overgrown with weeds and pitted with large holes." Nonetheless, given their youthful enthusiasm, it was the "perfect setting, for fun and for work." As a pleasant surprise, Luria also met his old pal Ugo Fano at the CSHL, who had been employed by the Institute's director, Misislav Demerec, as a biophysicist in radiation genetics. Demerec and Fano were characterizing bacteria resistant to phage infection, determining the rate of mutation underlying the development of resistant strains.

One of the first tasks that the European Duo carried out was to look at the phages for real. Such a task would settle once and for all the d'Hérelle-Bodet diatribe: if phages were corpuscular entities, as the former had always claimed, then it would be possible to see them with a microscope. They only needed a mighty lens. Delbrück and Luria were unaware that, in Germany, another researcher had already fulfilled such a quest. Ernst August Friedrich Ruska was born on Christmas Day 1906 and studied at the Technical University of Berlin, patenting magnetic lenses, the essential element of electron microscopy, in 1932. Already in 1931, he had placed two of these lenses in series, providing a magnification of 17× and inventing the first electron microscopy, which he named "hypermicroscope." In 1940, he described the infection of *E. coli* with bacterial viruses, noting that "the phage is a small round particle which, after cautious irradiation, appears homogeneous and rather dense." However, since the war was ongoing, the communications between German and Anglo-American scientists were severed. Thus, the German primacy of this achievement fell into oblivion.

Meanwhile, the Americans built their version of electron microscopy. In 1919, after the end of the First World War, Admiral H. G. Bullard and Commander S. C. Hooper of the United States Navy met *General Electric* president Owen D. Young. They urged him to acquire the *American Marconi Company* (a subsidiary involved in long-distance communications but not an American company) and use the assets to develop its own radio communications subsidiary to improve national security. Young smelled business and promptly agreed, promoting the formation of the *Radio Corporation of America* (*RCA*), which became a media giant. Within 10 years, *RCA* had purchased the Victor Talking Machine Company, the world's largest maker of records and phonographs, and obtained the rights to use the *His Master's Voice* trademark, which featured one of history's most memorable logos: a dog listening to a phonograph. In the early decades of the 20th century, *RCA* was one of the biggest world companies, similar in size to what *IBM* would become after the Second World War. *RCA*, just like *IBM* will do later on, could afford to pioneer high-risk projects. For instance, in 1933, *RCA* had the Rockefeller Center built, one of the most iconic landmarks of New York to this day.

Another high-risk endeavor was launched in 1939 by *RCA* president Vladimir K. Zworykin: the development of a TV set. Vladimir Zworykin was an energetic figure who pioneered the cathode tube application in TV sets. He was born in Murom, Russia, in 1888 and studied at the St. Petersburg Institute of Technology and the Collège de France in Paris. During the First World War, he served in the Russian Signal Corps, emigrated to the United States in 1919, filed a patent for a TV set in 1923, obtained US citizenship in 1924, and a Ph.D. in 1926 from the University of Pittsburgh. Later, he joined the *Westinghouse Electric Corporation* and was appointed director of RCA in 1929. Although the TV apparatus was presented at the 1939 New York World's Fair, it did not pay out in terms of economic gains.

Nonetheless, Zworykin pushed for an additional high-risk endeavor: the electron microscope, which was marketed in 1939 with a 100,000× magnification. To promote its use within the scientific community, *RCA* instituted a research grant and the very first winner was Thomas Foxen Anderson, the son of a Norwegian émigré, Anton Oliver. Anton was an electrical engineer who founded the *Oslo Power* and Light Company as well as the *Anderson Electric Company*. Thus, Thomas grew up with an outstanding knowledge of the electron and its applications. After obtaining his Ph.D. at Caltech in 1936, Thomas Anderson joined the University of Chicago and, in 1937, was offered a research position at the University of Wisconsin to investigate the effect of UV irradiation on yeast cells until, in 1940, he obtained the *RCA* fellowship, which Stuart Mudd, a bacteriologist at the University of Pennsylvania, managed.

One of the critical features discovered by Anderson was that specific phages (namely T4 and T6 but not T2) required a molecule to adsorb to their host (*E. coli* strain B). In particular, the amino acid tryptophan was essential for the proper adsorption of these viruses. Later on, Delbrück investigated this phenomenon in more detail, observing that the requirement for tryptophan was not absolute but raised by mutation. He isolated a strain that did not require additional factors for adsorption (T4), another strain that required tryptophan (T4,11), and another that required both tryptophan and calcium ions (T4,12). In addition, Delbrück reported that "tryptophan is a substance actively metabolized by E. coli," and the metabolic products of the metabolism of this amino acid "actively inhibit the adsorption of the cofactor-requiring phages."

Thus, pre-incubation of *E. coli* with tryptophan reduced the phage titers of T4,11 and T4,12 but not the wild-type T4. Indole, which derives from tryptophan's metabolism, was particularly active in inhibiting phage adsorption, leading Delbrück to remark that "indole plays the role of an antivirus drug, and to our knowledge it is the first drug on record that specifically protects a host cell from a virus invasion."

Anderson proved himself worthy of the *RCA* fellowship: he could tweak the newly invented microscope in ways nobody else could and obtain remarkable images. During the fellowship and the following years, Anderson collaborated with many scientists, achieving, among other things, the visualization of bacterial flagella and cell walls, the attachment of antibodies to bacteria and viruses, and the confirmation of the dimension of viruses (which were previously estimated by diffusion and ultra-centrifugation). Among the people who asked for Anderson's support was Luria, who visited Anderson in November 1941 to ask for microphotographs of phages. They started the joint work the day after Pearl Harbor's surprise attack.

On the 8th of December 1941, the Japanese forces led by Admiral Isoroku Yamamoto attacked the US naval base of Pearl Harbor in Hawaii without a previous declaration of war. Japan had started its colonization in the Far East in 1931, after staging a *casus belli* against China (the "Mukden incident") that prompted the invasion of Manchuria. Since then, the Japanese established an empire from China characterized by brutality against the conquered populations (*furia japonica*) that, if possible, was even worse than the Nazis'. The Japanese soldiers, in fact, routinely practiced massacres and mass rapes, the most notorious of all taking place in Nanking in 1937. After the surrender of France, French Indochina, which became part of the État français, was forced into the Japanese's economic sphere; both the UK and the US imposed a harsh embargo on Japan in May 1941. Thus, Japan and the Western countries were already at loggerheads well before the surprise attack on Pearl Harbor.

On the 6th of December, the US Navy had intercepted a message to the Japanese ambassador in Washington, anticipating a large-scale military operation in the Pacific. The intelligence was passed to the Chief of War Plans, General Leonard Gerow, but he did not take any action on the ground, as the Fleet was already in a state of high alert. Even more troublesome was that the Soviet spy Richard Sorge had passed intelligence to the Russians about an imminent invasion of the Pacific Island. Although the US was contributing heavily to the Soviet war effort, Stalin decided not to alert his ally, Franklin Delano Roosevelt. The Japanese, instead, had carefully planned. First, they signed a defense military pact with Germany and Italy (the "Pact of Steel") in May 1939. Then, in November 1941, they obtained assurances from the Nazi Minister of Foreign Affairs Joachim von Ribbentrop that Germany would have declared war on the US should Japan attack

America. The Japanese avoided alerting their allies that they intended to strike the US within a month. Göbbles commented that the attack came "like a bolt from the blue," hardly samurai's ethics, but practical.

The attack on Pearl Harbor prompted the entry of the United States into the war, a fact that was enthusiastically welcomed by Hitler, who thought a double front (in the Pacific and Europe) would have weakened Roosevelt's—and consequently Britain's—war force. On the contrary, the direct involvement of the United States in the conflict represented a historical watershed: without the US intervention, the United Kingdom would have never won the war. On learning of the engagement of the US in the conflict, Churchill declared that he "went to bed and slept the sleep of the saved." Ironically, Roosevelt authorized the institution of the Manhattan Project, which led to Japan's demise the day before the attack on Pearl Harbor.

Back in the day, though, the Japanese war machine, just like the Nazi's, seemed unstoppable. Simultaneously with the strike on Pearl Harbor, the Japanese launched the invasion of Formosa (today known as Taiwan and at the time an American possession) as well as the British colonies of Hong Kong and Singapore. Within a few weeks, as did Sumatra, Burma, Java, and the Philippines, the Dutch East Indies had fallen together with their petrol wells.

With the US at war, as citizens of enemy countries, both Delbrück and Luria became overnight suspected spies. "I found myself an enemy alien," Luria commented. Nonetheless, through the RF's involvement, they obtained security clearance to work in research institutions and continue their shared endeavor with Anderson. They were lucky: with "Executive Order 9066," Roosevelt ordered the deportation of over 100,000 Japanese people (even naturalized Americans) residing in the US in concentration camps ("relocation centers") for the duration of the war.

In March 1943, the first joint paper between Luria and Anderson was ready for publication. The stunning pictures of phages demonstrated, once and for all, that these viruses were not enzymes but much more complex entities. In the paper, they described the phages as made of "a round 'head' and a much thinner 'tail,' which gives them a peculiar sperm-like appearance," adding that "the 'head' is not homogeneous but shows an internal structure consisting of a pattern of granules." Furthermore, phages showed a variety of shapes that showed how these viruses were a monolithic entity, as d'Hérelle had believed, but a collection of organisms. For instance,

> Bacteriophage anti-coli PC … the diameter of the head appears to be about 80 mμ [nm]; the tail is about 130 mμ long.
>
> Bacteriophage anti-coli P28 … round particles are visible in the suspension of this phage which are somewhat smaller than those described for phage PC (about 50 mμ in diameter). An extremely thin tail … seems to be visible in many instances.

This publication granted Luria a Guggenheim fellowship and he used the money to join Delbrück at Vanderbilt. However, one of the requisites for the fellowship was to accept the first suitable position that could come over. The opportunity arose with an invitation in January 1943 for a research and teaching position at the Indiana University in Bloomington, "the smallest town I had ever been," said Luria. "It was an ugly town … a bit of depression town … a prejudiced town." Nonetheless, "the campus of the university was attractive ... the surrounding countryside agreeable." In addition, the university became a leading center in genetics due to the presence of Hermann Müller, who joined in 1946. Luria's very first Ph.D. student was James Watson, who joined Indiana University in 1947 and shared the office with another student, Renato Dulbecco, who had joined six months earlier—Watson remarked that "most of Luria's and Dulbecco's conversation was in Italian." While at the University of Indiana, "in Luria's flat," Watson met Delbrück, "who had briefly stopped over in Bloomington for a day." The meeting profoundly affected Watson's attitude toward science, for he always claimed that the German professor was "the model for what I wanted out of my own life."

Luria had always been keen on civil rights movements, a passion he acquired in Paris. He joined the University Teacher's Union, a small organization of a dozen members, even if that meant he was "looked upon with distrust by administrators and with embarrassment by colleagues." The distrust eventually resulted in his dismissal as a teacher. Thus, in 1950, he was forced to move to the University of Illinois in Urbana, a city, as he said, with "interminable hedges of cornfields" that made him and his wife "miss the lovely hills of southern Indiana." The new campus was "more citified, less interesting ... more sedate than Indiana's." Nonetheless, "the years in Urbana were fruitful."

Back in the day, when Delbrück joined Luria and Anderson, the scope of the research shifted. Instead of investigating the genetics of these viruses, as Luria wanted, the trio started to observe the process of infection itself, which was Delbrück's primary interest. The first thing that Delbrück, Luria, and Anderson noticed with their micrographs was that the capsids did not enter the infected bacterium. If the capsid was not the virus, what element carried the genetic information that produced it?

The paper summarizing these observations was written in a different mood than their previous ones, for the fortunes of the war had drastically changed. Having failed to secure a truce with Churchill, Hitler was mired in an attrition conflict. It was not a trench war like the First World War, but a novel warfare ("Bewegungskrieg") involving highly dynamic front lines. These fronts could stretch for thousands of kilometers, advancing and receding for hundreds of kilometers before their final collapse, and required massive support from sea and air. Thus, unlike the first conflict, the countries involved in the Second World War needed a well-equipped Army and state-of-the-art Navy and Air Forces. Consequently, the cost of the war was much higher than in the previous one. Just like in the First World War, though, the numbers would seal the sorts of the Second.

A simple audit shows that, in 1939, Nazi Germany could count on a population of 69 million people and 6.2 million soldiers against Britain, which had 48 million citizens and 5.5 million soldiers. However, the whole forces of the Axis combined could account for 259 million people against the 690 million of the British empire. More importantly, the Nazi empire extended to a single spot (Central Europe) and was essentially surrounded by the Allied countries. Japan was fighting its war in the Pacific, thus not backing Germany up. Hitler did not seize the French overseas possessions to avoid antagonizing Pétain and his control over the subjugated French territory; hence, the colonies remained at the direct disposal of the État français and not of the Nazis. As a result, the Nazi empire was hampered by a chronic shortage of raw materials and virtually no international financial support.

Despite its early gains and aura of invincibility, the economic system of Germany started to collapse pretty early in the war. Germany's gross domestic product (GDP) was equivalent to 839 billion euros in 1939, compared with 620 billion of Britain, but it shrank to 678 billion in 1945, whereas the UK's GDP increased to 724 billion by the end of the war. While Nazi Germany was investing 23% of its GDP in armaments in 1939, Britain had a weaponry expenditure of only 15%. In 1945, however, Britain spent 53% of its GDP on armaments compared with Germany's 70%, meaning that Britain could invest money in products and services besides military output, increasing the quality of life of its citizens. In contrast, Germany completely bowed down to the war effort. Things were not better for the rest of the countries entangled with Germany. The Axis' production almost halved during the war, from 1.6 trillion euros in 1939 to 1.0 trillion in 1945.

In contrast, the Allied countries increased their income from 3.8 trillion in 1939 to 5.2 trillion in 1945. Even the Axis' war machine shrank during the war years: in 1945, there were 15.6 million Axis soldiers compared with the Allied 28.6 million. Hence, in retrospect, it is not surprising that the German Army lost its initial momentum, and this Bewegungskrieg was sealed.

Things took a wrong turn almost immediately with the involvement of Italy in the conflict. Despite signing the Pact of Steel of mutual military backing in May 1939, Mussolini did not support Hitler's war campaign in September. Only when it was almost certain that the Nazis would conquer Europe, Mussolini decided unilaterally to enter the war on the German side by a surprise attack on France and Albania. Apart from disrupting the Nazi war plans, Italy was incapable of winning its fights. While the French Army melted in front of the Nazi war machine, the Frenchmen heavily beat the Italians, and it was only the French capitulation that saved Mussolini from losing the war within a few days from its onset. Also, the Italians could not defeat the Albanian Army, forcing Hitler to send a hefty contingent to the rescue. There was no need for Italy to be involved in the Second World War other than because of Mussolini's greed. Italy could have remained neutral just like Francoist Spain did. Francisco Franco died of old age in 1975, and his dictatorship peacefully transitioned back into democracy ("la Transición") within 2 years. Mussolini, instead, was executed in 1945 while trying to flee a disarrayed country dressed as a peasant, and his body was exposed hung from the feet in Piazzale Loreto (Milan), leaving Italy in tatters. All in all, the Italian involvement in the conflict meant that Germany had to shift essential assets from the east front against the USSR to the Mediterranean area.

After Pearl Harbor, the position of Nazi Germany worsened even still. By the end of 1942, with North Africa, virtually all of Europe, and vast swaths of Russia under his power, Hitler realized that his Juggernaut was already slowing down. Perhaps he could have gotten away if he had not signed the Pact of Steel and had fought the war alone. But his fascist arrogance against peaceful solutions and Churchill's stubbornness not to surrender meant that the war would be fought until its bitter end. The Japanese Emperor Hirohito shared the same tenet, but the US invasion of Midway Island in June 1942 meant that Japan's fate was also pronounced.

The combined British forces of the 8th Army led by Field Marshal Bernard Law Montgomery and the 1st Army led by General Kenneth Anderson managed to push the German Afrika Korp of Field Marshall Erwin Rommel out of Africa by the Summer of 1942. Having secured bases in North Africa, the Allied forces could implement Operation Husky, the invasion of Sicily, which started on the 10th of July 1943 and was led by Gen. Montgomery (British Army) and General George Smith Patton (US Army). The battle could be considered won when Patton entered Messina on the 17th of August. Even before the end of the Operation, on the 25th of July, Mussolini was arrested by King Victor Emmanuel and substituted by Gen. Badoglio; the catchphrase "Benito è finito" ("Benito is done") became a national favorite motto.

The invasion of the USSR did not go any better for the Germans. As part of Operation Barbarossa, the Nazi forces surrounded Leningrad in September 1941, beginning one of the cruelest sieges in history. The NKVD, responsible for securing the law and order in the city, arrested some 2000 people for cannibalism. This crime, born of the siege's desperation, contemplated both corpse-eating and body-eating (deliberate murder to secure meat) and it was rampant within families (parents eating their children). Despite these inhuman conditions, the city stood firm, owing to General Georgy Zhukov's merciless command, German skimpy supply lines, and the organizational delays due to the rescue of the Italian Army, unexpectedly fierce resistance from Soviet partisans, and the fact that Hitler diverted crucial resources from Leningrad to Moscow to begin a parallel siege there (Operation Typhoon, which also started in September 1941). During Barbarossa, German soldiers treated the Russian and Ukrainian populace in the conquered territories as "Untermenschen," or human vermins. Many soldiers lamented that there was nothing to loot in Soviet villages. In numerous cases, looting soldiers seized the little food available, leaving the civilians to starve to their deaths.

Like the siege of Leningrad, Operation Typhoon did not go as the Germans had planned: the siege of Moscow was broken on the 7th of January 1942, thanks also to the fact that Stalin charged Zhukov to organize the defense of the Soviet capital city. Despite these critical setbacks, Hitler decided to launch a third siege, this time of Stalingrad, with Operation Blau. The maneuver began on the 23rd of August 1942. Because of Soviet censorship, the population was unaware of the German whereabouts: they were picnicking when the German strategic bombing started. The German carpet bombing killed over 40,000 people by the time the state of siege was declared two days later. The city's defense was given to General Vasily Chikov, who was as ruthless as Zhukov. After the Germans entered the city, the battle became a building-by-building guerrilla. The Soviets initiated irregular warfare ("Rattenkrieg," the "Rats' fight") that unnerved the invaders: they moved through the sewer system to attack the Germans from behind. Snipers were also massively employed, the most famous being Vasily Zaitsev. In addition, the approach of winter brought the invaders to the brink of extinction since they were less prepared for temperatures well below zero than the Soviet citizens.

To make matters worse, an outbreak of cholera erupted in the city around mid-July 1942. To help fight the epidemic, the Russian microbiologist Zinaida Vissarionovna Yermolyeva was sent to the sieged city. Yermolyeva—who had also purified and produced the Soviet version of penicillin ("krustozin"), which could be used to treat infected soldiers during the conflict—decided to use bacteriophages by establishing a makeshift laboratory in the cellar of a bombed house, eventually producing

50,000 L of phage preparation per day. To obtain the bacteria required to isolate the viruses for the anticholera phage therapy, Yermolyeva ordered the snatching of infected German soldier corpses from behind the enemy lines. At the end of the year, Stalin made a direct phone call to Ermolieva, asking her, "Can the cholera epidemic interfere with the military plans?" She is said to have given the famous answer that she had already won a victory against the "invisible enemy;" now, it was time for the Red Army to win a victory against the invasion. Stalingrad was the second most crucial massive success of phage therapy; the first case was during the Soviet invasion of Finland at the beginning of 1940 to fight an outbreak of *Clostridium perfrigens* (the "Winter War").

In Leningrad, Zhukov launched a counteroffensive (Operation Spark) to break the Nazi encirclement of Leningrad on the 12th of January 1943. The Operation took about one week and 34,000 Russian casualties to succeed. However, the city was completely freed from the German invasion only on the 27th of January 1944, after 880 days of isolation. On the 2nd of February 1943, the siege of Stalingrad was also broken. From there on, the Soviet's advance became unstoppable: with a series of military operations (in particular, Uranus, Mars, Saturn, Ring, and Polar Star), the Nazis were pushed back toward Germany. The Soviets won a strategic battle at Kursk on the 7th of July 1943, representing the biggest tank fight in human history, but that also required massive support from the Air Force. Kyiv (Kiev), the Ukrainian capital, was freed on the 6th of November 1943.

5.2 THE MUTANTS

Back in February 1943, while attending a dance night at the Bloomington Country Club, Luria had a eureka moment: "during a pause in the music I found myself standing near a slot machine, watching a colleague putting dimes into it … it dawned to me that slot machines and bacterial mutations have something to teach each other." In other words, a bacterium that would become immune to phage infection would make a "jackpot:" it would gain access to the whole culture medium and replicate at will.

He imagined 100 plates with bacteria lawns sprayed with phages. If the mutations occurred after infection, all dishes should look the same. However, if the mutations were already randomly present in the plated bacteria, one might expect differences in the presence of bacterial colonies. A resistant bacterium already present in a lawn of sensitive bacteria would have the opportunity to spread throughout the plate, just like winning a jackpot, because the phages would have wiped out all other competitors.

Luria worked on Sunday and sent the preliminary results to Delbrück the next day; Delbrück replied that he would take care of the statistics behind such a genial idea. In the resulting paper, published in May 1943, the European duo—to use a moniker introduced by Delbrück's biographers Ernst Fisher and Carol Lipson—introduced the subject by noting that

> when a pure bacterial culture is attacked by a bacterial virus, the culture will clear after a few hours due to the destruction of the sensitive cells by the virus. However, after further incubation for a few hours … the culture will often become turbid again, due to the growth of a bacterial variant which is resistant to the action of the virus.

Since the initial culture was seemingly pure—composed of a clone of the same cell—the question they raised was where the mutant had come from. They summarized the hypotheses known at the time explaining such a paradox:

> D'Herelle [sic] (1926) and many other investigators believed that the virus by direct action induced the resistant variants. Gratis (1921), Burnet (1929), and others, on the other hand, believed that the resistant bacterial variants are produced by mutation in the culture prior to the addition of the virus.

That is to say, the alternative explanation for the rise of phage immunity was:

> First hypothesis (mutation): There is a finite probability for any bacterium to mutate during its life time from "sensitive" to "resistant".
>
> Second hypothesis (acquired hereditary immunity): There is a small finite probability for any bacterium to survive an attack by the virus.

How to discriminate between these two equally probable explanations? Following Luria's hunch, it can be assumed that, under the second hypothesis, there will always be the same number of mutants after exposure to the virus. Suppose a batch of *E. coli* strain B is cultivated in a liquid medium for a time t_x, with x ranging from zero to n time points. Luria and Delbrück chose the replication time of *E. coli* as the unit of time. At each time point, an aliquot of the culture is plated in a Petri dish containing phage T1. Only bacteria resistant to the viral infection would grow into colonies. The bottom line was that the original cultures only contained sensitive bacteria; thus, the resistant ones must have arisen at some point from mutants. If the mutation was triggered in response to the infection, in the Lamarckian sense, the number of cells resistant to infection would always be the same—or at least within a narrow margin—regardless of the t_x assayed because the resistant bacteria would come out only during the plating phase.

On the contrary, if the mutation was due to a random process preceding the infection, in true Darwinian style, there should be a high variability of results because the mutants would arise randomly in the liquid culture phase, not during the plating phase. If the first mutant had appeared at the beginning of the experiment (t_0), there would be one colony in the plate derived from such a bacterium. There would be two colonies at the second time point (t_1)—that is, after the time needed for the bacterium to duplicate. At the third time point (t_2), there would be four colonies, eight at t_3, and so forth. Considering only these time points, one would get 15 colonies overall. However, if the mutant had arisen at a different time point, say t_2, there would be no colonies at t_0 and t_1, but one colony at t_2 and two at t_3, for a total of three. On other occasions, the infection-resistant mutant might not raise at all. In other words, if the first hypothesis were true, the number of resistant colonies would "fluctuate." Figure 5.1 provides a visual exemplification of this concept. Delbrück and Luria provided a detailed statistical explanation of the differences between these alternatives.

The basis of Luria and Delbrück's statistical analysis was that a mutation is a rare event per unit of time that could be described by Poisson distribution. The period needed for the cell replication is chosen as the unit of time to obtain the number

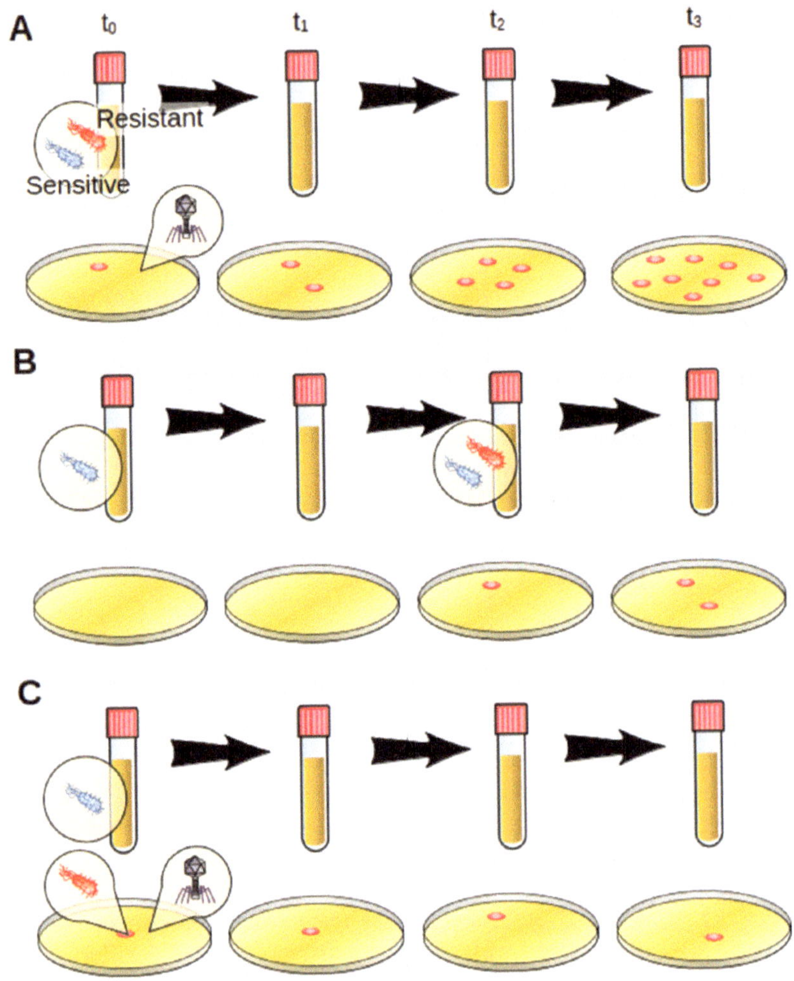

FIGURE 5.1 *The fluctuation test.* Idealization of the experimental setting needed to discriminate between Darwinian and Lamarckian models of phage resistance. There are two types of bacteria: those that are sensitive (blue) and those that are resistant (red) to phage infection. If mutations occur before infection (according to Darwinian theory), a liquid culture should have a mixture of these bacterial clones. Bacteria are left to grow over time, and the bacteria are duplicated at each time-point. At each time-point (t_0 to t_3) samples are plated on dishes containing the phage, showing colonies of resistant bacteria (dots). As shown in panel A, if a mutant arises at t_0, it will produce a single colony in the corresponding plate. There will be two colonies at t_1, four at t_2, and eight at t_3. Collectively, there will be 15 colonies. However, the mutant might rise at another time-point, for instance, t_2 (panel B), providing only three colonies overall. In other cases, there might be no mutants at all; thus, the colony count would be zero. The fundamental idea is that the number of colonies will be highly variable when the experiment is repeated multiple times. In other words, the number of colonies will "fluctuate." Conversely, in the Lamarckian context, the mutants will rise only after the infection (panel C). Thus, the number of colonies will remain extremely constant.

of cells at time t given by the equation $N_t = N_0 e^t$, where N_0 is the initial density of bacteria. Thus, "the number dm of mutations … during a time interval dt is … $dm = adtN_t$," where a is the mutation rate ("the chance of mutation per bacterium per time unit"). Since a mutation is rare, growing hundreds of thousands of plates to count the mutated bacteria would be impracticable. Instead, Delbrück and Luria solved the problem the other way around by calculating "the fraction of cultures showing no mutation," that is, $p_0 = e^{-m}$, where m is the mean number of mutants. From these premises, it was possible to demonstrate that statistically

> the mutation hypothesis, in contrast to the hypothesis of acquired immunity, predicts a distribution of the number of resistant bacteria with a long tail of high numbers of resistant bacteria, the determination of the *variance* of the distribution should be helpful in differentiating between the two hypotheses.

In other words, "the hypothesis of acquired immunity predicts a variance equal to the average … while the mutation hypothesis predicts a greater variance."

In statistics, the quantity variance (V) measures how far apart a bunch of numbers is from its mean; it is calculated according to the formula $V = \Sigma(x - x)^2$, where x is a given measurement, x is the mean of C measurements, and Σ represents the sum function of the C measurements. The square root of the variance is known as standard deviation: $\sigma = \sqrt{V}$.

Delbrück and Luria prepared many liquid cultures of *E. coli* B and plated an aliquot of these cultures in solid cultures containing phage α—now known as T1—at different times. Then, the European duo counted the number of resistant colonies showing off in Petri dishes. They provided the results from eight experiments by way of example. In the provided ensemble, they cultivated between five and 20 cultures, obtaining 3.8–62.0 colonies on average, with a variance ranging from 40.8 to 6620. The quantity σ/x was, on average, 1.48 ± 0.79. Since this value was higher than the unity—but only five times out of eight—the European duo concluded that the data were "in conflict with the expectations from the hypothesis of acquired immunity."

However, like the green pamphlet, even this work did not stand the test of time. The European duo's statistical explanation is hard to follow for our modern taste. For instance, despite focusing on data variance, Luria and Delbrück based their analysis on the quantity σ/x but did not define σ. Also, they did not provide a statistical test to support their choice: their essay was essentially a qualitative analysis. Later formalizations provided instead a solid quantitative approach, for instance, with the work of the British physicist Douglas Edward Lea and mathematician Charles Alfred Coulson.

Even in determining the mutation rate a—today indicated by the symbol μ—the European duo based the computation on the formula $p_0 = e^{-m}$, providing, as an example, the value of 0.33 (29 plates without colonies over 87 replicates) for the calculation; yet, $e^{-0.33}$ is 0.72 and not 1.10, as was reported. Almost four decades later, Adam Roger and colleagues at the Finders University in Australia provided an estimator of m, $\hat{m}_0$, defined as $\ln(C/z)$, where C is the number of replicates of an experiment and z is the number of tests giving no colonies. Using Luria and Delbrück's values, one obtains $\hat{m}_0 = \ln(87/29) = 1.099$, which is in line with the given value of 1.10.

In addition, the p_0 value is no longer used to estimate the mutation rates of bacteria; nowadays, the algorithm introduced by Sandri and Sarkar is the preferred method.

Today, it is also known that the resistance to T1 is due to mutations in the membrane protein FhuA, a surface porin—a channel connecting the extracellular environment with the cellular interior. FhuA is involved in the import of the molecule ferrichrome, which chelates ferric iron (Fe^{3+}) atoms, which are badly needed for cellular biochemistry. FhuA is also involved in the export of the antibiotics albomycin and rifamycin, providing antibiotic resistance. FhuA comprises a hollow tube closed on a side by a globular domain known as the "cork." The channel opens when FhuA comes in contact with the protein TonB on the inner cytoplasmic membrane. The ferrichrome binding facilitates the interaction between FhuA and TonB, opening the channel to permit the importation of iron. It turned out that phages T1 and ϕ80 can bind to FhuA directly, whereas phage T5 binds to the complex FhuA/TonB.

In addition, while the fluctuation test introduced by Luria and Delbrück could correctly discriminate between a Darwinian and Lamarckian hypothesis, it has been remarked that it could not distinguish between a pure Darwinian theory and a mixture of Darwinian-Lamarckian assumptions. It is now known that, while the bunk of genetic inheritance is based on the Darwinian theory, there are occasions where Lamarckian processes occur. One such instance is somatic recombination, which takes place in the lymphocytes. Twins are born with the same genetic repertoire for producing antibodies. Still, lymphocytes randomly edit some portions of these molecules, creating unique combinations in each sibling that are no longer within the Darwinian sphere.

Similarly, bacteria use CRISPR to fight viral infections. CRISPR stands for clustered regularly interspersed short palindromic repeats and was detected by chance during the sequencing of a phosphatase expressed by *E. coli*. CRISPR consists of several repeated elements formed by bracket sequences complementary to one another and separated by a spacer of random bases. For instance, in the sequence TCCCCCGCTGCGCGGGGGA, the underlined stretches are complementary and can complement each other, forming a hairpin structure. These CRISPR tandem repeats—meaning loci where the CRISPR elements are repeated hundreds of times—are present in the proximity of CRISPR-associated (Cas) protein genes responsible for processing these repeats. When a bacterium is infected with a virus, the viral genome is fragmented, and the Cas proteins add some of these fragments into the CRISPR clusters. This phase is known as "adaptation" and represents a somatic adjustment that fits into a Lamarckian context. In a later phase, known as "expression," the CRISPR sequences are expressed as single-stranded RNA and processed into hairpin structures known as CRISPR RNAs (crRNAs). During a subsequent infection with the same virus, the crRNA can bind to the phage genome due to its complementarity to viral sequences. Once bound to the phage genome, the crRNA guides the Cas nucleases to the foreign nucleic acid, causing the invading virus' genome degradation. The CRISPR-Cas system represents an adaptive immune response to phage infection and is used today to mutate eukaryote genes.

Despite the flaws pinpointed in later years, the fluctuation test was a widely influential paper, much like the green pamphlet a few years earlier. One of the first to be inspired by the work of Delbrück and Luria was Howard Borden Newcombe, a Canadian geneticist, in 1949. He replicated the fluctuation test in a slightly different manner. He plated *E. coli* B on two plates and left it to grow for a few hours; afterward, one series was additionally spread to redistribute the newly replicated bacteria. Both series were then exposed to phage T1. The idea was that, in the Lamarckian context, the number of cells would always be the same, as stated by the European duo six years earlier. In the Darwinian context, instead, the initial foci of resistant bacteria would be disrupted to generate new colonies immune to T1 infection. The results showed 76 colonies in the series without spreading and 13,001 in the series with additional disruption, confirming that mutants immune to T1 were formed before exposure to the virus.

The fluctuation test presented by the European duo was innovative because it introduced the concept of hypothesis testing based on rigorous mathematical grounds. Although there must have been a personal preponderance for the Darwinian concept over the Lamarckian theory, Delbrück and Luria left the choice between the two models to the numbers rather than to bitter discussions, severing the tradition that instead characterized the d'Hérelle-Bordet diatribe. As Anderson commented, Delbrück and Luria "formed a little green island of logic in the mud-flat of conflicting reports, groundless speculations, and heated but pointless polemics surrounding the Twort-d'Hérelle phenomenon." More importantly, the paper is today widely recognized as a founding stone of bacterial genetics. The path that leads to the production of recombinant insulin began from this seminal paper.

Other researchers influenced by the work of Luria and Delbrück were Demerec—essentially, the European duo's boss—and Evelyn Witkin. The former investigated the spontaneous generation of antibiotic-resistant bacteria, observing that "the variation in the number of resistant bacteria … is considerable … the results of this experiment, therefore, favor the assumption that resistance to certain concentrations of penicillin originates through mutation." The latter, who later delineated the mechanisms of DNA repair undergoing in bacteria, measured the frequency of generation of bacteria resistant to UV and X-ray irradiation, noting that the results show "fluctuations much greater than can be accounted for by sampling errors" with "the variance is significantly higher than the mean in every case." Thus, "change from sensitivity to resistance is a spontaneous mutation occurring in the normal culture at a rate of about 1×10^{-5} mutations per bacterium per generation." In addition, the mutations are derived from a "stable, heritable character." She established the *E. coli* strain *B/r*, noting that "resistance to ultraviolet radiation and x-rays is a stable, heritable character."

Furthermore, Witkin introduced Luria to New Yorker Zella Hurwitz, the daughter of a Jewish émigrée from Minsk; she became a feminist activist and a renowned psychologist who investigated how children's gender identities developed as well as the portrayal of sexuality in the media. Luria and Hurwitz married in 1945, a marriage that lasted for decades even if, as Luria admitted, "I started suffering from recurrent depressions, which did not hamper significantly my ability to work but made me tense and hypercritical … which caused much suffering to me and to my family." They had one child, Daniel, born in 1948.

5.3 MUTUAL EXCLUSION

Delbrück and Luria were among the first to report how infection with one type of phage inhibited the infection with another. By analyzing the range of burst sizes obtained during the infection of *E. coli* strain B with phage α, Delbrück and Luria concluded that "bacteria which are simultaneously infected with several virus particles of the same kind react as if only one of the virus particles were effective." In other words, if a phage has a burst size 180, a bacterium infected by a single phage will create 180 new virions. One would expect that a bacterium infected by two phages will produce 360 new virions, and so on. Instead, infected bacteria always produced the same burst size (180) regardless of the ratio of viruses to bacteria. Delbrück and Luria called such a phenomenon "self-interference" or "mutual exclusion."

"We soon discovered," Luria recounted, "that if two different phages attached the same bacterium, only one of them could multiply: there was a complete mutual exclusion." Working with phages α and γ, Delbrück said, "when a bacterium is simultaneously infected with two or more bacterial viruses ... only one of the viruses will grow." In other words, "in multiple infection ... virus γ displaces virus α." These observations were confirmed by taking advantage of the fact that a variant of phage γ (γ′) grew differently in *E. coli* strain B than in strain Bγ. For instance, the γ′ burst size was 95–130 in B and 40–60 in Bγ. Luria exposed Bγ to an excess of phage γ and, after one minute, added γ′. He then added antibodies to eliminate the free viruses, concentrating primarily on the viruses adsorbed to the bacterium. Afterward, he analyzed the plaques on solid media, observing that "about 90 per cent of the bacteria infected with both viruses γ and γ′ failed to liberate any virus γ′." He also noticed that "new virus particles arise by mutation in the course of the multiplication of normal virus inside the cell."

According to Delbrück and Luria, mutual exclusion was caused by a bacterial enzyme in small quantities inside the cells, and viruses competed for its employment. The virus that gained access to this enzyme obtained complete control over the bacterial biochemical machinery to the detriment of the other virus:

> the saturation may be due to the fact that among the bacterial enzymes which are necessary for virus synthesis, there is one "key-enzyme" which is completely engaged by one virus particle ... this means that the virus γ particle either succeeds in displacing virus α or does not succeed.

Phages α and γ were in a sort of total war with one another that could result in only one of them being victorious, a scenario that dramatically reflected the ongoing situation in Europe. Nonetheless, the key-enzyme theory was dropped relatively soon.

Using micrographs obtained in collaboration with Anderson, Delbrück showed that "most of the adsorbed virus particles do not penetrate the cell" and "several minutes may elapse between adsorption and penetration." However, there is a temporal window, specific for each type of phage, where additional phages of different origins are excluded: "it is less than 4 minutes for gamma, 8 to 9 minutes for delta, and 14 minutes for alpha." The scenario coming out from these data was that

> the entrance of the first virus makes the cell wall impermeable to other virus particles, just as the fertilization of an egg by one spermatozoon makes the egg membrane impermeable to other spermatozoa.

In support of the egg-membrane hypothesis, more formally known as the "penetration hypothesis," Luria presented data, in collaboration with other researchers, showing that phage ε could repress the expression of the receptor used in the infection of *Salmonella anatum*, making it impossible for other phages ε to infect the lysogenically converted host. The penetration hypothesis was later modified because other viruses could penetrate the cell. André Lwoff remarked in a paper that "the reality of the penetration [of the competing phages] was proven for the first time by Bertani in 1953." Thus, competing phages could enter the bacterium—hence, the membrane was not as impenetrable as a fertilized oocyte—but were prevented from developing normally.

Renato Dulbecco, in 1951, confirmed Delbrück and Luria's observation that the exclusion had a temporal dimension: the longer the interval between the first and the second infection, the lower the number of co-infected bacteria. Dulbecco had joined Luria's laboratory in 1947 with the intervention of Rita Levi Montalcini. Being of Jewish ancestry, Montalcini lost her job as an assistant professor in the Department of Anatomy of the University of Turin, where she was born. Nonetheless, she continued to study the growth of nerve fibers in chicken embryos in her bedroom. She joined the partisans fighting against fascists and, later, volunteered in the Allied health service. In 1946, she joined, with the help of Luria, the laboratory of Professor Viktor Hamburger at Washington University in St. Louis. Montalcini, who won a Nobel Prize in 1986 thanks to her research on tissue growth factors, was nicknamed "the queen" because of her "impeccable dresses and regal manner," remembered Luria. In turn, Montalcini helped Renato Dulbecco to join Luria's laboratory.

In the 1950s, a wealthy shingles patient was astonished that the medical community knew next to nothing about its etiological agent—chickenpox (varicella) virus, which is highly related to herpes virus. Thus, he established a hefty research grant through the RF to study herpes and other human viruses, shifting the RF's focus to human viruses. Dulbecco, sponsored by this grant, repurposed the plaque assay to grow eukaryotic viruses. Instead of having a lawn of bacteria, Dulbecco generated a single layer of eukaryotic cells, essentially a mini-tissue, in a Petri dish. Even if the viruses did not cause the lysis of the cells, analogously to the phages' plaques, the viruses would cause massive cellular alterations that would clearly show under a microscope. Dulbecco's technique is still a standard method in virology to this very day.

Dulbecco's phage studies took advantage of a feature of the plaques generated by *E. coli* co-infected with phages T2r and T2r$^+$: the doubly infected cells formed speckled ("mottled") colonies. As stated before, plaques obtained from T2r and T2r$^+$ could be clearly distinguished. Dulbecco infected *E. coli* growing in a liquid medium with either T2r or T2r$^+$ and added the other phage after a chosen amount of time. He then counted the resulting plaques on a solid medium. He observed, for instance, that the fraction of plaques derived from co-infected bacteria decreased from 100% when the viruses were administered simultaneously to 56% when inoculating T2r$^+$ first and

T2r after one minute delay and 4% when increasing the delay to 10 minutes. At the same time, the fraction of plaques infected only with the first given virus increased from 62% at t_0 to 78% after a 10-minute delay. In addition, if the first virus was irradiated with UV light, there was no significant decrease in the exclusion power of this virus. The phenomenon was not linked to viral UV-perishable molecules. However, the reduction of co-infected bacteria did not occur when bacteria were starved by not adding additional broth to the liquid culture. He concluded that "the exclusion we observe is dependent on metabolic processes requiring an external energy supply."

Later, Phoebe Fielding and Mary Lunt at the University of Oxford used *E. coli* strains that missed endonuclease to further explore the process of mutual exclusion. After a set amount of time, they infected *E. coli* with phage T5, added^{32}P-marked T4, and quantified the radio-labeled nucleotides released in the supernatant. In essence, Fielding and Lunt measured the amount of degraded phage DNA in endonuclease deficient (*E. coli* B.41 or K12.1100) and wild-type (*E. coli* B.31 or K12.1000) bacteria. They observed that the amount of degraded T4 DNA was similar in both strains of *E. coli*: 47% and 48%, respectively.

Thus, "bacteria infected with T5 were able to exclude and degrade superinfecting T4 phage DNA to acid-soluble fragments whether the cells contained an active bacterial endonuclease-I or not." However, infection of *E. coli* K12.1100 with T5 and simultaneous administration of chloramphenicol, an inhibitor of protein synthesis, "prevented breakdown of superinfecting T4 DNA." If chloramphenicol was administered five minutes after T4's primary infection, the degrading capacity of T4 was restored. These results indicated that phages could encode their endonucleases to actively exclude competitors from seizing the cell. Even more remarkable is the speed of the process: in just the time for preparing espresso, the phage could deliver its DNA inside the cell, unpack it, produce enough new proteins to modify the cellular environment, and replicate a 100-fold.

The process of exclusion from hijacking the host's biochemical features was not limited to phages. Delbrück cited an old paper about TMV, a plant virus, showing the same process. In that paper, written in 1929, McKinney reported two primary morphologies of mosaic in the tobacco plant *Nicotiana glauca*: yellow and mild dark green. Two separate TMV strains must have caused these two variants of the same disease. The author reported that "when tobacco plants with mild dark-green mosaic were reinoculated with the virus of … yellow mosaics, the former type became obscure and the plants showed the typical symptoms of yellow mosaic." Thus, the yellow strain of the virus had excluded the green strain from spreading into the cells. In other words, only one strain had prevailed in infecting the tobacco plant. Another early example of viral interference divulged by Delbrück included the cross-exclusion between influenza virus genotypes A and B, as reported by James Ziegler, one of the first graduates of the Phage Course, which will be established within a few months. Other instances included yellow fever virus, excluding subsequent infection with Rift Valley fever virus and among different strains of poliomyelitis virus.

Today, we also know that all viruses have evolved molecular mechanisms to prevent other viruses from infecting the already hijacked cell. These mechanisms are diverse. *Streptococcus thermophilus* phage TP-J34, for example, encodes a lipoprotein known as ltp (lipoprotein of temperate phage), which is thought to interfere with the injection of DNA from viruses adsorbing to the infected cell. Another example is that of phages targeting *Pseudomonas aeruginosa*, which modify the structure of surface receptors—namely the pili and the so-called O-antigens—used by other phages for infection, thus preventing the competing viruses from adsorbing to the hijacked cell.

5.4 HERSHEY HEAVEN

"One of the tragedies of my scientific life is that I make too many discoveries." The biology class students laughed as these introductory words were uttered, but the lecturer remained stoic. "This is no laughing matter but a tragedy," he continued, "because new discoveries interfere with completing each previous line of research." Alfred Day Hershey, the man who pronounced these lines, was a man of very few words, but when he spoke, people listened. James Watson described him with these words: "Al had no trace of Delbrück's almost evangelical charisma or Luria's assertiveness" but "he could enjoy what he called Hershey Heaven—doing experiments that would give the same answer upon repetition."

Jun-Ichi Tomizawa, who became a renowned Japanese geneticist and worked with Hershey at the CHSL, highlighted how Hershey could easily find the heart of the matter of a problem and focus on its solution. For instance, he commented that "we can use B to investigate A," but often, "this fails because we find a new phenomenon C, that we were unaware of." Benzer even declared that there is a "Hershey's razor," in direct association with Ockham's razor: "when one has a word, phrase, or sentence in a manuscript that somehow doesn't fit, regardless of twisting it every which way you can, just cut it out." Another of Hershey's aphorisms was: "there is nothing more satisfying to me than developing a new method: ideas come and go, but a method lasts." It also remarked that experiments rarely returned a clear-cut response: when "distinguishing between two hypotheses," results often "fall just in between." These maxims might sound silly, but they provide zen-like concepts to experimentalists who spend their days in the laboratory trying to generate valuable data.

Hershey was born in 1908 in Owosso, Michigan. He obtained a Ph.D. in bacteriology in 1934 from Michigan State College with a work focused on characterizing the pathogenic bacterium *Brucella*. Afterward, he received a position as an instructor at Washington University in St. Louis, where he collaborated with Jacques Bronfenbrenner, a Ukrainian Trotskyist émigré who participated in the 1905 upraise in St Petersburg. Forced to leave Russia to avoid being arrested by the Tsarist police, Bronfenbrenner went first to France, where he eventually joined the PI, and then moved to the US after obtaining a research position at the RF to study the serodiagnosis of infectious diseases. Bronfenbrenner obtained a Ph.D. at Columbia University in 1912 and was naturalized as an American citizen the year after. He became an assistant professor at Harvard in 1917, returned to RI in 1923, and obtained a chair in Bacteriology and Immunology at Washington University in 1928. He managed the public health response against what was later called the St. Louis encephalitis virus. Around 1940, Bronfenbrenner gave Luria two phage strains (P28 and PC) that were later named α (T1) and γ (T2)—β indicating the bacterial host.

From the late 1930s, Bronfenbrenner started investigating the nature of phages, together with Hershey, assuming they would be much smaller than reported in the literature. Hershey was struggling to provide experimental support to Bronfenbrenner's hypothesis and, to get through his assignment, contacted Delbrück in 1942 in the hope of obtaining valuable insights for the experiments. Delbrück invited Hershey to attend a seminar in Nashville in January 1943, and the student reciprocated by inviting the German professor to St. Louis in the spring of the same year. In February, Delbrück met Hershey, and in April, both Delbrück and Luria went to St. Louis to visit Hershey, marking the birth of what became known as the "Phage Group." Delbrück termed this newly established partnership "two enemy aliens and one social misfit," referring to Hershey's disenfranchised manners: he "drinks whiskey but nor tea ... likes living in a sailboat for three months [in a raw]," wrote Delbrück to Luria, adding that he "likes independence." Excited by the meeting, though, Hershey wrote back to Delbrück in July saying that he was hoping he "could come here sometime in the fall, maybe we can do some experiments ... my guess is that there will be a great rush on phage work after the war."

Waclaw Saybalski, another Phage Course graduate who worked at the CSHL for a while and accompanied him on sailing excursions—Hershey's chosen pastime in addition to classical music—described him as a dedicated worker who completed "every experiment by himself and in perfect silence." Hershey's assistant Martha Chase was also not chatty, and sometimes "they seemed to communicate only by 'sign language'." Naville Symonols, a colleague and friend, remarked that Hershey was "always the first" to arrive in the laboratory and "often the last to leave." Saybalski noted also that Hershey could become so absorbed by a project to compromise his own life. In 1953, Hershey neglected eating and "lost weight throughout the whole year." He only recovered when his wife Jill forced him to take an extended vacation on a boat and fed him generously. Another Phage Course alumna, Ann Skalka, had been Hershey's student and described how Hershey—who was referred to as "Dr. Hershey" by his team members—provided his students with complete research freedom. When the students had a problem, Hershey would listen silently and deliver a written suggestion a few days later.

Bruce Wallace, one of Hershey's collaborators who later became University Distinguished Professor of Biology at Virginia Tech, reported an answer that Hershey gave to one student who did not understand a topic. This response provides a hint of Hershey's mentoring style, which was oriented on self-awareness:

> No, I won't explain. If I did, you would see how it works. If I don't and you think about it, you will see how it works and you will have done it yourself. I think that's best.

Hershey joined Delbrück and Luria in organizing and running the Phage Course at CSHL. When the course started, the war had already ended. In the south of Europe, the Allies began a massive assault against Italy, which had capitulated following Mussolini's imprisonment. Instead of alleviating the Italians from the miseries of war, Victor Emanuel III's decision enraged Hitler, who declared the erstwhile ally a traitor and ordered the occupation of the Italian peninsula. As a result, Italy, like northern France, had become a Nazi colony. As retaliation to Italy's change of side, Himmler even devised one of the first acts of biological warfare in modern times: he ordered the flooding of the marshlands around Rome, which the fascist regime had dearly reclaimed just a few years earlier, and reintroduced mosquitoes carrying malaria germs. As a result, more than 50,000 people acquired this disease in the following months.

The Allied advance prompted a riot in Naples (27–30 September 1943) that pushed away the German occupiers. The Anglo-American troops landed in Anzio, near Rome, on the 22nd of January 1944, and the capital was liberated on the 4th of June. Two days later, Operation Overlord, the largest invasion in human history since King Agamemnon attacked Troy, took place in Normandy, northern France, followed by Operation Anvil (15th of August), a landing in southern France. It was the beginning of the end of the Nazi empire. As in Italy, the arrival of the Allies prompted the revolt of the subjugated people: riots in Paris pushed away the Germans, and the city was freed on the 25th of August. Although the Germans fought back more than the Allies had expected—for instance, at Arnhem (during Operation Market Garden, led by Gen. Montgomery), in the Hüregen Forest, and in the Ardennes—the Allied troops swept through Europe like a Juggernaut. On the 3rd of September 1944, Brussels was freed, followed by Aachen—already within German territory—on the 18th of October, a feat that prompted a Communist uprising in Cologne at the end of the month. Trier was liberated on the 1st of March 1945, followed by Hannover (8th of April) and Magdeburg (11th of April).

On the East front, the advance of the Red Army instigated a riot in Warsaw (from the 1st of August until the 2nd of October 1944). However, in this case, the Nazis responded with bloody brutality, devastating the city. Stalin cunningly ordered the Red Army to slow down its advance purposely to allow the massacre of the Poles. The Soviets liberated Vienna on the 11th of April 1945 and met the Anglo-American troops at Thurgau on the Elbe on the 25th. However, the march of the Red Army was poisoned by the brutality shown by the Russian soldiers on the freed population, regardless of their nationality, and even fellow communists were brutalized. The war correspondent Natalya Gesse commented that "the Russians soldiers were raping every German female from eight to eighty," adding that "it was an army of rapists." It is estimated that Russian soldiers raped over two million Germans women during their march, resulting in countless waves of suicides—as well as fathers killing their daughters to avoid them ending up in Russian hands—and pressuring the Germans to keep fighting to avoid being captured by the Reds. Even Hitler did not want to be imprisoned in Russia. The Red Army began attacking Berlin, where the Führer had sought refuge, on the 16th of April 1945. He committed suicide on the 29th, together with his mistress Eva Braun, to avoid being captured by Soviet troops. On the 5th of May, Russian secret service agents from the SMERSH Department recovered his remains, which had been burned to escape detection.

Even on the Pacific front, the savagery of the Japanese Army could not halt the Americans' relentless advance. The Emperor had abandoned the Japanese soldiers, who received orders not to surrender at any cost but provided no support to fulfill the order. Several troops, particularly in New Guinea, organized hunting expeditions to kidnap and devour people, including Allied prisoners of war. While cannibalism was a desperate deed of isolated families in Leningrad, it was institutionalized and carried out systematically by Japanese soldiers.

While an armistice between Germany and the Allied forces was signed on the 9th of May, Japan continued its belligerence despite the evidence that it could not win the conflict. The Imperial Army surrendered on the 15th of August only after atomic bombs, developed through the Manhattan Project, were dropped on Hiroshima (on the 6th) and Nagasaki (on the 9th), killing about 160,000 people either directly from the blast or indirectly from radiation poisoning within four months. Despite the tragedy of the attack, the destruction of Hiroshima and Nagasaki was not qualitatively different from other so-called "strategic bombings," a concept introduced by Sir Arthur Harris—nicknamed "Bomber Baron"—Air Chief Marshal of the Royal Air Force, as a retaliation for the accidental bombing of London that took place on the 24th of August 1940 during Operation Sealion. Harris argued that civilians were as good targets as military installations because their deaths, combined with the destruction of urban monuments, would have broken the German morale, easing the victory for the Allied forces. The Germans labeled such air power as plain terrorism and developed, later on, "Vergeltungswaffen" ("vengeance weapons") 1 (an autonomous flying bomb) and 2 (a rocket) in retaliation for Harris' strategy. The Nazis, though, had little moral ground to criticize the British for their areal incursion: even though the first bombing of London had been an accident, the following Battle for Britain killed, just in London, 41,000 and injured another 137,000. The Americans initially refused Harris' scheme with disdain, but after losing hundreds of pilots during their first engagements in the war, they kowtowed to its effectiveness.

The Allies knew that the surrender of Japan through conventional means would have been a lengthy affair that would have cost hundreds of thousands of lives, both combatants and prisoners of war. Even more, intelligence reported that the Japanese Army was developing biological warfare. General Ishii Shirō's Unit 731 (officially named Kwantung Army Water Supply and Epidemic Prevention Department) had weaponized the etiological agents of plague, anthrax, cholera, and typhoid, testing them on thousands of prisoners of war, including American soldiers. The human guinea pigs in this massive production endeavor used for biological warfare were not referred to as subjects or, least of all, patients but as logs ("maruta" in Japanese). The term came from a joke made by one of the doctors employed in the project, who once commented that the unit was a saw-mill where "the people are the logs." The maruta were typically infected and then vivisected to study the spread of pathogens in real time. They were also used to study frostbites and as incubators for germs in ways akin to horror movies. Unit 731 was based in Pingfang in Manchuria and was complemented by several other branches, particularly Unit 100 (Anti-Epizootic Protection of Horses Unit), which developed veterinary and agricultural pathogens and was based in the Manchurian city of Xinjing, and a poison gas production complex near Hiroshima in Japan.

Shirō was a social climber and managed to embed his undertaking well into the academic system of the Japanese empire. Thousands of medical students were sent for internships within the Unit. Top professors were too compliant to publish empirical results from human experimentation, including vivisection, in high-ranked medical journals. Similarly, wicked doctors such as Josef Mengele and Carl Clauberg were conducting human experiments in the concentration camps scattered around Nazi Germany, albeit not focused on biological warfare. After the war, the US recruited many of Unit 731's top scientists instead of prosecuting them. A similar approach was taken with rocket scientists (Operation Paperclip), such as Wernher von Braun, who had developed the German "vengeance weapons." The use of war criminals to advance American scientific endeavors was made in the context of a confrontation with the USSR. Shirō, for one, despite all the suffering he caused, was granted legal immunity by the US government; he died of neck cancer in 1959, aged 65.

Unit 731's germs were weaponized within dedicated warheads produced by the finest Japanese porcelain maker, even inside fountain pens and chocolate bars. In 1938 and 1939, Unit 731 dropped bombs containing plague and cholera in Guangdong, China. The plague bacteria (*Ye pestis*), the pathogen of choice of the unit, were inoculated into rats bred by children, and prisoners of war were forced to feed fleas on the mice and then collect them back once infected. During the brief Soviet-Japanese Border War in 1939, the Halha River in Manchukuo was infected with salmonella and typhoid bacteria. In 1940 and 1941, Shirō's Unit tried another strategy by dropping plague-infested fleas and rice—which would attract rats, facilitating the spread of the disease—and pouring vials of typhoid bacteria into freshwater sources around the Chinese cities of Ningbo, Quzhou, and Changde. The resulting outbreaks killed thousands but were fortunately contained—although, in some locations, these diseases are still endemic to this today.

Shirō also devised a daredevil plan to attack the USA directly: a submarine would carry an airplane that, after reaching San Diego, would have spread plague-infested fleas over the city. The plan, dubbed Operation Cherry Blossom at Night (or Operation PX), was scheduled for the 22nd of September 1945. Hurling airplanes from a submarine might seem a ludicrous plot even today, but it was not far-fetched. Two weeks after the surrender of Japan, the US Navy captured and then sculled an enormous submersible, I-400, capable of carrying and launching three purpose-built dive bombers. Luckily for the Americans, this stealth aircraft carrier could not be employed in the conflict. The capitulation of Japan in August prevented the execution of Operation PX. In other accounts, chief of Staff Gen. Umezu Yoshijiro, who was imprisoned for war crimes after the capitulation, well aware of the risk of igniting a plague pandemic, vetoed Shirō's scheme because "Japan will earn the derision of the world." Hence, the Allies were conscious that if the Japanese had succeeded in striking a large-scale biological attack, the consequences might have been much more excruciating than the destruction of a couple of cities. There was high pressure to end the war as soon as possible. The atomic bomb provided such a shortcut.

The bombing of Hiroshima, thus, was not a cruel action that came out of the blue but simply another Monday in the 509th Composite Group of the United States Army Air Forces. The only difference was that only one B-29 Superfortress—the *Enola Gay*—was involved in hitting the city. The attack was only more effective. The destruction of these cities was carried out by only one bomber, compared with the 279 employed to attack Tokyo on the 9th of March. However, about 41 km^2 of the city were destroyed in Tokyo compared with 12 in Hiroshima and five in Nagasaki. The number of people killed was about 80,000 in Hiroshima, 40,000 in Nagasaki, and more than 83,000 in Tokyo. The bomb on Hiroshima (dubbed "Little Boy") generated 15 kilotons of energy. In comparison, the bombing of Berlin between August 1942 and March 1944 unleashed 17 kilotons of TNT (plus 16,000 tons of incendiary bombs), killing over 9000 civilians.

Despite the sorts of the conflict, in the mid-1940s, Luria teamed up with Raymond Latarjet, a French radiation researcher he had met in Paris. Latarjet visited Luria's laboratory and proposed a research plan to measure phages inside bacteria. They observed that the sensitivity to irradiation changed during the infection. Such an observation goes under the name of the Luria–Latarjet effect. They carried out an experiment that consisted "in infecting a homogeneous bacterial population with a known average number of particles of a given virus, and irradiating it with increasing doses at a given time during the growth process (latent period); then in determining, for each dose, the survival of the infective centers." Using *E. coli* B and phage T2, Luria and Latarjet reported that the former had a survival of 10^{-4} (1 CFU/mL of output for every 10,000 CFU/mL of input) and the latter 10^{-5} (1 PFU/mL of output for every 100,000 PFU/mL of input). The viability of T2 inside B is slightly higher due to the cell's shielding property.

Already in the 1930s, thanks to the work of Eugène and Elisabeth Wollman at the PI, it was understood that phages—and viruses in general—were no longer infectious soon after infection. It was, in fact, possible to lyse the bacterial cells at different time points and count the number of infectious phage particles; plaques were generated only several minutes after the primary infection. The interval between infection and the appearance of intracellular infectious particles was called "eclipse." Viruses were infectious before release from the cell, but why not before? It appeared that viruses entered a vegetative phase that removed their activity. Luria postulated that, once inside the cell, phages were broken down to their essential components, which were then replicated and recombined before reassembly in newly infectious particles.

However, the viability of the phage was not constant. Before five minutes post-infection (p.i.), Luria noted, "the survival curves are almost exponential." After five minutes p.i., the "survival for high doses increases more rapidly than for low doses." For instance, with 400 ergs/mm^2 of UV irradiation, the T2 survival was about 0.1 after 5 minutes of exposure and five at 10 minutes; with 800 ergs/mm^2, the survival was less than 0.02 after 5 minutes and about four at 10 minutes. Thus, doubling the dose caused a 0.2-fold increase at 400 ergs/mm^2 and 1.25-fold at 800. Furthermore, the situation changed after 11–12 minutes: "resistance to high doses drops progressively with time." For instance, the survival at 18 minutes to 1200 ergs/mm^2 was lower than that at 9 minutes.

Luria and Latarjet hypothesized that "the increase in resistance could be brought about would be the accumulation, around the particles, of some ultraviolet-absorbing material." Such a "'screening' hypothesis" might indicate the encapsulation of the viral genomes inside the newly synthesized capsids. Bacteriophage replication ceased to be a straightforward process but a series of steps, a feature that upset Delbrück's sense of harmonic simplicity.

5.5 THE PLEASURES OF COLD SPRING HARBOR

Delbrück and Luria started a course on phages in the summer of 1945, just a few days after the conflict's conclusion. Luria and Delbrück continued their work in idyllic peace. Delbrück molded this new Phage Course to the cast he had experienced in Bohr's department. The atmosphere was friendly and wholly informal. Luria described the routine of this summer school: "we worked, swam, square-danced, and held daily seminars."

James Watson, who attended the 1948 edition of the course, described a typical week at the camp:

> we played baseball next to Barbara McClintock's cornfield, into which the ball all too often went ... Practical jokes dominated the mood during the late summer in Cold Spring Harbor, culminating in an evening when Gordon Lark, Victor Bruce, and his sister Manny Delbrück, and I let the air out of the tires of several friends' cars parked before Neptune's Cave. Afterwards buckets of water were poured over our beds. On another evening, Visconti interrupted a staid Demerec social evening with an attack with a toy machine gun.

Delbrück would interact with the participants on many occasions, such as during tennis matches or at lunch in the canteen. On the other hand, Luria not only was not a good player—"why don't you learn enough to play with grownups?" Delbrück reproached him once—but he was "reluctant to enter the relation of almost intimacy with the students ... which Max evidently enjoyed."

The first edition of the course lasted nine days, from the 23rd of July to the 11th of August and was advertised as:

> the purpose of the course is to acquaint the student with some of the techniques used in bacteriophage research, and with the recent results of such work.
>
> The course will consist of about nine half-day laboratory periods and nine half-day study periods for the evaluation and discussion of the experiments performed by the students. No outside work should be planned by the student, since it may be expected that his full time will be occupied by this course.

An admission test was required to ensure the mathematical understanding of the underlying calculations necessary to determine the titers of bacteria and phages. August Dörmann was the laboratory assistant, along with one J. Reynolds. The first group of students was formed by Martha Baylor, Rollin Hotchkiss, Herman Kalckar, Phyllis Margaretten, Stuart Mudd, and Thorbjorn Sigurgeirsson (sponsored by Niels Bohr). The following year, the course was extended to three weeks, a format that remained unchanged for the next quarter of a century. Each edition had 14 pupils on average. Delbrück taught the course's first three editions and, after that, participated by giving individual speeches on various topics. Teachers and professors changed significantly between editions, but one constant was that they were all highly regarded phage researchers. An attendee commented that "nearly all of the instructors throughout the history of the phage course were either former students in the course or had other connections to Delbrück."

This summer boot camp covered the theory and practice of phage biology and drew some of the world's best scientists. Among the high-ranking participants was Leo Szilard, the same physicist who had started the Manhattan Project. In the last days of the atomic undertaking, Szilard had tried to convince Harry Truman, who had succeeded Roosevelt to the Presidency of the United States, to drop the bomb on an uninhabited island in the Pacific Ocean in front of an international delegation to show the devastating capabilities of the new explosive device without killing innocent people. His plea was refused because the military echelons believed that the Japanese would surrender only in front of a massive human hecatomb: Harris' strategy had been fully embraced and brought to its final consequences.

Franklin Stahl, one of the early students, commented that the course was colloquially known as the "Phage Church," which "was led by the Trinity of Delbrück, Luria, and Hersley … Delbrück's founder status and ex-cathedra manner made him the Pope." Eduard Kellenberger also gave a similar description. Segrè agreed, saying that Delbrück was the "joyful leader of the Phage Church," but also noted that 'the Pope's' "directness and brusqueness could be hurtful, and many did not approve." Others compared Delbrück to the *pater familias* of the group. James Watson commented that Delbrück "could be ruthlessly dismissive in his criticism and wounded many young people." Nevertheless, he added that "many students were afraid of Luria who had the reputation of being arrogant toward people who were wrong," although "I saw no evidence of the rumored inconsiderateness towards dimwits."

One student remarked that "the shorts-and-sandals atmosphere promoted an easy familiarity … the camaraderie made it easier to maintain the style of ruthless critique for which they [mostly Delbrück and Szilard] became known." In fact, despite his friendliness, Delbrück was also harshly mannered in criticizing someone else's research. His comments during many meetings, such as "this was the worst seminar I ever heard" or "I don't believe a word of it", became his trademark. Nonetheless, it has been remarked that "it was a friendly challenge … he or anyone else who had seemed just as harsh, might invite you for a swim or buy you a beer." There were also seminars and, in some cases, unclear presenters were interrupted in the middle of the talk and asked to come back the following day better prepared.

The Phage Course significantly impacted the attendees' scientific attitudes: it required them to take a scientific approach to their investigations, which meant developing and proving a theory through experiments. Waclaw Szybalski, who attended several editions of the happening, commented that the seminars "taught us to think in quantitative terms, to propose only such hypotheses as could be experimentally tested." He added that "our creativity was also stimulated by many bull sessions, Max's unrelenting criticism, and the many questions asked by Leo Szilard." In addition, the course was a place to advance science, not merely a training avenue of established protocols—even if state-of-the-art at the time. One student recalled the scientific process that took place during the course:

> a person gave birth to an idea, but then cast it out like a balloon into the surrounding atmosphere. At the point it ceased to belong to any one person; the other members poked and pushed at it, tried to see if it would pop. It wasn't you they criticized, it was the idea.

Another crucial aspect of the course was the graduation ceremony: an attendee recalled that it was an "evening of merriment" that "involved skits, silly costumes, and considerable amounts of ethanol-fortified drinks" as well as food fights, which "became an annual tradition at Cold Spring Harbor." Once graduated, the attendee became an official member of the Phage Group, which was thus an extension of the

Phage Course, an exclusive circle that, like many brotherhoods in American universities, maintained such engagement through annual activities that supported the sense of belonging to such an "extended family." One graduate recalled that "new recruits came into the group every year ... instructors and lecturers ... returned each summer ... scientists brought their families and everybody stayed on the bucolic campus ... much science was done on the tennis courts and beach as well."

Millard Susman, who attended the course in 1957 and later became an instructor, gave a vivid account of the atmosphere surrounding this prestigious happening:

> Graduation parties were part of the tradition ... probably because Delbrück liked parties ... The party was at the Prufrock House ... The celebration began with organized silliness; for example, Ilga [Lielausis], blindfolded, had to use her toes to determine the agar content of the medium in several petri dishes, and Ann [Roller] and John [Cairns] had to step into a closet and recombine five of their "phenotypes" (articles of clothing) ... The party ended with a water fight...
>
> Students ... felt that they had been initiated into the community of biologists. The course featured seminars by leaders in the field, and the students could sit and drink beer with them ... or chat with them on the beach.
>
> The pleasures of Cold Spring Harbor come not only from the science, which is wonderful, but also from the place, which is enchanting ... We would schedule labs in the very early morning in order to be able to dig clams in the afternoon. On a hot day, when the beach was particularly inviting, we would delay an experiment until late afternoon and head for the outer harbor.
>
> That is what the Phage Course and Cold Spring Harbor are all about — the wonder and the elegance of life.

The graduation meant the official introduction to the Phage Group as well as the right to subscribe to the "Phage Information Service," a mailing list curated by Delbrück to provide relevant information on the trends in phage research. This newsletter was a typed letter summarizing meetings and experimental results. The first number was released in March 1947 and was distributed by mail for 14 issues until 1959. On the other hand, the membership also implied that the associates were required to send drafts of their scientific papers to Delbrück for approval.

Even before starting the course, Delbrück was invited to join the University of Manchester in early 1945. This position would have helped him to stay closer to his family in Germany. While his mother had died during the war, he still had siblings and friends in Berlin he wanted to support in the critical years following the end of the conflict. Delbrück delivered four lessons in Manchester in preparation for employment. However, in December 1945, he received a proposal from George Beadle—the geneticist who shared the Nobel Prize with Tatum and chairman of the Division of Biology—to establish his chair of Biology at Caltech. Delbrück wrote to Bohr that he was "very happy about this, because it signals the completion of my metamorphosis into a biologist." Delbrück requested that, to take the position, he obtain the endorsement of a biochemist, Mark Adams, one of the instructors of the Phage course at CSHL and part of the Phage Group, whom Delbrück highly regarded. Albeit Beadle promptly accepted the request, Adams kindly refused because he preferred other avenues of research—sadly, he died prematurely in 1949.

Thus, Delbrück joined Caltech without a proper biochemist, knowing biochemistry would be the key to unlocking the phages' secrets. He was given a windowless office in the basement of the Department of Biology. Despite the humble location, Delbrück shaped the atmosphere of his laboratory to match that of the Phage Course at CSHL: apart from the usual informal environment, the Delbrücks organized frequent excursions with colleagues and students in the Joshua Tree National Park desert—tours that Luria shunned. Unsurprisingly, under Delbrück's guidance, Caltech became the world reference center for the research.

6 New York

6.1 THE MUSKETEERS OF THE BACTERIOPHAGES

"You are working for the knowledge and for the good of mankind," said Hershey to the audience attending the Nobel ceremony. He, Delbrück, and Luria were awarded the prestigious prize for their experiments that "first of all imply a deeper insight into the nature of viruses and virus diseases" and "bring about an increased understanding of inheritance" of genetic characters. Hershey's Nobel Lecture recalled the experiments that revealed the DNA structure. Luria talked about the alteration of the bacterial biochemistry carried out by phages. For instance, the genomes of T-even phages contained "hydroxymethyl cytosine (HMC) instead of cytosine in their DNA and determine … the production of an enzyme, deoxycytidine triphosphatase, that destroys dCTP, a specific precursor of host DNA," causing the degradation of the host genome and assuring the focus of the biochemical machinery to the synthesis of viral DNA. He added that "an even more intriguing situation is that of phages Oe of *Bacillus subtilis…*" which contain "hydroxymethyluracil (HMU) instead of thymine in their DNA and, upon infection, determine a series of enzymatic changes directed at converting the path of DNA synthesis from bacterial type to phage type DNA." Consequently, "host-DNA synthesis stops a few minutes after infection with phage Oe."

On the other hand, Delbrück's lecture provided more general terms, emphasizing the point of the irreconcilability between physics and biology (which was the fulcrum of Bohr's complementarity principle) and combining his high regard for Kepler. In antiquity, he declared, the difference between living and non-living organisms was blatant. For "Aristotle … it was obvious that things developed according to plans" because "every animal and plant is generated in some definite way, runs through a cycle of development in which it unfolds its inherent plan, and succumbs to death and decay." Instead, "astronomy is the exception and offers the contrast of an eternal periodic system subject to neither generation nor decay." Even in modern times, "life seemed to have unique properties quite irreducible to the world of physics," and "many physicists even today look upon biology as something outside their domain." The emergence of molecular biology—which he helped pioneer—had answered the "riddle of life" (that is, the duality of living and non-living) because "the ideas of information storage, of the replication of the stored information and of its programmed readout have become commonplace … the marvel that the mechanical and chemical machinery underlying all these affairs can, in fact, be worked out is keeping a host of scientists very happy and very busy." Nonetheless, he still had reservations about neurobiology.

Delbrück also addressed the distinctions between science and art:

> the books of the great scientists are gathering dust on the shelves of learned libraries. And rightly so. The scientist addresses an infinitesimal audience of fellow composers. His message is not devoid of universality but its universality is disembodied and anonymous.

DOI: 10.1201/9781003657712-6

> While the artist's communication is linked forever with its original form, that of the scientist is modified, amplified, fused with the ideas and results of others and melts into the stream of knowledge and ideas which forms our culture.
>
> The scientist has in common with the artist only this: that he can find no better retreat from the world than his work and also no stronger link with the world than his work.

Delbrück, Luria, and Hershey shared the Nobel podium because they pioneered the development of bacterial genetics, implying that this developing field of biology had become a mainstream topic. The work started by these three Nobel laureates was pushed forward by a whole group of people, some of whom will become Nobel laureates in their own right. Such a gathering of researchers became known as the "Phage Group."

The Phage Group met officially in Nashville in March 1947. It was formed by the European duo Hershey and Anderson, as well as by Mark Adams, Stanley Cohen, August Dörmann, and Max Zelle. Within a couple of years, Leo Szilard joined and started organizing monthly meetings attended by Joshua Lederberg, Herman Kalckar, and James Watson. In 1952, André Lwoff organized an International Phage Symposium at Abbaye de Royaumont, near Paris, decreeing the international nature of the Phage Group.

Meanwhile, the so-called Cold War between the West (the United States plus the countries that the Allies had freed during the conflict) and the East (the USSR plus the countries that the Soviets had occupied during the war) had erupted even before the Second World War was over. During a visit Churchill made in Moscow in October 1944, Stalin showed the British Prime Minister a note of how Europe should be split between themselves after the defeat of the Nazis:

> Romania: Russia 90%; the others 10%
> Greece: Britain (in accord with USA) 90%; Russia 10%
> Yugoslavia: 50% 50%
> Hungary: 50% 50%
> Bulgaria: Russia ~~75%~~ 90%; the others ~~25%~~ 10%.

Churchill felt uncomfortable facing the Russian dictator, and he found himself in an awful situation afterward because the worst conflict the human race had ever experienced began, on British insistence, to support Poland's freedom and ended with the incorporation of Poland within the Soviet empire. After 60 million deaths, Poland had simply swapped one tyranny for another. Stalin, just like Hitler, considered the Poles as second-rate people and started to deport Poles to make space for Russians. For instance, he forced the entire population of Lwów to move to Breslau, Germany, which became known as Wrocław. The British Field Marshall Brooke commented, "there is no doubt that from now onward Russia is all powerful in Europe."

Well aware of the risks the USSR would cause to international stability, Churchill had a military plan (Operation Unthinkable) drawn to push the Russians out of Poland. It was never implemented because it was well understood that the British veterans had no intention of starting another war for the Poles.

Even before Churchill had thought of Operation Unthinkable, still in 1944, Stalin had designed a similar plan involving the invasion of France and Italy with the support of the national communist parties. Beria dissuaded Stalin from executing the plan when his spies—chiefly among them a German émigrée, Klaus Emil Julius Fuchs, who worked in sensitive areas of the Manhattan Project—reported that the Americans had developed the atomic bomb: the retaliation would have been devastating for the USSR. Truman later commented that, had not been for fear of nuclear retaliation, "the Russians would have taken over Europe a long time ago."

The leak of atomic secrets to Russia allowed the Soviet scientists to develop an atomic bomb in 1949, well ahead of American expectations, testing it on the 19th of August. Truman had hoped to use nuclear dominance to relieve the civil society from the strain of military spending. Instead, he had to maintain a large military force in action, with a considerable contingency stationed in Europe as a preventive measure. In addition, he had to authorize the development of a weapon even more devastating than the atomic one: the thermonuclear (or hydrogen) bomb. This bomb, dubbed the "Super" back in the day, was initially designed by Edward Teller but had been opposed by many scientists, including Oppenheimer—who collaborated in its development, although not in a prominent position as in the Manhattan Project—and Szilard.

Teller, in collaboration with Stanislaw Ulam, completed the design of this novel nuclear device in March 1952; it was tested with the codename Ivy Mike in November on the Eniwetok Atoll, a small island in the Pacific Ocean and released an energy equivalent to over 10 million tons (Mt) of TNT. On the 1st of March 1954, the US tested the most potent bomb of its arsenal ever: Castle Bravo. The device was expected to release 5 Mt; instead, it released 15 Mt. The island of Bikini in the Marshall archipelago, where the test took place, was vaporized with two other adjacent islands. The radioactive fall-out covered an area of 130,000 km^2 (corresponding to the surface of Nicaragua) and hit several islands in the Marshall archipelago, as well as one Japanese trawler. Over 600 people were directly contaminated—and thousands suffered from the long-term effects of radioactive poisoning—and radioactive material ejected from the bomb was later recovered in Japan, Australia, the US itself, and even in Europe.

To compare the differences between the atomic and thermonuclear bombs, let's consider these examples. The detonation of Little Boy (15 kt, the one used in Hiroshima) in the center of Berlin would wreck everything from the Tiergarten to Volkspark Friedrichshain, some 4 km away but still well within the city limits. However, if the detonation had used Castle Bravo (15 Mt), the devastation would have extended from the village of Falkenrehde (West of Berlin) to that of Rehfelde-Dorf (East of Berlin), some 40 km away. If Fat Man (23 kt, the one used against Nagasaki) were to explode in the center of Rome, everything from the Vatican to the San Lorenzo district (5 km away) would have been destroyed. Nevertheless, suppose a NATO-issued thermonuclear device from the 1960s, such as the B28 (1.5 Mt), would have exploded instead, not only the immortal city would have been completely obliterated but also everything from Castel di Guido (West of Rome) to Torre Gaia (East of Rome), 27 km apart, would have been brushed away. The radioactive fall-out would have rendered inhabitable thousands of hectares for thousands of years. If the atomic bomb had brought Harris' strategic bombing to its most drastic consequences, the hydrogen bomb went well beyond even Göbbels's concept of total war into an outright nuclear holocaust.

Despite the massive investment in economic and social terms required to develop the thermonuclear bomb, though, the US supremacy in armaments did not last long. The Soviets detonated their version of the Super (RDS-37) on the 22nd of November 1955, exactly two years after Ivy Mike. After losing the dominance due to the sole ownership of nuclear weapons, the military high brass, instead of dismissing this newly developed and exceedingly lethal weapon as however useless in combat, opted for a Samson doctrine—"Let Samson die with all the Philistines," a concept that corresponds to the more Anglophone "cut off one's nose to spite one's face." The military strategists advised the President of the US—and in turn, the leaders of the USSR, UK, France, Pakistan, India, Israel, and North Korea—to stockpile nuclear arsenals (estimated 12,492 warheads thus far) under the assumption that it is better to destroy the whole world rather than lose a war. Churchill introduced the term "equality of annihilation" to describe this novel historical situation; Robert McNamara, the national security advisor for US President John Fitzgerald Kennedy, coined the term "mutually assured destruction".

Back in the day, Truman's decision to go ahead with the development of the thermonuclear bomb was primarily based on a strategic analysis provided by George Kennan, a diplomat stationed at the American embassy in Moscow. In a very long telegram (8000 words), he described how Stalin wanted "more": more land, more influence, more power. Kennan remarked how the Russians needed an external enemy (the USA) to justify their dictatorship because "without which they do not know how to rule." Kennan urged the President to exert a "patient but firm and vigilant containment of Russian expansive tendencies." Truman agreed, officially setting what became known as the Cold War in motion.

The term 'Cold War' has been used since the 19th century, but it has gained its contemporary meaning thanks to articles written by Graham Hutton in The Atlantic in August 1939 and George Orwell in The Tribune in October 1945. The name emphasized no proclaimed war and no overt military actions between countries. Still, there was a *de facto* confrontation between the two blocks of influence—the capitalist West and the communist East. In more modern times, such confrontation on a planetary scale has been called the Third World War outright. The post-war conflict between the USSR and the US/UK blocks was never declared but was fought at the political, economic, and intelligence levels. Nonetheless, several regional conflicts were supported by either block for their interest.

The first of these regional conflicts that had the potential to destroy the whole world through nuclear holocaust was the Berlin blockade. Since Stalin had refused to release the territories that the Red Army had occupied during the war, half of Europe was in the hands of the USSR. Specifically, Germany and Austria were divided into sectors: the eastern part of thc countries was under Russian occupation, and the western part was further subdivided among the US, UK, and France. Berlin was a case on its own because it was located entirely in the Soviet section of Germany but was split into the four occupation zones (East Berlin was in the Soviet area, West Berlin in the US/UK/France area). In June 1947, Truman started implementing the containment doctrine by providing vast economic support for European countries (the European Recovery Program, better known as the Marshall Plan) to prevent these countries from relying on the USSR. Stalin indignantly refused to accept what he saw as a way to sway people away from the USSR, preferring instead to leave the Eastern Europeans to starve and live on the rubble of their destroyed cities.

At the beginning of 1948, the US, UK, and France proposed to unify their sectors and create a single German state, an idea against Stalin's propositions. When, in June, the Western Allied unexpectedly introduced a new currency in West Germany (the Deutschmark), Stalin went berserk and imposed a blockade on all goods reaching West Berlin. The US and UK organized a massive aid operation to support the almost three million West Berliners from the risk of starvation: they carried over two million tons of items from food to fuel by air, with about 300,000 flights for almost a year. This support maneuver was an enormous logistic and economic burden for countries that had just ended a world war but proved successful: Stalin ordered the removal of such a useless blockade in May 1949.

The second of these early regional crises that risked triggering a thermonuclear exchange occurred in Southeast Asia just a few months later. Korea, like Europe, had been divided into occupation zones (American and Soviet), each controlled by nominal leaders: the pro-American Syngman Rhee in the South and the communist Kim Il-sung in the North. Each wanted to invade the other's territory. While the Americans had no interest in supporting another conflict, Stalin gave implicit consent to the invasion of the South. Thus, Kim Il-sung promptly raided South Korea in June 1950, igniting a bloody conflict that South Koreans could not manage to win on their own. Even with the US intervention, the North Koreans managed to conquer large swaths of the southern state. However, Gen. Douglas McArthur, who had led the US forces against the Japanese during the Second World War, organized a successful raid behind the enemy lines that cut the invading forces from the mainland. Following the American counteroffensive, Stalin promptly abandoned the North Koreans to their fate. Luckily for Kim Il-sung, though, another communist dictatorship had been established in China by Mao Tsetung, who decided to support his fellow comrades in the South. The Chinese managed to repel the US army. McArthur requested the use of atomic bombs, but—luckily for humanity—Truman categorically refused, even if that meant that the Americans were almost pushed out of the Korean peninsula. Eventually, the US-backed troops organized yet another counteroffensive, and the conflict ended in July 1953, essentially re-establishing the pre-war borders.

There were numerous instances of attrition between the United States and the Soviet Union during the Cold War, where a nuclear exchange was narrowly avoided. As a result, while scientists worked to improve the world, politicians and the military brass sought to destroy it. The terror for mutual annihilation was the background against which the Phage Church and the world were working during the Cold War years.

6.2 THE REVENANTS

One day in 1946, Delbrück and William Thomas Bailey observed that UV-irradiated phages could generate plaques again. Irradiated phages were considered too damaged to be infective; instead, they have been somehow brought back to "life"—since viruses do not have an independent metabolic apparatus, they need to hijack that of their host; thus, the idea that viruses are alive is still today hotly debated. Delbrück

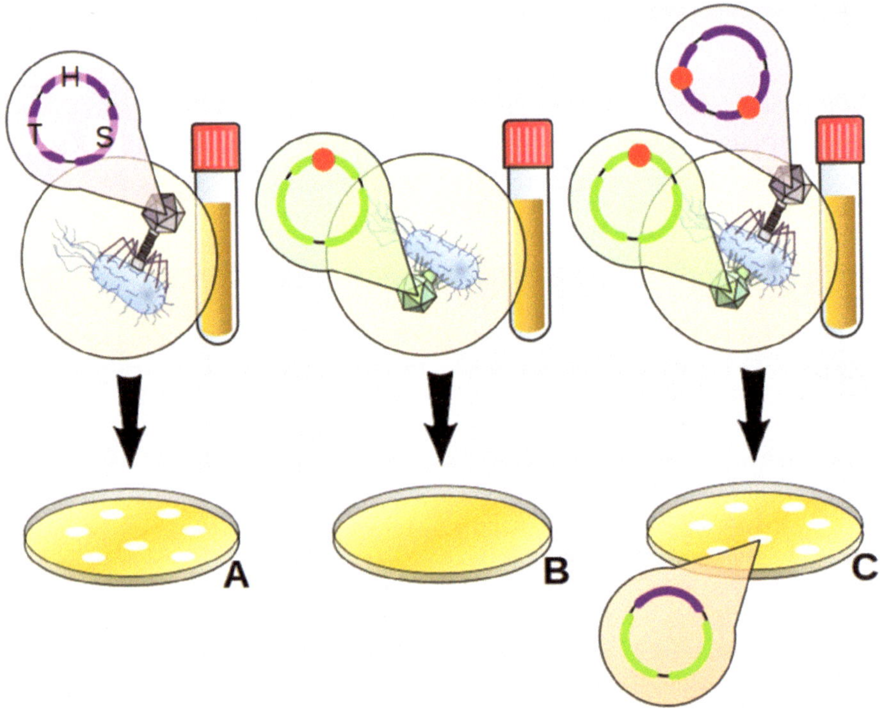

FIGURE 6.1 *Phage reactivation*. Idealization of the recombination experiment. To generate an infectious spawn, a phage needs functional genes (or "units") encoding the proteins forming the head (H), tail (T), and sheath (S) (panel A). UV irradiation might cause mutations that generate non-functional versions of these genes (dots). Phages carrying at least one non-functional gene cannot replicate (panel B). When a bacterium is infected by multiple irradiation phages (panel C), its genome may recombine, forming a chimera virus with functional genes originating from either progenitor. However, for recombination to occur, the progenitors must be genetically related.

did not follow this path, but Luria did. In a paper dated 1947, Luria noted that "an irradiated sample can be diluted 1:10 into a heavily [highly concentrated] bacterial suspension, and after a few minutes—during which no phage liberation occurs—diluted again 1:10^4 in broth and plated" will contain more phages than one sample that "had been first diluted 1:10^4 in broth, then 1:10 in bacterial suspension." He hypothesized that there should be a factor "capable of causing the reactivation of inactivated phage particles" (Figure 6.1).

Reactivation was not a universal process; it worked only for large phages (T2, T4, T5, and T6) but not for the small ones (T1 and T7). More importantly, reactivation occurred when a single bacterium was infected by multiple phage types simultaneously. For instance, Luria reported that "T6 can reactivate T2 in presence of strain B, sensitive to both, but not in presence of B/6, sensitive to T2 and resistant to T6." It was possible to determine the fraction of bacteria that were infected by more than one phage:

> if the average number of phage particles adsorbed per cell is x, the proportion of cells with more than one phage particle is given by the expression: $1-(x+1)e^{-x}$.

When x—which today is better referred to as multiplicity of infection, MOI—reaches zero, the expression becomes $1-e^{-0}=1\ -1=0$. This means that one needs a large number of phages to obtain reactivation. Luria remarked that

> we shall assume that a phage particle contains a certain number of different self-reproducing "units" (loci), each capable of undergoing a lethal mutation under the action of radiation. Reactivation depends on transfer of these units, the requirement for reactivation being that a given locus does *not* carry a lethal mutation in *all* of the particles that infect the same cell.

It is possible to estimate the number of units within a viral genome that can give rise to lethal mutants using the formula $y=[1-(1-e^{-r/n})^k]^n$, where y is the probability that n units are mutated in k particles adsorbed to the name cell, r is the number of mutations per particle. The latter quantity can be calculated as $r=\ln(P_0/P)$, where P_0 and P are the initial and final phage titers. For low doses of radiation, the probability of having two particles adsorbed to the same bacterium becomes $y=e^{-r^2/n}$.

Luria noted that "the probability of reactivation increases with increasing values of k, that is, of the number of phage particles adsorbed per bacterium." He also reported that it would be possible to increment the number of mutations by increasing the intensity of radiation on the phages. With T2 phages irradiated with UV at 400 erg/mm^2, Luria obtained y values of 0.385 for $x=0.39$ and 0.606 for $x=7.8$. Doubling the radiation's strength, the y values were 0.017 for $x=0.39$ and 0.189 for $x=7.8$. In essence, the results showed that the UV radiation dose was directly proportional to the number of mutations per particle r. In addition, the higher the number of particles adsorbed to a cell (x), the greater the probability of recombination. Luria concluded that "a virus particle may rather be comparable to a gene complex than to an individual gene … the virus particle might better be compared to a chromosome."

Renato Dulbecco and Albert Kelner discovered independently that UV-irradiated phages could become fully infective simply by photo-reactivation. In particular, Dulbecco observed that plates on top of the stacks left on the bench gave higher phage titers than those on the bottom. It turned out that visible light was enough to remove the mutations that had inactivated the phages. Such an explanation also solved the problem of high variability in phage counts: different light conditions were enough to hamper the reproducibility of the results. Delbrück later expanded on this observation by introducing the "principle of limited sloppiness," which stated that a bit of carelessness in the experiments is required to discover novel events by serendipity.

6.3 LINKAGE DISEQUILIBRIUM

Since Mendel's time, genetic information analysis has been grounded in a simple concept. Consider the phenotypes α and β, which may correspond to the color and texture of a pea seed, the eye color and wing shape of a fruit fly, or the metabolization of an amino acid and resistance to an antibiotic in a bacterium. These phenotypes correspond to the specific genotypes in various flavors or alleles, namely *a*/*A* and *b*/*B*. The combinations *aa*, *aA*, and *AA* will generate the phenotype α and *bb*, *bB*, and *BB* will determine β. Eukaryotic organisms such as pea plants, fruit flies, and humans carry two alleles—they are known as diploid from the Greek for "duplex-shaped"—that can recombine with specific frequencies. For instance, parents with genetic profiles such as *aa*/*bb* and *AA*/*BB* generate progeny with genotypes *aA*/*bB* 100% of the time, but parental profiles are observed 0% of the time. Parents with a genetic profile *aA*/*bB* will generate a progeny with a genotype *aA*/*bB* 25% of the time and *aa*/*bb* or *AA*/*BB* only 6% of the time. This sort of recombination—known as dihybrid crossing because it considers the inheritance of two genetic traits—is visually rendered by a table (square) introduced by Reginald C. Punnett in 1905. In the Punnett square, each column and row will contain the combination of alleles derived from each parent, and the cells will contain the combination inherited by the progeny (Table 6.1).

Such a simple inheritance pattern is based on the alleles passed to the progeny as statistically independent units. In other words, the transfer of these alleles corresponded to the toss of two coins: the first coin represents the phenotype α and the second the phenotype β. If, on the other hand, certain traits are inherited simultaneously, one would conclude that these entities were not independent, much as always getting tails when tossing a coin suggests the coin is not fair. Such a phenomenon is known as "linkage disequilibrium." Let us consider four traits: α, β, γ, and δ. The mating of parents both with heterozygous genotypes (*aAbBcCdD*) will generate a prole with parental genotype 6% of the time and with homozygous genotype *aabbccdd* only 0.4% of the time. Instead, the mating of homozygous parents *aabbccdd* and *AABBCCDD* will generate heterozygous prole with genotype *aAbBcCdD* 100% of the time. Variations from these well-defined frequencies will indicate that the traits are not independent but in linkage disequilibrium.

Genetic mapping, a process introduced by Morgan's student Alfred Sturtevant, takes advantage of the linkage disequilibrium of specific traits to measure their

TABLE 6.1
Example of a Punnett Table for a Dihybrid Cross[a]

		♂	
		ab	***AB***
♀	*ab*	⚥ *ab*/*ab*	⚥ *ab*/*AB*
	AB	⚥ *AB*/*ab*	⚥ *AB*/*AB*

[a] Parental strains are indicated with the symbols ♀ and ♂; the resulting spawn is indicated by the symbol ⚥. The parental strains are both heterozygous (they contain both *a*/*A* and *b*/*B* alleles); 25% (one cell out of four) of the spawn will be *ab*/*ab* homozygous; 25% *AB*/*AB* homozygous; 50% (two cells out of four) will have the parental genome.

distance. This procedure considers the genotype traits physically linked on a string of some sort, an entity called a "chromosome." Thus, it is possible to consider a string with the phenotypes α—β—γ—δ— ... , assuming the phenotypes *a*—*b*—*c*—*d*— ... and A—B—C—D— ... on the homozygous parents. When the prole is generated, it is assumed that these strings of genotypes will physically interact, forming a new string with mixed traits, a process known as "crossing over." For instance, the recombination between parents with genotypes, say, *a*—*b*—*c*—*d* and *A*—*B*—*C*—*D* will produce recombinants via the reaction:

$$a—b—c—d \times \boldsymbol{A—B—C—D} \rightarrow a—b—\boldsymbol{C—D} + \boldsymbol{A—B}—c—d,$$

through an intermediate that can be represented as

$$\begin{array}{c} a—b\ \boldsymbol{C—D} \\ \times \\ \boldsymbol{A—B}\ c—d. \end{array}$$

The more frequently two traits are passed together in the progeny, the closer they are in this statistical map. For instance, in the example above, one could redraw the map as: *a*—*b*———*C*—*D*. Distances on Sturtevant's map—more formally, linkage map—are measured in "centimorgans" (cM), corresponding to a 1% chance that two traits will end up on different chromosomes during crossing over. As blunt as this instrument might seem, it was instrumental in unleashing the secrets of genetic structure.

John Cairns measured the length of the phage T2 genome using a typical physicist's approach. Knowing the properties of radioactive atoms, Cairns assessed that "DNA containing [^{3}H]thymine of specific activity 10 c/m-mole will have roughly one disintegration per micron of double helix per week and should produce a near-continuous line of grains along its length after a few weeks' exposure." Thus, he grew phage-bearing bacteria in a medium containing radioactive thymines, extracted the phages' DNA, placed the nucleic acids on gel-coated microscope slides over a photographic plate, and left them there for a couple of months. What he got after this lengthy experiment, which he dubbed "autoradiography," were pictures of dots arranged in strings, whose length was proportional to the size of the viral genomes. He obtained a mean length of 52 μm corresponding to a weight of 110 million Daltons. Using autoradiography, Hershey and colleagues assessed that the T2 genome consisted of a single molecule of DNA and obtained a weight of 130 million Daltons (77 million for T5 and 31 million for phage λ). Cyrus Levinthal and colleagues at the Massachusetts Institute of Technology (MIT) employed several techniques aside from autoradiography—including zone sedimentation and fraction chromatography, the latter devised by Hershey—to confirm that T2 virions contained a single molecule of DNA. George Streisinger and Victor Bruce at the California Institute of Technology (CIT) confirmed that phages T2 and T4 had a single "continuous genetic map, and therefore, presumably, a continuous genetic structure."

Later on, the genetic map showed that some phages have a circular genome. George Streisinger, Robert Stuart Edgar, and Georgetta Denhardt traced a series of markers on T4's genome and measured the recombination frequencies in the viral spawn. They observed that "*am54* is closely linked to *r67* ... and *am85* is closely linked to *tu42b*;" thus, these markers constituted a genetic cluster, whereas the markers *tu41*, *h42*, and *ac41* formed another cluster, a relation can mapped as:

tu41—h42—ac41————————tu42b—am85———am54—r67.

The authors observed that "the linkage map may have ends [linear] ... or the linkage map could be a circle." However, "the cross *r67 h42 ac41*⁺ × *r67*⁺ *h42*⁺ *ac*41 distinguishes between these alternatives," where the '+' superscript indicates the presence of a functional allele of a genetic locus (trait). The results showed that "the *r* marker is more closely linked to *h* than to *ac*." Thus, the trio concluded that "the linkage map of phage T4 is circular." Later, it was even discovered that phages—as viruses in general—can have single-stranded genomes. Robert Sinsheimer, an American biochemist, was the first to report about these organisms, showing that phage ϕχ174 has a genome formed by circular single-stranded DNA molecules.

Delbrück tried to stay in this field by publishing a paper in 1952 on population genetics together with Niccolò Visconti di Modrone—who attended the phage course in 1950 and had been described as a "very aristocratic, best looking, and extensively brilliant geneticist." The paper tackled a critical problem in recombination analysis: the higher the number of genomes in a cell, the higher the recombination outcomes. Delbrück and Visconti commented that "one had been forced to assume that the parental particles break up into subunits which multiply independently of each other and are later reassembled into complete particles ... after an interval of time in which each particle has mated on the average *m* times, there will be a Poisson distribution, $p_r = m^r e^{-m}/r!$, of the number of particles that have mated," where *r* is a chosen time interval. The situation is way more complicated than the four-way Punnett table derived from the example of the above two alleles/two genes. For example, the probability pr that a genome could recombine once in a unitary time unit is 37%, while pr for recombining twice is 27% and 15% for triple recombination.

The complexity of the recombination process escalates rather quickly. For instance, Delbrück and Visconti reported that, for a set of genetic markers (or factors) 1–3,

> this segregation can occur in four possible ways, namely, (0) into the parental types, (1) such as that factor 1 has been segregated from 2 and 3, (2) such that factor 2 has been segregated from 1 and 3, and (3) such that factor 3 has been segregated from 1 and 2. We will designate the probabilities of these four alternative events with c_0, c_1, c_2, and c_3, the sum of which is equal to unity. The probabilities c_r are related in a simple manner to the probabilities prs of recombination between any two factors, namely, $c_r + c_s = p_{rs}$.

If all factors were independent, the probability of obtaining the parental allele c_0 would be 25%; if the markers 1 and 2 were linked, the probability of c_0 would be $\frac{1}{2}(1 - p_{12})$; if all markers were linked, c_0 would be $(1 - p_{12} - p_{23})(1 - p_{12})$.

Delbrück and Visconti covered many other combinations, but it was clear that phage recombination was not an easy issue.

However, this avenue of research did not go anywhere; Frank Stahl even commented that the Delbrück-Visconti paper was "one of the worst ever written," and provided an alternative theory in a work published a few years later. Nonetheless, the theory provided by Stahl (in collaboration with Edgar and Jacob Steinberg) was even more complicated than Delbrück and Visconti's. For instance, the authors presented a switch model where the probability of recombination per mating process for two markers separated by a distance D and for a synapsis of length K was $\frac{1}{2}(D-K)(1-e^{-2x})+K\{1-[(1-e^{-2x})(2x)^{-1}]\}$ when D was larger than K and $\frac{1}{2}(K-D)[1-\exp(-2DxK^{-1})]+D\{1-[1-\exp(-2DxK^{-1})(2DxK^{-1})]\}$ when D was smaller than K, with x being the average number of crossovers for the selected region of the viral genome. Punnett could simply determine the probability for each allele by counting the number of cells in his tables. For instance, a two-way table (two alleles, two loci) would have four cells; thus, 1/4 = 0.25 (25%); a three-way table (three alleles, three loci) would give 1/9 = 0.11 (11%); and so on.

On the other hand, Stahl and colleagues had to use an "electronic computer" to calculate the different probabilities. Given that in the 1960s, when the paper was written, computers were a fancy luxury used chiefly in engineering and physics problems, it is possible to guess how labor-intensive phage recombination was. Such a shift in computational power shows that Sturtevant's method had reached its limits, and it was high time that a more fundamental way of analyzing genetic information was found.

6.4 BREAKING GLASS

In 1952, Luria stumbled across another fundamental aspect of virology:

> I came upon a peculiar class of bacterial mutants … When infected by a certain phage the mutants were killed but seemed to produce no phage … one day, the test tube containing the phage-sensitive bacterial culture I was going to use happened to break. I have never been a very neat laboratory worker … I got from my colleague Gio Bertani a sample of completely different bacteria called Shigella … it worked only too well … My mutant bacteria had not failed to produce phage; they had produced a modified phage that refused to grow in its usual bacterial host but grew perfectly well in Bertani's bacteria … I had discovered the first instance of the phenomenon of *restriction* and *modification* … My finding was completely serendipitous.

Working with Mary Human, Luria reported that although it was recognized that "the properties of virus particle are unaffected by the host in which they grow," they "have encountered a novel situation: the genotype of the host in which a virus reproduces affects the phenotype of the new virus." They observed a mutant T2, named T*2, that was morphologically similar to the parent strain since it was recognized by antibodies raised against T2. T*2 could produce plaques in the same manner as T2 when the host was *S. dysenteriae* strain Sh. Still, the production of plaques in either *E. coli* B or *S. dysenteriae* Sh depended on the age of the original B strain. There would be no plaques if the *E. coli* B cells where T*2 was adsorbed were fresh.

Conversely, if the *E. coli* B cells were old, there would be a progeny of T2 ("without admixed T*2") at an output of one CFU for every 100 inputs. In other words, "the T* form of phage … fails to multiply in young, actively metabolizing cells but multiplies in an appreciable fraction of old, starved bacteria." Such observation represented "a hitherto unsuspected plasticity of virus properties … controlled in part by nongenetic, 'cytoplasmic' components of the virus particles."

The following year, another of Luria's students, Giovanni Bertani, in collaboration with one of Delbruck's collaborators, Jean Weigle, reported something similar. Bertani was born in Como (north of Italy) but was raised in Naples (in the South of Italy). He obtained a degree in zoology at the University of Milan, Italy, just after the end of the Second World War and then extended his studies in Naples and Zurich before joining Demerec's group at CSHL and then, in 1949, Luria's laboratory at the University of Indiana. He recalled that "Luria, Jim Watson, and I were sharing a smallish laboratory, from which a corner had been cut out for Luria's desk"—in pure d'Hérelle style. Bertani attended the 1949 edition of the Phage Course and joined Delbrück at Caltech in 1954. Weigle was a professor of physics at the University of Geneva but retired early due to health conditions and then started to cooperate with Delbrück at Caltech.

Bertani and Weigle used a non-lysogenic derivative of *E. coli* K-12 (that is, a K-12 without λ), *E. coli* B, another *E. coli* strain (C), and phage λ (which had become the workhorse of phage research). They noticed that "phage λ forms plaques with equal efficiency" on both B and C. They isolated the phage progenies produced from these infected cells and re-infected the bacteria. "The phage produced in C cells infected with λ, however, is different from that produced in S cells: it has lost its ability to grow in S cells while retaining the ability to grow in C cells." In other words, λ phages produced from B cells could always re-infect B and C, while those obtained from C could always re-infect C but could infect B only 0.01% of the time. Thus, in the space of a single generation, λ had drastically changed. The speed in genetic alteration made it "very unlikely that the change … is due to a mutation."

One of the pitfalls of the plaque assay is that it fails to account for bacteria affected by some essential genes, such as DNA replication. An alternative assay was based on the production of phage-encoded lysozyme, which occurs late in the infection and can be associated with completing the replication cycle. It was shown that the levels of lysozyme were unaltered when irradiation was given before a 4-minute threshold (instead of the 5 minutes observed with the plaque assay) but decreased in a time-dependent (or, rather, dose-dependent) manner after that. Dörmann confirmed this observation using a method that liberated virions from bacteria. Recombination could take place during this eclipse phase.

Werner Abner, at the University of Geneva, summarized the effect by saying that "if phage λ·K, grown on the bacterial host *Escherichia coli* K-12, is used to infect a new host strain, two alternative results are possible … (1) the infecting phage DNA is accepted by the host … [or] (2) [the phage] grows only exceptionally." Abner observed that the phage DNA was degraded within the recipient cell by marking the viral nucleic acid with ^{32}P, allowing ¼ of an hour for adsorption on starved cells, and giving 15 minutes of growth in a fresh medium. Afterward, he spun the suspension and observed that the radioactivity was limited to the supernatant. That meant that the viral DNA was fragmented and ejected outside the bacteria. On the other hand, when he used the K-12 derivatives C600 and W3350, most of the radioactivity still resided with the bacterial cells (that is, it was present in the precipitate). With the collaboration of Matthew Meselson, Werner Arber could isolate the restriction factor, identifying it with an enzyme called restriction enzyme.

Today, we know that almost 20,000 restriction enzymes are classified into four main types (I–IV). The first type I restriction enzyme was isolated from K-12 by Meselson and Robert Yuan at Harvard University in 1968. Hamilton Smith and Kent Wilcox isolated the first type II restriction enzyme at the Johns Hopkins Medical School from *Haemophilus influenzae* serotype d (*Hind*II, originally called endonuclease R). Kathleen Danna and Daniel Nathans demonstrated that R could fragment the genome of the simian virus 40 (SV40) into fragments of specific length.

The possibility of digesting the same genome with different restriction enzymes opened the possibility of obtaining "specific fragments useful for the determination of nucleotide sequence." In other words, restriction enzymes allowed the generation of a "cleavage map ... a physical map based on specific sites susceptible to endonucleases." In a cleavage map—today better known as restriction map—a genome is digested with a combination of endonucleases, the lengths of the generated fragments to an arbitrary zero point that usually cut the genome only once (in the case of SV40, such an enzyme was *Eco*RI), the lengths of the fragments are recorded and then aligned to the origin. For instance, one might digest a genome of an arbitrary 10,000 bp genome with a combination of endonucleases (EN) and obtain the following fragments:

Combination	Endonucleases	Length
1	EcoRI	10,000
2	EcoRI+HindII	1000, 9000
3	EcoRI+HpaIV	2500, 7500
4	HindII+HpaIV	1000, 1500, 7500

The fragments are recomposed as follows:

```
EcoRI          HpaIV
  ⇓              ⇓
  |— — — — — — — — — — — — — — — — — — — — — 1
  |— —|— — — — — — — — — — — — — — — — — — — 2
  |— — — — —|— — — — — — — — — — — — — — — — 3
   — —|— — —|— — — — — — — — — — — — — — — — 4
      ⇑
    HindII
```

Restriction mapping was a significant step up over the linkage map, providing a direct observation of the microbial genomes.

Allan Maxam and Walter Gilbert brought the method to its limits in the late 1970s. In the Maxam–Gilbert method, the DNA is digested either As/Gs, Ts/Cs, Cs, or Gs and the fragments are arranged by length as in the restriction map. This approach allows us to read the nucleic acid directly rather than large chunks of them, as in the following example. Let us assume a DNA fragment of ten base pairs and a digestion pattern of this kind:

Combination	Target	Length
1	A+G	1, 3, 5, 8, 9, 10
2	T+C	2, 4, 6, 7
3	G	3, 5, 10
4	C	4, 7

The fragments are aligned using the molecular weight (the length of the fragments) as a reference (here represented by *):

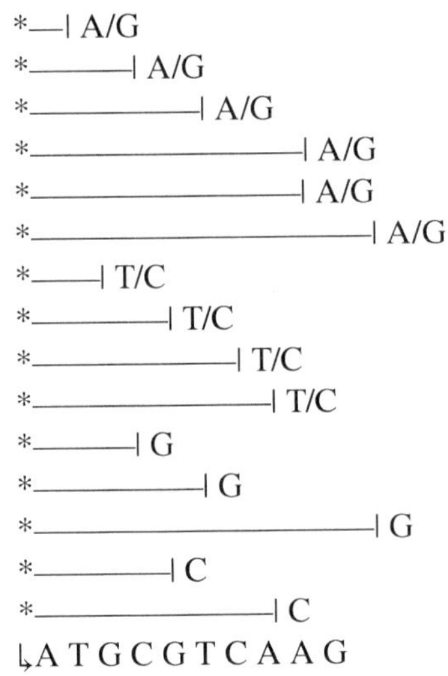

The Maxam–Gilbert method lost popularity when Frederick Sanger introduced the dye-terminator method just a few years later. This method, unlike the others, does not rely on DNA fragmentation but on the use of special nucleotides (dideoxynucleotides) that block additional polymerization of the chain. By adding a terminator specific for each base, it is possible to immediately determine the sequence of nucleotide in a DNA fragment faster and cheaper than with the Maxam–Gilbert method. For instance,

A-terminators:

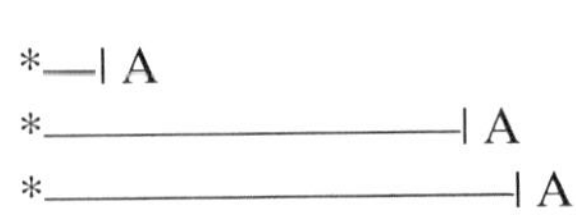

T-terminators:

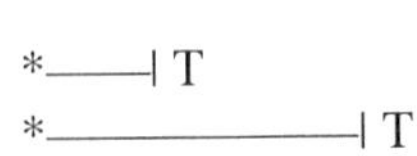

C-terminators:

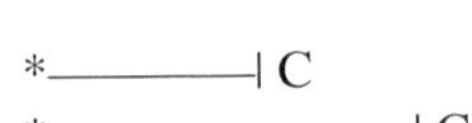

G-terminators:

```
*——| G
*————| G
*——————————| G
↳A T G C G T C A A G
```

In addition, the possibility of fragmenting a specific section of the genome combined with the capability of separating and recovering the fragments by electrophoresis made it possible to isolate desired pieces of DNA, opening the gate for recombinant DNA technology.

6.5 THE BLENDER EXPERIMENT

Despite the progress in understanding the process of genetic inheritance, in the mid-20th century, it was still unknown what was the carrier of the genetic information. Luria wrote in 1945 that

> we should not forget the lack of any direct evidence of the presence, in bacteria or viruses, of "genes" in the sense of discrete material units, whose existence in higher organisms is provided by linkage studies.

In 1928, Frederick Griffith reported a curious occurrence while working with two morphologically different strains of pneumococcus (*Diplococcus pneumoniae*, now known as *Streptococcus pneumoniae*). Strain S formed smooth colonies on Petri dishes and could kill mice injected subcutaneously; conversely, strain R developed into colonies with rough surfaces but was non-pathogenic.

Griffith injected R bacteria with heat-inactivated S bacteria and observed that he could isolate live S bacteria. The dead S bacteria had been resurrected; more accurately, the R bacteria had been transformed into the S form. Later on, Oswald Theodore Avery at RF determined that the transforming factor could be circumscribed in the DNA: cleavage of this acid hampered the transformation. The Austrian Erwin Chargaff—another Jew who flew to America following the Nazi repression—reported that the amount of adenine (A) in a cell equaled that of thymine (T); similarly, the amount of cytosine (C) was the same as guanidine (G)—this equality now goes under the name of Chargaff's rule. However, he also noticed that the relative amounts of ATs and CGs varied in different animals.

Rollin Douglas Hotchkiss, who had taken part in the inaugural edition of the Phage Course, joined Avery's laboratory in 1946 and took up the work after his boss retired in 1948. Hotchkiss substantiated once and for all that DNA was the carrier of genetic information. In 1949, Hotchkiss reduced the percentage of protein contaminants in the bacterial extracts to less than 0.02%. Thus, he proved that proteins were not responsible for transferring phenotypic tracts from one bacterium to another. Hotchkiss showed that the chromatographic DNA patterns differed in pneumococcal strains, an observation also recorded by Erwin Chargaff. Afterward, he also demonstrated that highly purified DNA extracts could transmit resistance to penicillin and multiple phenotypes.

The chromatography principle works a little like this: imagine an airplane, a train, a car, and a bicycle. Then, let them go off at the same time. Within a few minutes, they will split: the airplane will be farther ahead, followed by the train, then the car, and finally the bicycle. Each vehicle had moved in a medium (the air) but at different speeds. In chromatography, molecules move through a medium—in this case, paper but, usually, a gel made of agarose, acrylamide, or other suitable porous materials—according to properties such as volume and electrical charge of the molecule. In paper chromatography, given a time window of about 24 hours, thymine will move farthest, followed by adenine, cytosine, and guanidine. These nucleotides will form specific bands that can be individually cut, releasing their content into a solution. UV-spectrometry can then evaluate the amount of nucleotides inside each band: the higher the decrease of incident light passing through the sample, the higher the concentration of nucleotides.

Today, it is known that S bacteria encode a polysaccharide that is transferred to the cell's surface, where it polymerases, forming a protective shell (capsule) surrounding the cell. Such a shell masks the rugosity of the bacterial outer membrane, interspersed with proteins, giving it a smooth appearance. The capsule is also a virulence factor because it protects from phagocytosis, allowing the bacterium to replicate unmolested until the host succumbs. Different enzymes are involved in synthesizing the capsule polysaccharide; one is *cap3*, a gene about 780 base pairs long. The bacterial cell bursts open during heat inactivation, and the streptococcal genome is fragmented. Bacteria have evolved to internalize DNA fragments from the environment, a process known as "transformation." With the transformation, bacteria can import foreign DNA, process it, and insert it into their genome, another instance of somatic adaptation. If the inserted fragment contains a properly structured coding sequence, the bacterium has acquired an additional gene and extended its genetic repertoire. Otherwise, it has added junk to its nucleoid, and little harm is done. Griffith observed precisely this phenomenon of transformation: some R-type bacteria imported properly structured S-type genome fragments, and after inserting them into their nucleoid, they started to encode the capsule virulence factor. Such an observation revolutionized the field of microbiology and the nascent field of bacterial genetics.

Stimulated by these seminal experiments, Alfred Hershey understood that DNA was the carrier of genetic information. In contrast, Delbrück considered it "a stupid molecule" that left a vital question unresolved: "we are thus confronted with the paradox that phage enters its … cycle containing all DNA and no protein, and seems to emerge from the cycle all protein and no DNA." The DNA had been discovered by the Swiss physician Johannes Friedrich Miescher in 1868. Only afterward it was realized by the scientific community that there was an additional nucleic acid, ribonucleic acid (RNA); Hershey was one of the first researchers who differentiated RNA from DNA.

Hershey married Harriet Davidson in 1945, but in 1950, he re-married Martha Cowles Chase, his research associate at the Carnegie Institution in Washington (which he directed from 1962). The couple split a few years later; Chase married another colleague, Richard Epstein—but they also divorced. Although Martha Chase obtained a Ph.D. under the tutelage of Giuseppe Bertani, she always felt depreciated—regarding herself as "just an underpaid technician"—and eventually left academia for good.

In 1951, Hershey and Chase attempted to discover the essence of the gene using only a centrifuge and a blender. They infected *E. coli* strain B with phage T2 using a medium containing radioactive isotopes. As a result, the viral offspring included the radioactive isotopes of phosphorous (^{32}P) in its DNA chain and sulfur (^{35}S) in the capsid proteins. They marked the phagial components with either ^{32}P or ^{35}S. They ran the experiments in parallel with a single detection system—since the former isotope releases beta-radiation at 76 keV and the latter at 49 keV, it could be possible to detect their emissions at 16 and 25 nm, respectively, allowing to run the experiment at once. Hershey and Chase ensured that the isotopes were adequately incorporated into the viral components by lysing the phages with osmotic shock (diluting the phages in a high concentration of sodium chloride) and centrifuging the suspensions. The pellets emitted a signal when ^{35}S was added, indicating that the capsid had been marked as expected. Conversely, the supernatants, which contained nucleic acids, emitted a signal in the presence of ^{32}P.

Having established their experimental procedure, Hershey and Chase allowed T2 to infect their host for specific periods and then vortex the suspension in a Waring blender. After 2 minutes, most virions were detached from their hosts, showing that about 80% of ^{35}S was present in the pellet compared with 30% of ^{32}P. As a result, it was possible to infer that the capsids were removed from the bacterium while most of the phosphorous remained within the bacteria. The bacteria were further cultivated, and a viral spawn was produced, with "little or no S^{35} [sic] is contained in the mature phage progeny." Hershey and Chase observed that "the bulk of the phage sulfur remains at the cell surface during the infection and takes no part in the multiplication of intracellular phage."

Conversely, "P^{32} [sic] shows that phosphorus is transferred from parental to progeny phage to the extent of 30 percent at yields of about 30 phages per infected bacterium." They concluded that DNA is the carrier of genetic information. In other words, the whole virion, composed of a chain of nucleic acids and several proteins arranged in a complex manner, can be re-generated by the DNA alone without the need for any protein. Such a relatively simple experiment had the effect of a bombshell in the scientific community (Figure 6.2).

6.6 SUDDEN DISAPPEARANCE

In the summer of 1944, Delbrück decided to standardize the work on phages, proposing what became known as the "Phage Treaty" (Table 6.2). There were now several working groups in this field, but it was difficult to compare the results. Thus, he decided to focus only on *E. coli* strain B—which he had isolated—and seven types (T) of phages (1–7), chosen because they were all lytic, produced clearly visible plaques, and could be handled easily in the laboratory. Working with a predetermined set of microorganisms is not as far-fetched as it might sound. The direct comparison of experimental results requires standardization of protocols, and, in general, geneticists must work with organisms with an established genetic pattern. In the last decades, great care has been taken to generate "pure breed" organisms—that is, with a precise set of genes—such as C57BL/6 mice.

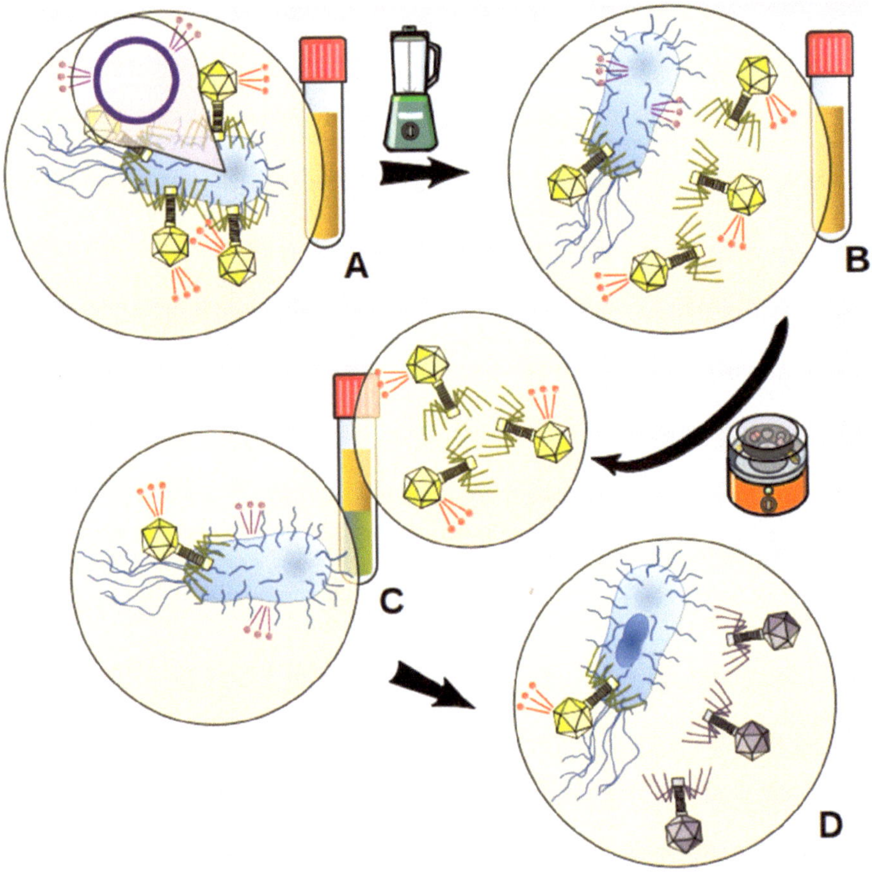

FIGURE 6.2 The blender experiment. (A) Phage T2 substructures were radioactively labeled with isotopes in the capsid proteins (orange lines) and the genome (purple lines). (B) A blender separated the phages from the *Escherichia coli* B a few minutes after adsorption. (C) The suspension was centrifuged, resulting in two distinct radioactive phases: the supernatant contained particles of low molecular weight, including unattached empty virions, while the pellet contained the marked bacteria. (D) The exposed *E. coli* were able to develop additional radioactive phage T2, showing that the genetic entity encoding the phages was not part of the proteinaceous substructure.

He was utterly uninterested in viruses different from those he had chosen. For instance, when Hershey discovered a mutant T2 (T2r) that produced halo plaques, Delbrück did not care to understand more. In contrast, Hershey went even further by observing that such mutations could be reversed. Based on the fact that phage mutants of the type r (for rapid) produced a halo that was missing from the wild type, Hershey could calculate the rate of forward mutation (T2 to T2r) and backward mutation (T2r to T2): 10^{-3} in the former case, 10^{-8} to 10^{-9} in the latter case.

In 1953, André Michel Lwoff reported that bacteria irradiated with a small amount of UV light generated plaques without incident phages. Such an observation

TABLE 6.2
Characteristics of the Viruses of the Phage Treaty[a]

T-Code	Old Names	Plaque Type	Plaque Diameter[b]	Halo[c]	Ring	By[d]
1	α/P28	Large	0.5–2.0	Large	None	L/D, B
2	γ/PC	Small	0.1–0.2	None	None	L/D, B
3	—	Large	1.0–2.5	Large	Yes	R
4	—	Small	0.3–1.0	Small[e]	None	R
5	—	Medium	0.5–1.0	Narrow	None	R
6	—	Small	0.1–0.2	None	None	R
7	—	Large	1.0–2.5	Large	Yes	M
—	Λ[f]	Small/large	—	—	—	Z
—	φχ174[f]	Large	—	Large	—	Bu

[a] Based on Demerec, M. and Fano, U. (1945) "Bacteriophage-resistant mutants in *Escherichia coli*". *Genetics*, vol. 30, pp. 119–136.

[b] In mm; the plaques develop after around 5 hours of incubation.

[c] The halo develops after around 18 hours of incubation.

[d] Isolated by: B, Bronfenbrenner; Bu, Bulgakov; L/D, Luria and Delbrück; M, MacNeal; R, Rakieten; Z, Zimmer.

[e] Subsequent purification separated T4a (without halo) and T4b (with halo).

[f] Later additions to the group; data recovered from the German Collection of Microorganisms and Cell Cultures (DSMZ).

extended the phenomenon of lysogeny first reported by Bordet in 1925: the spontaneous development of plaques in otherwise pure bacteria cultures. It was well established then that temperate phages produced turbid plaques in opposition to the clear ones associated with virulent phages. Lwoff and his collaborator Gutman isolated individual particles of *Bacillus megaterium* within microdrops and observed them over time. Lwoff noted that "a bacterium is there, and suddenly disappears," adding that "when this happens, phage is found in the droplet, around 100 phages per lysed bacterium." Lwoff summarized previous experiments where antibodies against temperate phages P2 had been exposed to lysates of lysogenic strains of *dysenteriae*. The exposed antibodies had the same neutralizing power as non-exposed ones, indicating that the lysates did not contain antigens that could engage the antibodies. "It must be concluded," commented Lwoff, "that this specific protein [the antigen] is synthesized only when the lysogenic bacterium produces bacteriophages." Lwoff advanced the hypothesis that "prophage is a specific structure anchored on a defined chromosomal locus." The droplet experiments also showed that the proportion of bacteria spontaneously producing phages varied greatly: sometimes, it was almost 100%, and in other cases, 1/200. "This strongly suggested that phage production was controlled by external factors." Indeed, Lwoff found that irradiation with UV, X, and γ rays, as well as exposure to compounds such as ascorbic acid, could activate dormant prophages. "It appears that inducing agents … are mutagenic."

If mutagenic conditions could induce dormant prophages, "one of the most efficient ways to decrease aptitude is starvation," noted Lwoff. Through observations and experiments on environmental microbial communities, it turned out that starvation is the key to maintaining the balance between phages and their prey. Since the energy sources available in the environment, such as in rivers or the oceans, are limited, bacteria are mainly in a state of hibernation: bacteria grow—or, more appropriately, bloom—in bursts only when energy sources become sporadically available. Bacteriophages can reproduce only during these blooms. With the phages prevented from multiplying at will, these viruses can coexist with their hosts without obliterating them.

Lwoff introduced the term "aptitude" as "the quantitative specific response [the development of a prophage into a phage] of a lysogenic system to an inducing agent." Before 1921, the word lysogenic strain referred to any bacterium carrying a bacteriophage; after that, it became more precise, referring to a clone of bacteria bearing prophages. Lwoff also coined the term "prophage," defining it as "the form in which lysogenic bacteria perpetuate the power to produce phage." Years later, Lane Barksdale and Sheldon Arden, both at the New York University School of Medicine, gave a striking definition of lysogeny ('lysogenic conversion') in one of their review on the subject:

> In the change from the nonlysogenic to the lysogenic state every bacterium becomes converted to the capacity to liberate phage and to an immunity to that liberated phage. The acquisition of these two properties is the *sine qua non* of all lysogenic conversions.
>
> Lysogenic conversions, then, involve the integration into the bacterial replicating system of a particular linkage group of genes which in the process of integration and in the process of excision act as a unit—that unit is a viral genome.
>
> Each group of temperate phages exhibits some distinctive properties regarding the clustering of related gene functions about their chromosomes, the number of and/or the places they sit in the lysogenic cell, and the mechanisms operating to insure their stability in that cell.

The term "curing"—again one of Lwoff's neologisms—could instead be defined as the opposite of lysogenic conversion: the "elimination of prophages from lysogenic bacteria." This concept is distinct from "carrier strain," another term introduced by Lwoff, which refers to a mixed collection of genetically related bacteria that are not all harboring prophages.

In microbial communities, the carrier state is a critical concept. Even though a lysogenic bacteria carrying, say, an α prophage is impervious to re-infection with α, a temperate virus still needs a host to replicate. If all of the bacteria in a community carry a prophage, the wave of infectious phages generated by the activated prophage—"gonophage," to use another of Lwoff's terms—will die out due to a lack of hosts. The presence of prophage α will eventually burden the lysogenic strains due to the increased metabolic demand for the synthesis of the viral integrant and the mortality of the bacteria undergoing spontaneous prophage induction. Such a burden will place lysogenic strains at a competitive disadvantage against bacteria immune to phage α infection.

Conversely, things change when a population X comprises bacteria of the same genetic stock but with a sub-population X^{α} lysogenically converted (meaning it carries prophage α) and a naïve sub-population X^{-}. The gonophages will find hosts for their

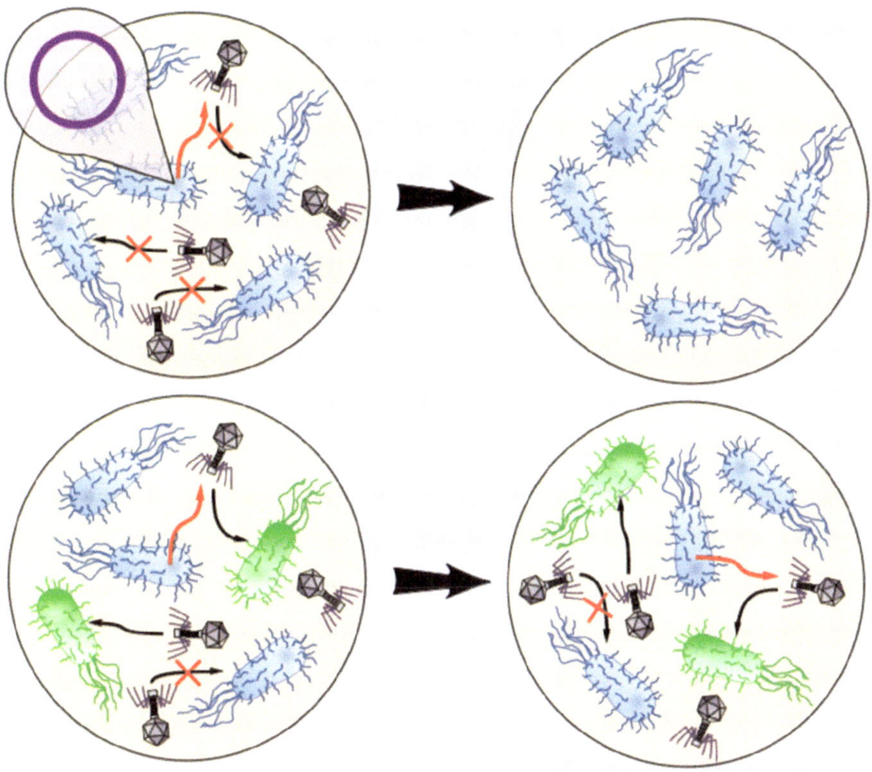

FIGURE 6.3 *Idealization of the carrier state concept.* (Upper panel) A community of lysogenic strains is formed only by bacteria carrying a given prophage. The prophage can naturally be induced at a specific rate (arrows), producing a wave of infectious phages (gonophages). However, the new spawn of viruses cannot infect any bacteria in the community because the lysogenic bacteria are immune to infection. As a result, the gonophages are wiped out from the community. In the long term, carrying the prophage will become a burden to the bacterium, and it can be expected that a new strain will rise that will maintain the immunity to the phage but without generating gonophages. Nonetheless, the permanence of the virus in this community is jeopardized. (Lower panel) A carrier state community is formed mainly by lysogenically converted bacteria (in blue) and a small fraction of naïve bacteria (in green). When the prophages are induced, the newly generated gonophages can find hosts for their infection cycle. The ensuing community will be stable if the infection is kept under certain limits. Lysogenic bacteria will benefit from the loss of bacteria caused by induction by expanding in the spaces left by the infected naive bacteria. The naïve bacteria will remain as a subpopulation within the community. The virus will be maintained in the community as a regulator of genetic variability.

replication, X^{α} will benefit from the prophage induction by replicating in the space freed by the infected bacteria, and even X will be granted their replication, albeit at low density, as long as phage α will not exterminate all naïve bacteria. Such a community formed by α, X^{α}, and X^{-} will be based on a tight equilibrium, but it will be more genetically varied than in the previous scenario. Figure 6.3 illustrates such a concept.

We now know that prophages are widespread in bacterial populations. It has been shown that up to 20% of the bacterial genome is composed of prophages depending on the strain analyzed. By comparison, the human genome contains about 8% of viral integrants. Each bacterium usually carries at least one prophage, and there are cases where it is possible to observe up to 20 prophages in one bacterial genome. Phage integration can occur at a single spot or multiple spots, with or without sequence specificity.

6.7 λ

Another unintentional event in New York was the next step in comprehending phage biology. The discoverer, Esther Miriam Zimmer, was born on the 18th of December 1922 in the Bronx, New York, to David (a Romanian émigré) and Pauline Geller, an American working part-time at the local general practitioner. She was very bright, particularly with languages, but refused to follow the convention to become a school teacher. Instead, she enrolled in biology at the university by convincing Bernard Ogilvie Dodge to admit her to his laboratory.

Dodge came from an impoverished family and struggled to gain a college degree due to financial problems. During his thesis, he discovered by chance that it was possible to germinate spores by heat shock, which was quite challenging to achieve back in the day. This feat, along with his determination and ingenuity, granted him a teaching position at Hunter College in New York. Dodge became *the* expert in *Neurospora crassa* biology and cultivation. This microbe is an orange-tinted common mold observed on bread, albeit mainly found in tropical and subtropical regions, which caused an outbreak among French bakeries at the end of the 19th century. One of the first researchers to use *Neurospora* was Francis Ryan (Columbia University), who asked Dodge for a strain of his mold when he joined Beadle and Tatum's laboratory in 1942.

After joining Dodge's unit and becoming an expert in *Neurospora* manipulation, Zimmer visited Alexander Holländer at the Industrial Hygiene Research Laboratory in Bethesda to generate mutants of this microbe by UV irradiation. Around this time, she was contacted by Joshua Lederberg, one of Ryan's students, for a sample of *Neurospora*. Later, she joined Tatum's laboratory at Stanford University, where she developed *E. coli* strain K-12, one of this bacterium's most widely used strains. Zimmer obtained her Master's degree in 1945. Without additional positions, she returned to New York and attended the 1946 CSHL Symposium, where she met Lederberg in person. It was love at first sight. Both had read *Microbe Hunters* by Paul De Kruif, an influential biography of the early microbiologists from Antonie van Leeuwenhoek to Paul Ehrlich published in 1926—incidentally, Witkin had read *Microbe Hunters* alongside *Arrowsmith*, and she also produced *E. coli* mutants by UV irradiation. Zimmer "was not eager for the stresses of marriage," Lederberg recalled, but "she gave in to my impetuous courtship." After just three months of engagement, Lederberg and Zimmer married in December 1946, moving to Wisconsin, where he obtained his academic position at the School of Agriculture. Zimmer received a Ph.D. studentship in the Laboratory of Royal Brink, a geneticist at the University of Wisconsin who had first contacted Lederberg, offering him the professorship.

In 1951, during her Ph.D. at the University of Wisconsin, Zimmer noted that the UV-treated K-12 sub-strain W518 developed plaques "showing turbid centers" after being in contact with a parental K-12. Studying this observation more systematically, Zimmer and Lederberg remarked that "the lysogenicity of strain K-12 had remained unsuspected despite its maintenance for over 25 years and close study as the subject of mutation and recombination experiments since 1944" because a lysogenic strain requires a susceptible strain to be exposed. "The novelty," they noted, "consisted of two exceptional mutant sub-strains W-435 and W-518 which were sensitive to the phage." Zimmer and Lederberg concluded that "*Escherichia coli* strain K-12 carries a symbiotic phage, λ," so-called after the κ plasmid present in Paramecium. However, λ turned out to be a virus and not a plasmid. Zimmer and Lederberg also noted that "mutation of λ to a more virulent mutant λ2 has been observed," where λ2 could infect strains that were immune to λ.

Soon after, λ was extensively investigated by many researchers worldwide, including Hershey. Electron microscopy was promptly employed to visualize radioactively labeled λ molecules. It was shown that native λ genomes—obtained from the lysis of λ virions—were 17 µm in length—today, we know that this corresponds to a 48 Kb—that could spontaneously circularize at 60°C due to the presence of cohesive ends—or "sticky ends," a term introduced by Hershey—meaning stretches of complementary single-stranded DNA in an otherwise double-stranded chain. The circularization was not covalent since obtaining linear genomes from circular ones was possible. Thus, the circularization must occur upon annealing by the following Watson–Crick model's complementarity. The λ genome could also form concatenamers *in vitro*. Interestingly, the presence of cohesive ends was demonstrated by the λ inactivation in the presence of DNA polymerase since the enzyme would fill the single-stranded gaps, hindering the circularization of the chain. Without circularization, the virus was no longer infective, highlighting the importance of this step in λ replication.

Although λ was named after a plasmid (*κ*), it became clear that it had unique characteristics, chiefly that it could integrate into the host's nucleoid. The first to put forward such an idea was Allan McCulloch Campbell—another Phage Course attendee who worked with Hershey at the CSHL for about one year—who had graduated from the University of Illinois and then obtained a research position at the CSHL under Demerec's supervision. Even Campbell observed that λ obtained from one *E. coli* K-12 sub-strain (C600) could not grow in a similar but distinct sub-strain (W3350). It was also understood that plasmids could integrate; in particular, François Jacob and Élie Léo Wollman at the PI had coined the term "episome" to describe a plasmid—essentially an independent mini-chromosome—that could integrate into the host's chromosome. Furthermore, Jacob and Wollman believed that λ could strongly associate with *E. coli*'s chromosome while remaining an independent entity. Today, it is known that some viruses could do just that: for instance, herpesvirus can form synapsis with human chromosomes, making it challenging to separate the viral DNA from the human's physically, and yet the two entities are independent. However, there was a problem with the Jacob-Wollman model.

As Campbell stated, "at the time, one thing that was known for sure about the λ genome was the order of genes." Genetic mapping performed by Enrico Calef and

Giuseppe Ricciardello at the University of Pavia, Italy, established that λ virions had linear genomes with the gene markers *m5* and *mi* at the extremities and *co* in between. On the other hand, Ray Wu (Cornell University) and Armin Dale Keiser (Stanford University) had shown that λ had 20 bases at the extremities of the genome protruding as single-chain. These extremities were complementary, allowing for the circularization of the genome that was later covalently linked by the host's enzymes.

It was also established that λ integrated between the *E. coli*'s genes *gal* and *trp*, but "the gene order in the prophage had changed from the phage order (*m5 co mi*) to *co mi m5*." Jacob and Wollman did not contemplate any genomic rearrangement in their model. Campbell "noted that, if the phage genome were to circularize and then to insert by reciprocal crossing over between *m5* and *co*, their [Calef and Ricciardello's] result would be obtained." That is,

[to *mi*] ← —***m5***————***co***——***mi***— → [to *m5*]

×

[to upstream bacterial genes] ←——*gal*——*trp*—— → [to downstream bacterial genes]

↓

——*gal*——***co***——***mi***————————***m5***——*trp*——.

The Phage Group did not readily accept Campbell's model. Campbell recounted that "Luria … regarded the model as beautiful but wrong … Max Delbrück … proclaimed that the idea was 'very ugly'." Nonetheless, Wu and Keiser demonstrated that the λ genome could indeed circularize.

The demonstration of the actual insertion of λ into *E. coli* genome, rather than simple synaptic association, was obtained by employing the full range of the state-of-the-art techniques of the time: genetic mapping, restriction mapping, electron microscopy, and the newly developed Maxam–Gilbert sequencing. Gel electrophoresis showed that the mass of the nucleoid of lysogenic *E. coli* was higher than that of the naïve bacterium, and the increase corresponded to the mass of the λ genome. In particular, genetic analysis and direct microscope visualization indicated a core homology region between λ and *E. coli* genomes. The region used by λ to recombine was composed of a core element (O) about 15 bases long enclosed in less specific nucleotides; the region on the phage genome is known as *attP* (or *attϕ*, originally *POP*, att being short for "attachment"), and that on the bacterium is *attB* (*BOB*). While λ has a single att site, the prophage has two because *attP* literally splits into two, termed left (*attL*) and right (*attR*). The core region *O* allows the homology recombination between virus and bacterium. *O* was also known to be bracketed by restriction sites; in particular, recognition sequences for the Hind enzyme were the closest to this region. It was possible to determine the location of O using phage mutants missing these restriction sites. This process highlighted the power of induction associated with this approach.

To put it simply, let us represent the core sequence with ◍ and that of restriction enzymes with ▲/■, whereas the absence of the site can be represented with △/□; the segment containing ◍ is highlighted below in boldface. Thus, one possible conformation for a wild-type λ might be:

————▲————■—◍—■————▲————

which will generate the following fragments:

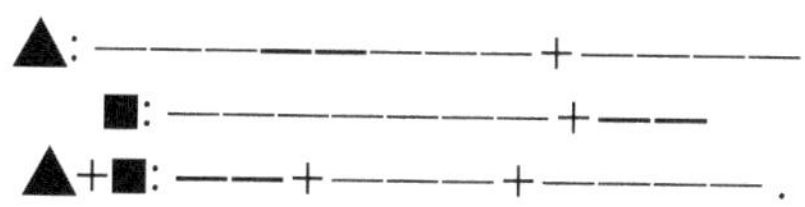

Mutant phages might have the profiles:

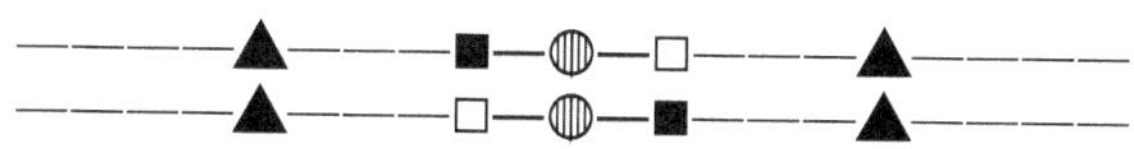

which will generate the fragments:

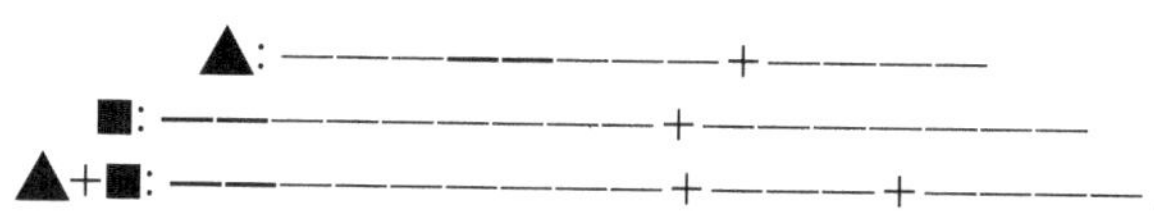

The mutants miss the small fragment containing ◍. This observation was the basis for obtaining a series of mutants with peculiar restriction profiles. Simplifying, one might obtain the following series:

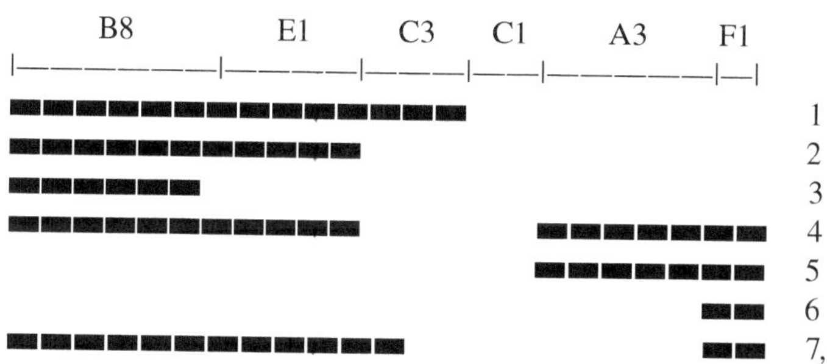

where ■ indicates a mapped restriction fragment. In all profiles (1–6), the segment C1 is missing. Since the restriction sites are external to *O*, C1 must contain the core region. It would have been impossible to obtain these profiles in the presence of synapses because one would have gotten completely different fragments' lengths that would not have fit the expectations.

Sequencing served as the ultimate confirmation means. It was shown that the sequence of λ genome around *O* was:

… CGTTTCTCGTTCAGCTTTTTTATACTAAGTTGGCATTATAA … ,

where *O* is underlined. On the other hand, the sequence of naïve *E. coli* provided the following string:

… TCCGTTGAAGCCTGCTTTTTTATACTAACTTGAGCGAAACG … ,

where *O* is underlined and the bacterial sequence is highlighted in boldface. The λ lysogens, instead, could be shown to have two *O* regions:

*att*L **… TCCGTTGAAGCCTGCTTTTTTATACTAA**GTTGGCATTATAA …
*att*R … CGTTTCTCGTTCAGCTTTTTTATACTAA**CTTGAGCGAAACG …** .

In other words, it was evident that the bacterial sequence had been split in two, with the phage sequence placed in the middle. Given the straightforward answer given by sequencing, the rigorous logic required to generate a restriction map became obsolete. Remarkably, not only has sequencing become the gold standard for genetic analysis, but we have accumulated so much genetic data—albeit enough is never enough—that it is possible to predict the structure and function of unknown genes and their products.

Later on, Zimmer devised the technique named "replica plating." To determine whether a microbe could grow on one medium but not another—a feature critical for the genetic analysis underway at the time—a small amount of one colony (a clone with a specific genetic profile) from one master plate was picked and placed in another plate, with great care taken to keep track of the various colonies. She realized that she could use the powder pad of her cosmetic set to pass the colonies from the master to other plates. Nonetheless, according to different sources, Lederberg developed the process after studying the features of many materials and deciding that velvet was the best option for carrying germs. Since the layout of the plates would be the same, it would have been easier to track the clones, speeding up the procedure. The protocol was published in a joint paper by Zimmer and Lederberg in 1952 and remains a standard procedure.

7 West Coast

7.1 THE BACTERIAL SEXUAL REVOLUTION

"I was startled—and privileged—at age 21 to have made a surprising discovery that involved merging bacteriology and genetics," said Joshua Lederberg in an interview with *The New York Times* in 1983. In the early 1950s, it was essentially established that the essence of the gene was the DNA, a concept unknown just a decade earlier. Two pieces of the genetic mosaic were still missing: how bacteria evolve their genes and the DNA's structure. The completion of this genetic mosaic will be carried out by scientists involved in phage research.

Back in the day, bacterial genomes were thought to be relatively stable. According to Darwinian theory, all creatures must evolve their genomes to adapt to their environment. For instance, let us assume an organism that requires an enzyme *z* to digest a specific nutrient, say cellulose for a protist or cholesterol for a human. Such an organism would die if the source of cellulose or cholesterol is exhausted. However, suppose the gene encoding *z* evolved through random mutation into an alternate form *Z*, an enzyme capable of catabolizing other substances such as lignin or phosphatidylcholine. In that case, the organism may thrive in environments lacking cellulose or cholesterol but containing lignin or phosphatidylcholine. Even better, if an organism could carry both *z* and *Z*, it would have an evolutive advantage over competitors bearing either *z* or *Z*. This process is known as genetic evolution, and all organisms, from viruses to people, are compelled to fulfill it.

Genetic evolution might occur by direct mutation. Changes in the DNA's nucleotides—for instance, due to UV irradiation or in inherent error-prone characteristics of the enzyme processing the DNA—cause a shift in the gene sequence. The *Z* gene could arise if the mutation occurred within the *z* gene and was placed correctly. Of course, the probability of obtaining a functional new gene just by random mutation is low. Consequently, genetic evolution through this mechanism is also extremely plodded. Eukaryotic cells developed ways to modify their genes more extensively and quickly, speeding up the evolution process.

From Mendelian and cytogenetic studies, it was known that animals, including people, used a process known as sexual mating. Essentially, this process requires two types of cells—conventionally called mating types or simply female (♀) and male (♂)—with a different genetic profile. For example, let us assume ♀ with a genetic profile *abcd* and ♂ with a profile *ABCD*. The mating is carried out by direct interaction between ♀ and ♂ that literally fuse to generate a zygote ⚥ with a genetic profile *abcd/ABCD*. The organism ⚥ has two alleles for genetic locus, doubling its probability of adaptation to the environment. While unicellular organisms such as protists exhibit sophisticated breeding patterns through this mechanism, mating occurs in animals, including humans, primarily in reproductive cells (sperms or egg cells, collectively known as gametes).

DOI: 10.1201/9781003657712-7

Moreover, mating provided an additional feature for adaptation: recombination. Inside ⚥, the chromosomes *abcd* and *ABCD* could shuffle their genes by crossing over to produce, for instance, *abCD* and *ABcd*. The resulting gametes will have the genetic profiles *abCD* and *ABcd* instead of the initial *abcd* and *ABCD*. Thus, in the span of a single generation, a genetic reshuffling can increase the zigote's chances of environmental adaptability.

The replication of *Neurospora crassa* provides a prototypical example of this process. *Neurospora*'s ease of manipulation and haploid genome, which becomes diploid during mating, make this microbe perfect for genetic investigations. Bernard Dodge even told Morgan that this mold "would be more important than *Drosophila* one day." Later events fulfilled this prophecy. *Neurospora* has two mating types, A and a, which are haploid cells. These cells are known as ascospores and can replicate by direct scission to produce the mycelia that can be observed, for instance, on old bread, spoiled flour, or even on the bark of trees. The mycelia are colonies of these haploid cells, and some of the cells can differentiate to generate reproductive cells known as germinating conidia that correspond to ♀ and ♂. The conidia fuse to form a diploid ascus (⚥) that later generates haploid ascospores.

On the other hand, bacteria are strictly haploid; thus, crossing over cannot occur. As a result, it was believed that bacterial genomes were relatively stable. This conventional wisdom was shaken by Griffith and his work on the phenotypical transformation of pneumococci. A complete paradigm shift occurred through the work of pioneers such as Edward Tatum and Joshua Lederberg.

Tatum had worked first at Stanford, where, in collaboration with Beadle, they had characterized the genetics of *Neurospora*. He moved to Yale in 1945 but returned to Stanford in 1948, remaining until 1957, when he joined the RI. Joshua Lederberg was born in Montclair, New Jersey, on the 23rd of May 1925. His father was a rabbi, and his mother was a housewife whose ancestors had been rabbis. Despite his father's desire to raise Joshua as a rabbi, the young boy became fascinated by the lives of Luis Pasteur, Robert Koch, and Albert Einstein. Lysenko and his denigration of genetics had also galvanized him—a feeling also shared by Evelyn Witkin—thus, he favored genetics, which at the time was, he said, "a battleground of political ideologies." Lederberg enrolled at Columbia University, obtaining a degree under the supervision of Francis Ryan. Despite his remarkable intelligence, Lederberg was not an outstanding laboratorist. Ryan commented, "You could tell that Joshua was in the lab because you could hear the tinkle of breaking glass."

In 1943, Lederberg volunteered in the US Navy Medical Corps and obtained a bachelor's degree in zoology the following year, studying Plasmodium infections. Afterward, Lederberg continued his path into microbiology by obtaining a position as a Ph.D. student in the laboratory of Edward Tatum, who had also been Ryan's doctoral supervisor at Stanford University. The collaboration aimed to investigate the details of bacterial genetics, indirectly promoted by the European duo: "the principal encouragement to think about genes in bacteria," Lederberg recalled, "had come from Luria & Delbrück experiments on the statistics of mutation in *E. coli*."

Lederberg and Tatum started from the assumption "that bacteria have no sexual mode of reproduction" and sought to present "evidence for the occurrence in a bacterium of a process of gene recombination, from which the existence of a sexual stage may be inferred." Lederberg used *E. coli* K-12, initially supplied by Zimmer, and developed mutants that required additional growth factors and, therefore, could not be grown in a minimum medium as the parental strain. While the parental strain, which can produce all the required biochemical molecules from the raw components present in a minimal medium, is known as a prototroph, the mutants are dubbed auxotrophs. They remarked that "a mutant strain can be signified by suffix a '–' sign to the initial of the substance in question" and "particular emphasis can be placed on the ability of a strain to synthesize a particular growth factor by suffixing a '+'." Thus, they obtained a Y10 strain that required threonine, leucine, and thiamine but was capable of synthesizing biotin, phenylalanine, and cystine (T–L–B1–B+ϕ+C+) by X-ray irradiation, and a strain Y24 with opposite profile (T+L+B1+B–ϕ–C–) by UV irradiation. They added that "the minus sign is justified by the *a priori* consideration that a mutation establishing a growth factor requirement generally represents the loss of a function," in this case, the biosynthesis of a molecule.

If mutations could knock down an enzyme, causing the switch from a prototrophic to an auxotrophic phenotype, it was assumed that mutations could also restore the activity of an enzyme, transforming back an auxotrophic strain into a prototrophic one. Lederberg and Tatum observed that one cell out of ten million could undergo this recovery process; thus, the rate of back-mutation was 10^{-7}. Nonetheless, the cells could recover only one locus, requiring two growth factors. The probability of recovering three loci by random mutation would have been $(10^{-7})^3 = 10^{-21}$, a rare event indeed. Remarkably, Lederberg and Tatum could not use *Neurospora* because the rate of back-mutation was much higher than in bacteria and precisely because of the sexual recombination involved in the reproduction of this mold.

Lederberg and Tatum recorded no colonies after plating Y10 or Y24 alone on a minimum medium. Instead, "about 100 colonies developed for each billion (10^9) cells inoculated" after plating Y10 and Y24 together. In other words, they obtained cells with the parental phenotype T+L+B1+B+ϕ+C+. Therefore, the rate of back-mutation was $10^2/10^9 = 10^{-7}$, way too high to be explained by random mutation. They concluded that

> to account for these recombinants, we postulate a sexual phase in this strain of *E. coli*: a cell fusion which allows the segregation of genes in new combinations into a single cell … The alternative to this conclusion is that transforming substances, capable of inducing mutations *in both directions*, are present in the medium … Bilateral transformation cannot be distinguished from a sexual process by the genetic analysis of a population.

However, as stated before, the probability of complete back-mutation (even due to "transforming substances" present in the medium) would have been 10^{-21}, too rare to be statistically significant. Lederberg and Tatum leaned toward the physical merge of Y10 and Y24 to form a fully revertant spawn cell, just like eukaryotes can do.

Lederberg's findings were confirmed by Bernard Davis a few years later. Davis used Y10—he employed a more modern symbology than Lederberg's, annotating that this strain had a genetic make-up Thr$^-$, Leu$^-$, Thi$^-$—and an additional K-12 mutant 58–161 incapable of producing methionine (Meth$^-$). He then used "a U-tube, 25 mm in diameter, partitioned by an 'ultrafine' fritted glass disk" that was "impermeable to the organisms under investigation." He added Y10 to one extremity and 58–161 to the other, then "a vacuum was applied alternatively to each side," assuring that the liquid medium would pass through one extremity and then the other. This apparatus allowed for a homogeneous medium: if there were a soluble mutagenizing factor (capable of passing through the filter), it would cause bacterial recombination without direct contact among the cells (incapable of passing through the filter). After 4 hours, the medium was transferred to a minimal medium, and the colonies were counted the next day. While cultures without a separating filter produced 58 colonies, those separated by the glass mash gave none. Therefore, bacterial recombination requires intimate contact with the cells.

Electron microscopy provided the ultimate verification of Lederberg's theory of bacterial connections during conjugation. For instance, photographs of *E. coli* plasmid E1 demonstrated that these elements were circular dsDNA strings independent of the main bacterial nucleoid. While visiting the PI, Anderson worked in collaboration with André Lwoff, Francois Jacob, and Elie Wollman, and he took photographs of bacteria connected while exchanging genes. The picture was published in a scientific outlet (the PI Annals) but was so groundbreaking that it was reprinted on the front cover of the weekly news magazine Paris Match.

Essentially, Lederbeg revolutionized the field of microbiology while still being a Ph.D. student. His debut to the scientific community was also dazzling. It occurred at the 1946 CSHL Symposium, which Luria had organized. The CSHL Symposia had started in 1933 but were put on hold in 1943 due to the war. The 1946 annual edition, titled *Heredity and Variation in Microorganisms* and held between the 2nd and the 12th of July, was attended by 130 scientists from all over the US and Europe (including many fellows of the Phage Group: Szlizard, Lwoff, Demerec, Hotchkiss, but also Raymond Laterjet, Curt Stern, Howard Newcombe, Evelyn Witkin, Jaques Monod, Alexander Holländer, and John Haldane), with 25 talks scheduled. One of the speakers was Tatum, who gave a seminar on Neurospora genetics; Lederberg was only attending the conference as a student. During the talk, Tatum mentioned that bacteria *apparently* lack a sexual stage, and one of the attendees asked why "apparently" since the conventional wisdom was that bacteria did not have any sexual stage at all. When Tatum answered by mentioning his student's results, there was turmoil in the audience, and Lederberg was asked to give an impromptu talk by popular acclaim. The improvised seminar lasted several hours due to the volley of questions the audience threw at Lederberg and that he systematically answered.

Among the tricky questions, one of the most challenging came from Lwoff. He pointed out that the generation of prototrophs might be because one strain could have produced a growth factor used by the other strain. Since the growth factor is a soluble molecule, it could easily pass the filter. In Lwoff's opinion, the recombination was simply a case of cross-feeding. Unmoved, Lederberg answered that the new colonies of prototrophs, once formed, did not require other species to grow on a minimal medium. The genetic transformation was stable and heritable and did not fit with cross-feeding. Among the people unimpressed by Lederberg's work was Delbrück: he quite curtly expressed his disinterest in the phenomenon for the lack of a kinetic analysis," as Lederberg recalled.

Lederberg acknowledged that his discovery had been fortuitous—"just one more example of Ed Tatum's notorious serendipity," he later commented. Lederberg's experiment was not due to the direct recombination of bacterial genomes, as in the case of *Neurospora* or other eukaryotic organisms, but to the exchange of mobile genes. The lucky strike was that *E. coli* K-12 carried such plasmid—a term coined by Lederberg from "plas" (as in cytoplasm) and "id," the Latin for "thing"—in the first place: not all strains of this species do. For instance, Lederberg recalled that "Luria promptly tried to emulate the findings, using phage resistance markers in *E. coli* strain B," but "he failed" because "only about one strain in twenty would have worked with the protocol used." Instead, "Aaron Novick and Leo Szilard ... had no trouble reproducing my results with the K-12 strains I furnished them." Following such a lucky strike—"we weren't looking for plasmids, we bumped into them," he later said—Lederberg obtained (in only one year, thanks to Tatum's engagement of a "fast track") his Ph.D. in 1947 and (thanks again to Tatum's involvement) a position as an assistant professor at the University of Wisconsin, where Tatum had studied.

Bacterial recombination remained a curious feature in the microbiology field without much follow-up. As Cavalli-Sforza commented, "three years after ... the results obtained by Lederberg with the strain known as K 12 [sic] still stand alone, no other bacterial strain having been found to show a similar behaviour." William Hayes, a geneticist from Dublin at the University of London, demonstrated that plasmids moved unilaterally from a donor to a recipient and called the plasmid he was working with "fertility factor" (*F*). The *F* factor was shown to be a fraction (about 1/40) of the length of the nucleoid and carried specific genes not observed in the main bacterial chromosome. Hayes first reported his results at a scientific conference in Oxford, where he was mostly ignored, in April 1952. In September, however, he presented them again at a conference held in Italy, and he was incensed. Even Delbrück, who initially was very skeptical about Hayes' report, invited him to give a talk at the 1953 CSHL symposium.

Moreover, Hayes serendipitously observed that in some old *E. coli* F^+ colonies, the fertility factor could be lost and recovered, at least in a fraction of them. Nonetheless, the genes associated with the *F* plasmid were not lost, implying that the plasmid could integrate into the nucleoid and then excide once again. Thus, *F* turned out to represent a specific subclass of plasmids known as "episomes" because they could insert themselves on the bacterial chromosome and then move out again. The genomes of viruses, such as λ, can also be considered episomes because they can exist either as independent molecules or as integrated elements of the main chromosome. Using the same blender approach used by Hershey, Hayes could determine the time required to transfer the whole plasmid and the order of the genes on it.

Furthermore, Hayes isolated *E. coli* F^+ strains that showed recombination frequencies a thousand-fold higher than Lederberg's reported. Cavalli-Sforza also obtained some strains that could recombine at a rate at least ten times higher than Lederberg's K-12. These strains were named high frequency of recombination (*Hfr*) and further subdivided into H (from Hayes) and C (from Cavalli-Sforza).

7.2 SERENDIPITY, AGAIN

Although it had become established within the microbiological community that bacteria could transfer genetic material upon close contact, one of Lederberg's students at the University of Wisconsin, Norton Zinder from the Bronx, New York, provided the exception to the rule—another momentous breakthrough that, as Zinder remarked, was "discovered by accident and good luck." While repeating Davis' approach to strains of *Salmonella* he had produced by UV irradiation, Zinder observed transformation even in the presence of the separation between parental strains. There was indeed a soluble and filterable mutating agent, after all.

Zinder obtained dual auxotrophs of *S. typhimurium* and applied the Davis' process to the strains LA-22 (a collection of sub-strains with different mutations) and LA-2 (Meth$^-$, His$^-$), then plated in a minimal medium the two strains separately. In theory, there should have been no colony (no prototroph formed), but he obtained colonies in the LA-22 inoculum but not with LA-2. A filterable agent must have passed a gene from LA-2 to LA-22 to restore the latter to the parental genotype, but the agent could not pass genes from LA-22 to LA-2. In other words, "LA-2 produced a filterable agent (FA), under stimulation from LA-22, that could elicit prototrophs from LA-22."

Zinder and Lederberg repeated the experiment with an LA-22 sub-strain (SW-435) that required galactose and xylose, but that was streptomycin resistant (Gal$^-$, Xyl$^-$, S^r) and the parental prototroph LT-22 that was streptomycin sensitive (S^s). They obtained strains that recovered one of the metabolic pathways (but not both) above the background level of back mutation. They also managed to transfer the streptomycin resistance trait from *S. typhimurium* to an S^s strain of *S. typhi*. "We conclude," they remarked, "that an FA filtrate has many activities, producing many different transductions (but no more than one per cell) that result in singly transduced clones."

Nonetheless, Zinder and Lederberg were not wholly in the dark in understanding the origin of this filterable mutating agent. Firstly, they knew that "LT-22 is lysogenic for a virus (hereafter referred to as PLT-22) active on LT-2." Secondly, FA resisted enzymatic digestion with trypsin, pancreatin, ribonuclease, and deoxyribonuclease. Thirdly, FA was released from old colonies but not young ones (a characteristic of lysogenic bacteria). Fourthly, cell-disrupting agents such as penicillin, benzene, and heat did not produce the filterable agent. "These results," they concluded, "indicate that FA is not released by mechanical, chemical, or biological disruption of the cell." Instead, FA was released when the cells were "only slightly inhibited." In addition, by employing progressively smaller filters, Zinder and Lederberg assess that FA had "a particle size slightly less than 0.1 μm" (100 nm). They also applied electron microscopy, observing round particles of roughly the same size.

The experiment was later repeated with *E. coli* K-12, which was known—thanks to the work of Esther Zimmer—to carry a prophage. The gene encoding galactosidase (but not that of lactase) was transferred from W-435 (gal$^+$, lac$^-$) to W-518 (gal$^-$, lac$^-$) sub-strains (both sensitive to infection with the K-12 phage). Through genetic mapping, it became clear that galactosidase was transferred because it was located near the insertion point of the prophage. In contrast, lactase was not transferred because it was located further away. Somehow, the virus was packing a portion of the bacterial genome inside the virion with the viral genome, albeit at a low frequency.

Lederberg presented the preliminary results regarding the genetics of *Salmonella* at the 1951 CSHL Symposium—the same where Barbara McClintock presented her discovery about transposons—winning, as Zinder recalled, "all competition for incomprehensibility," speaking non-stop from 7 PM to 1 AM," as Charlotte Auerbach, another attendee, reported. Nonetheless, Lederberg and Zinder talked with Harriet Ephrussi-Taylor, an American geneticist participating in the meeting, who suggested that the transformation might have been due to "a virus with some DNA stuck onto it." Primed by the tip, Zinder and Lederberg theorized that the genetic exchange occurred due to a virus moving across bacterial strains. In essence, as Zinder later explained:

> one of the two strains (the recipient) [LA-22] carried a temperate virus ... the second strain (the donor) [LA-2] was especially susceptible to this virus. When the two were mixed, an occasional infective virus developed in the recipient strain and invaded a cell of the donor strain ... Some of the incorporated particles of *Salmonella* DNA unchanged in their genetic cores ... [which] instead of killing their host [LA-22], they simply disappeared. The DNA they brought with them took its place in the chromosomes of the recipient cell and modified the nutritional or other characteristics of the cell's offspring.

Lederberg coined the term "transduction" from the Latin "trānsdūcō" ("lead across") to describe this novel genetic process. John Parkison summarized the process from the University of Utah in this way: a virus "must have moved from the LT-22 culture through the filter to the LT-2 side of the U-tube, where it induced FA activity, which then moved back across the filter to produce prototrophic recombinants in the LT-22 cells." This virus, PLT-22 (now called simply P22), is a member of the *Podoviridae* family and has a relatively short tail—unlike the phages employed by Delbrück and Luria that Anderson photographed—giving it the roundish form noted by Zinder and Lederberg.

Based on the data shown by Luria, a lysogenized bacterium could not have been infected by another phage. The transduction process described by Zinder and Lederberg required a phage (P22) to be produced by a lysogenic strain (LA-22), infect a naïve strain (LA-2), load some genes, and then re-infect LA-22 (Figure 7.1). However, to infect a lysogenic LA-22 containing P22 prophages, P22 needed to overcome the latter's superinfection exclusion characteristics.

It turned out that P22 contains four systems to accomplish superinfection exclusion. First, P22 modifies the bacterial O-antigen to prevent other phages from adsorbed to the lysogenic bacteria. However, this system only reduces but not eliminates the adsorption of P22 to P22-lysogens. Second, P22 encodes for a protein (*sieA*) that hampers the injection of viral DNA inside the infected cell. However, according to Miriam Susskind (University of Massachusetts) and David Botstein (Massachusetts Institute of Technology), "the *sieA* exclusion system can be saturated by high multiplicity of infection." Given how many bacteria inside the U-shaped tube became infected with P22, releasing thousands of phages into the culture medium, a high multiplicity of infection is highly likely to have occurred in the Zinder–Lederberg experiment. Third, P22 encodes a second exclusion protein (*sieB*) that, as Susskind and Botstein highlighted, "does not act on wild-type P22." Fourth, P22 encodes for a repressor (C2) that inactivates lytic genes but does not interfere with generating a new prophage.

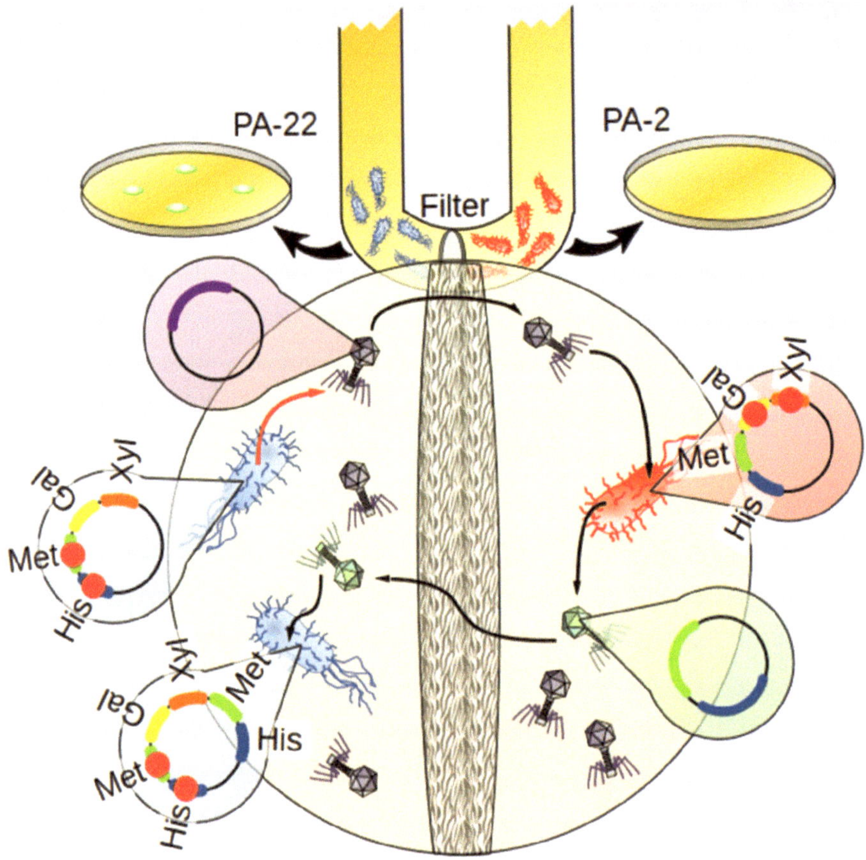

FIGURE 7.1 *The Zinder–Lederberg experiment.* Upper panel: Overview of the experimental apparatus. This consisted of a U-shaped glass tube separated by a porous membrane (filter). The tube was filled with culture medium, and different bacterial strains were placed at the two extremities. The filter allowed only small particles to cross over, not bacteria. Samples of the media could be withdrawn at specific times and plated on minimal medium. It was observed that *E. coli* PA-22 formed colonies but not PA-2. Lower panel: overview of the molecular process underlying the experiment. The filter is represented by a grey mash in the middle of the panel. The original *E. coli* PA-22 (blue cell) was deficient for *his* and *met* genes (dots) but was lysogenic for a phage containing only viral genes (purple virion). The phage was small enough to pass the filter and infect *E. coli* PA-2 (red cell), which was instead deficient for the *gal* and *xyl* genes. Occasionally, phage replication errors introduced bacterial genes (*his* and *met*) into the capsid (green virion). This so-called transductant phage could cross the filter again and infect PA-22, introducing complementary genes for *his* and *met*. The resulting lysogenic bacterium could grow in the minimal medium.

Therefore, P22 coming from LA-2 bacteria could very well infect LA-22 lysogens. The high titers in the culture vessel would have compensated for the reduced infection efficiency. In fact, not only could P22 infect bacteria pre-infected with P22, but the whole transduction process required P22 prophages as an insertion point for the newly acquired genes (the "transducing material"). In other words, the machinery that allows the integration of the viral genome into the bacterial nucleoid will use the sequences of the prophage already present in the bacterial genome as an anchor to the integration process. Not only that, but it was even possible to distinguish two different modes of transduction.

In normal conditions, the phage DNA processing machinery recognizes signals in the viral genome known as packaging (pac) sequences, cuts the DNA at these loci, excites the viral genome as a linear chain that is then circularized and used as a template for a concatenamer through θ replication. The concatenamer is then cleaved into unit-size elements packaged inside the phage capsid before release from the infected cell. After release from the donor cell, the aberrant viral particles will inject their DNA into another recipient cell, and recombination will occur at the homology location between donor and recipient DNA. In "specialized transduction," which Larry Morse and the Lederbergs discovered in 1956, viral particles ("transductants") primarily contain viral DNA with some extra bacterial DNA flanking the prophage. Such an addition is conceivable because the capsid may hold more DNA than the viral genome. The addition of bacterial sequences is because the phage machinery fails to recognize the end of the viral genome and continues pulling out DNA from the nucleoid. In the so-called "generalized transduction," the phage machinery recognizes bacterial sequences similar to the bona fide packaging signals (*ψpac*) but present in the bacterial nucleoid. Some phages use an alternate sequence (*cos*), which, like *pac*, is randomly represented in the bacterial nucleoid. However, *cos* is far less common than *pac*, so it is considerably less likely to play a role in generalized transduction. In this case, the transductants will contain only bacterial DNA. Recently, a third transduction type has been discovered: "lateral transduction." In this process, packaging starts on the *pac* site of viral genomes, but instead of ending with the viral segment, it continues into the bacterial DNA. Figure 7.2 visualizes the differences between these processes.

As for conjugation, the discovery of transduction had been a pure stroke of luck. Not only did Zinder and Lederberg work with a carrier bacterium, but if P22 had carried out specialized transduction instead of generalized transduction, the two researchers would not have seen the rise of a prototroph. They probably would have explained the results as simple back-mutation and forget about it. Furthermore, if the mutated genes (gal, xyl) had been located at very distant loci, they would not have been loaded on the same transductant, and, again, there would not be the development of a prototroph. Instead, with just two experiments, Lederberg had caused a complete paradigm shift in the conception of bacterial genetics. No wonder he was considered a prodigy and became a Nobel laureate in 1958.

7.3 RECRIMINATIONS

Lederberg founded the Department of Medical Genetics in 1957, then moved to Stanford University the following year. He also received a Nobel Prize with Edward Tatum and George Beadle that year. The Prize caused a deterioration of the relationship between Lederberg and Zimmer. One of the first drawbacks of their marriage came during her Ph.D. studies. Her spouse recommended she should not insert her chance findings on the F plasmid and phage λ in her thesis to speed up the conclusion of her doctorate. However, Lederberg's Ph.D. thesis had been boosted precisely

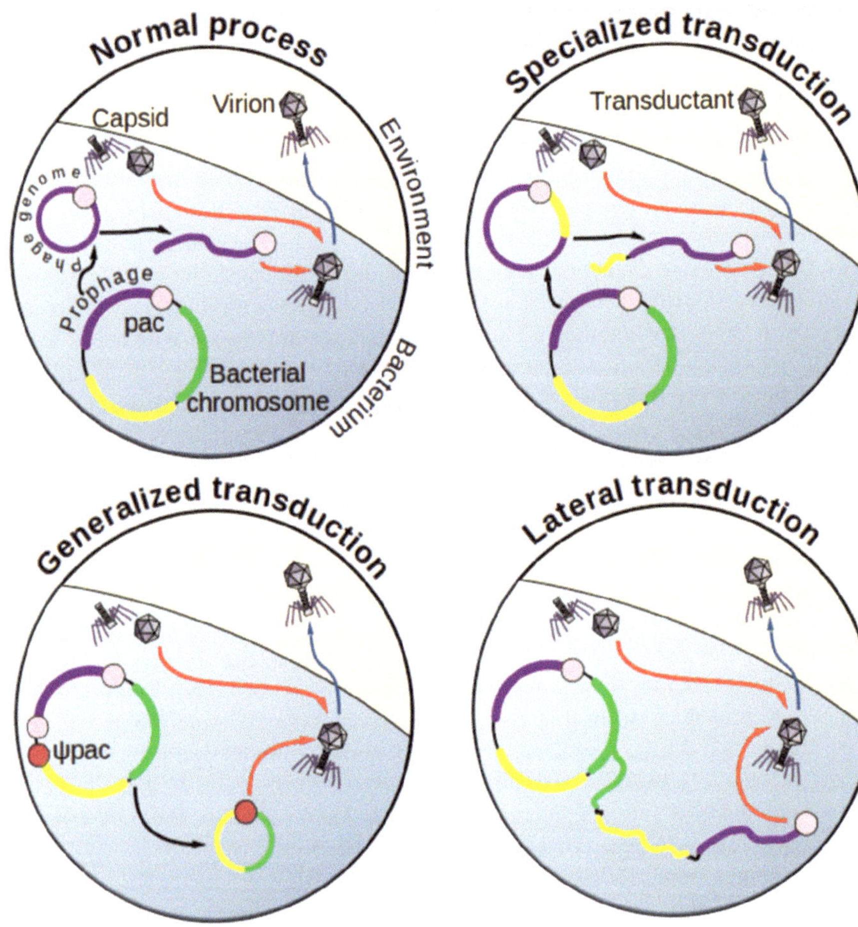

FIGURE 7.2 *Different types of transduction.* Normal process panel. The replication of a temperate phage involves inserting the viral genome (purple) into the bacterial nucleoid, which contains bacterial genes (green/yellow circle), to produce a prophage. Upon release, the phage genome is excited from the nucleoid in linear form after recognizing packing sequences (pac) by the viral machinery. The genome is packed into the capsid, and the virion is released from the cell. Specialized transduction panel. In some occasions, errors in the excision of the viral sequences cause the presence of bacterial genes close to the prophage to be included in the phage genome. The bacterial sequences are then packed into the virons to produce transductant phages. Generalized transduction process. Errors in recognition of the pac sequence induce the excision of bacterial genes at pseudo-pac sequences (ψpac) contained in the nucleoid. This process causes the production of transductants containing no viral genomes but high levels of bacteria genes. Lateral transduction panel. In this case, errors in the termination of the excision process produce a concatemer that starts with the phage genes but continues with the bacterial genes, resembling a theta replication. The concatemer is then cut and packed into the transductant phages.

by this kind of serendipity. Thus, while her husband became a professor thanks to his thesis, Zimmer's dissertation, lacking crucial data, was accepted but soon forgotten. She remained a research associate all her life. Conversely, other researchers such as William Hayes, Francoise Jacob, Ellie Wollman, and Jean Weigle promptly explored—and exploited—her discoveries in detail.

Another reason for the attrition between Zimmer and Ledergberg was the assignment of the Nobel Prize itself. While Pierre Curie refused to get the Nobel Prize unless the Committee gave one to his wife as well—because both husband and wife had collaborated closely and equally on the work that granted the 1903 prestigious recognition—Lederberg did not quarrel with the Committee. To be fair, neither did Thomas Morgan (1933 Nobel Laureate in Genetics; his wife and collaborator Lilian was neglected), Alfred Hershey (he shared the Prize with Delbrück and Luria, but Martha Chase was never considered), or André Lwoff (1969 Nobel Laureate in Microbiology; his wife Marguerite was excluded). Thus, the 1958 Nobel Prize for Medicine went to Joshua Lederberg jointly with George Beadle and Edward Tatum, but Zimmer was utterly ignored. After the award ceremony, people saw just the husband and overlooked the wife, no matter what she did.

Zimmer experienced, like Meitner, Skłodowska, Franklin, and many others before and after, what is called the "Mathilda effect," named after Mathilda Gage, a 19th-century feminist who asserted that women are disregarded in scientific professions. At the same time, their accomplishments are handed down to male colleagues. A few years later, a similar problem arose when Francis Crick, James Watson, and Maurice Wilkins shared the 1962 Nobel Prize for Medicine despite Rosalind Franklin's fundamental contribution to uncovering the DNA structure.

The Nobel Prize Committee charter allows for a maximum of three yearly *ex aequo* winners for each class. This clause was not an issue when the Prize was established in the late 19th century because research was generally carried out single-handedly. Thus, Pasteur devised the experimental procedure, designed the instruments, and worked in the laboratory; Einstein conceived his theories alone (albeit his wife helped him write some papers). In more modern times, research is conducted by a team; thus, assessing the paternity—and maternity—of the results becomes a bitter affair. Usually, professors are preferred for the Prize under the assumption that they have conceived the experiment and contextualized it in the current scientific paradigm. Instead, research technicians are only considered mindless executioners of the professor's idea and automatically excluded from the tender.

The problem with Zimmer was that she remained—unlike Zinder, who became a professor of Molecular Genetics at RI and later discovered the first phage with an RNA genome (f2)—an untenured (that is, unpaid) research associate even when working in the laboratory she established with her husband. She endured such a strain—just like Meitner did before her—for sheer love of science. Nonetheless, the ever-growing shadow of his spouse, which increased considerably after obtaining the Nobel Prize, relegated her to oblivion, and she could no longer tolerate such a stigma. While Lederberg once reproached Ryan for publishing an account on the replica plating that neglected Zimmer, Lederberg bypassed his wife's role in subsequent articles on the subject. For instance, in a retrospective on the discovery of bacterial recombination fate in 1987, he remarked that "Tatum had acquired K-12 from the routine stock culture collection in Stanford … the same strain was then in hand when he set out to make single, and then double, mutants in *E. coli*." No mention of his wife, not even in the acknowledgments.

The relationship between Lederberg and Zimmer went south; he remarked, "I found myself under a malignant spell where Esther's disparagement of me became all too believable." The couple divorced in 1967. Lederberg quickly remarried to Marguerite Stein, who was 12 years his junior; Zimmer remarried many years later (in 1993, when she was 70) to Matthew Simon. While Lederberg expanded his involvement in academia, Zimmer started a career in classical music, co-funding the Mid Peninsula Recorder Orchestra.

Afterward, Lederberg collaborated with NASA during the Saturn Project, which sent men to our natural satellite in the 1960s and 1970s, raising the problem of microbial contamination of celestial bodies such as the Moon. He also developed the microbial detection system mounted in the Viking probe that landed on Mars in 1975. Such a work prompted him to be used as a model for the character of Dr. Stone in Michael Crichton's novel *The Andromeda Strain*, the story of a scientist fighting a microbial pathogen from outer space. Lederberg also became an advisor on biological warfare for the US Department of Defense. He pioneered computer analysis for biological purposes, contributing to designing the programming language *DENDRAL*, an early example of artificial intelligence applied to biochemistry. Well aware of the practical application of plasmids and transducing phages, Lederberg became a consultant to *Syntex* and *Cetus*, two pioneer biotechnological companies.

Another example of the Mathilda effect within the Phage Church is that of Margaretha' Grete' Gujer, born in 1919 in Rümlang near Zurich. Like Meitner, she chose a career in academia rather than that of a housewife, as was customary at the time. Thus, in 1942, she enrolled in the Polytechnic of Zurich (ETH), but she soon dropped out for economic reasons and started working in a financial firm. Nonetheless, she met Eduard Kellenberger at the ETH, who she married in 1945. The year after, Kellenberger obtained a PhD studentship in Jean Weigle's laboratory in Geneva. Gujer helped her husband with his job, focused on characterizing bacteria by electron microscopy. Kellenberger and Gujer co-authored several papers together.

When Weigle joined Delbrück in Caltech, Kellenberger became director of the Center for Electron Microscopy (CEM), which had a good start with the support of Thomas Anderson—who visited for a while. Afterward, Weigle would bring news and research topics he gained at Caltech and stimulate the work in Kellengerber's group. Thus, despite limited funding, CEM became the reference center for the electron microscopy study of phages. Gujer contributed significantly to this work but did not raise much more than a research assistant. Her modesty made her fade compared with her husband, who had a vivid character. In addition, her poor mastery of the English language made her unsuitable to participate in international conferences, making her even more marginalized. For instance, she taught a young Ph.D. student, Werner Arber—who obtained the Nobel Prize in 1978 for discovering the restriction enzymes and worked with the Lederbergs and Luria—the laboratory techniques for phage analysis, but the official tutor remained Weigle. Gujer collaborated with Ludovica Zichichi, daughter of the Italian physicist Antonino, for the characterization of the mechanism of recombination using an abortive lysogen λb2 as a model.

Like Zimmer, she had a blow on his academic track due to poor supervision. Weigle suggested she delay the publication of her work on recombination to have experimental confirmation from Matthew Meselson. It turned out that the paper by Meselson and Weigle was published before Gujer's, so she lost primacy.

In 1965, Kellenberger moved to the Kansas State University for a sabbatical in the laboratory of one of his former Ph.D. students, Gordon Lark. Gujer went with him, but their relationship deteriorated, and they divorced the following year. She obtained a position in Lucien Caro's laboratory at Oak Ridge but returned to Geneva in 1971. Despite her substantial contribution to phage characterization, she was not recognized as a prime figure within the phage community. Shortly before her retirement, she only obtained an honorific title, the Nessim-Habif Prize. "I was nothing, as I never finished my studies," she described herself. Luria, instead, portrayed with the words: "she is well known and well appreciated for her contribution to molecular genetics," adding that "she was responsible for training in bacteriophage research some of the best European geneticists." Despite her involvement, though, she remained in the background while others, including her husband, went under the spotlight.

7.4 THE ESSENCE OF GENES

At the end of the 1950s, bacteria had ceased being static entities. They had become lively organisms with dynamic genomes shaped by the insertion of foreign DNA (transformation), direct exchange of plasmids (conjugation), and indirect acquisition of genes through viral infection (transduction). Indeed, it turned out that bacterial genomes displayed more plasticity than eukaryotes'. The puzzle of genetics was mainly solved, but nobody had any clue about the actual structure of this molecule, the primal question that led Delbrück to move into genetics. The DNA composition (carbon, hydrogen, nitrogen, and phosphorus) was known from chemical studies, but how these atoms were arranged in space was still a mystery.

Luria understood biochemistry and was instrumental in solving the DNA enigma but recognized that he was too old to learn this discipline. Luria recalled that Pauling had advised him "not to do biochemistry unless I knew lots of chemistry." Thus, Luria decided to prompt his "first and most successful" graduate student at Indiana, James Watson, on this quest by sending him to the laboratory of Herman Kalckar in Copenhagen. Watson's Ph.D. thesis was about the measurement of phage recombination after X-ray irradiation, and his pursuit of the structure of the gene from the biochemical point of view made him an external figure within the Phage Group, remaining constantly in contact with Delbrück (rather than Luria), providing updates on the subject. However, even the biochemistry path was not enough to solve the mystery of the inner structure of the gene, so Watson moved to the Cavendish laboratory to learn X-ray crystallography and teamed up with Francis Harry Compton Crick, another Ph.D. student to solve the riddle of the genes.

During his time in Cambridge, Watson had the opportunity to attend Maurice Wilkins' Laboratory at King's College in London in January 1953. Wilkins and his team, which included Rosalind Franklin, used X-ray diffraction to investigate the intimate structure of DNA, just as Watson and Crick were doing. Wilkins innocently showed Watson, unbeknownst to Franklin, some unpublished pictures (in particular, the so-called "photograph 51") of DNA diffraction patterns she had recently taken. "Rosy"—how Watson called her in his memoirs—"of course did not directly give us her data," he admitted years later; in fact, "no one at King's realized they were in our hands." Showing visiting researchers the experiments conducted in the hosting laboratory is a standard practice—otherwise, what would be the point of visiting a lab? However, instead of acknowledging the discovery and complimenting the work done in pursuit of collaboration, as customary, Watson began ruminating about the shown data to find his own interpretation. A further unintentional help to Crick and Watson's work was given by Max Perutz, another of Wilkin's collaborators, who handed over to Crick a draft of a paper containing Franklin's latest calculations on the structure of the DNA.

Primed with these insights, Crick and Watson could proceed speedily to generate a DNA structure model published in March 1953. The model they provided was summarized in a few words:

> the structure has two helical chains each coiled round the same axis … The novel feature of the structure is the manner in which the two chains are held together by the purine and pyrimidine bases. The planes of the bases are perpendicular to the fibre axis. They are joined together in pairs, a single base from one chain being hydrogen-bonded to a single base from another chain … The hydrogen bonds are made as follows: purine position 1 to pyrimidine position 1; purine position 6 to pyrimidine position 6 … These pairs are: adenine (purine) with thymine (pyrimidine), and guanine (purine) with cytosine (pyrimidine)

Peter Lawrence, one of Crick's colleague at the University of Cambridge, gives a clear account of how the structure of the DNA was discovered:

> [it] occurred on a Saturday morning, February 28th, 1953. It was then that Jim [Watson], fiddling with cardboard cut-outs of the bases, saw that an A-T pair filled exactly the same space as a G-C pair.

Such a complex riddle was solved by such an elementary method. The discovery needed a celebration, so the two researchers went to their favorite pub, *The Eagle* in Cambridge, where a plaque now commemorates that "it was here on February 28th, 1953, that Francis Crick and James Watson first announced their discovery of how DNA carries genetic information." "The key discovery was Jim's determination of the exact nature of the two base pairs (A with T, G with C)," Crick remarked. "He did this not by logic but by serendipity."

After the Nobel Prize for Medicine in 1962 had been assigned to Crick, Watson, and Wilkins, Crick and Watson's approach to the investigation—that involved peeking into someone else's results—sparked a controversy where it looked like the researchers from Cambridge had stolen the priority of the discovery from the Londoners. More precisely, the whole situation appears to be a clear illustration of the Mathilda effect, in which the Nobel committee overlooked Franklin because of misogyny. Francis Crick commented on this subject that

> feminists have sometimes tried to make out that Rosalind was an early martyr to their cause, but I do not believe the facts support this interpretation … I don't think Rosalind

> saw herself as a crusader or a pioneer. I think she just wanted to be treated as a serious scientist. Rosalind's experimental work was first class. She was less at home, however, in the detailed interpretation of the X-ray photographs.

Moreover, the Committee did not neglect the King's College group: Wilkins shared the Prize with Crick and Watson. Both Watson and Crick always acknowledged the importance of Franklin's contribution to the discovery. Another aspect that is often neglected in this dispute is, to put it in Crick's words, that "Rosalind Franklin, who had done such a good work on the X-ray diffraction patterns of DNA fibers, had died in 1958"—even if the Committee had decided to give her an award, she would not have been present to accept it.

Soon after the Nobel celebration, Watson wrote a memoir about this tremendous discovery. Despite its popularity, Watson's account was not unanimously praised; *Harvard University Press*, for instance, refused to publish it. Written with "Pepys like frankness," as Lawrence Bragg put it—Samuel Pepys was the Chief Secretary to the British Admiralty in the late 1600s whose diary, published in the 19th century, contained many details of the Great Fire of London occurred in 1666—it did not favorably portrait the people engaged in this historic discovery. Francis Crick complained about the invasion of his privacy. Such a reaction was perhaps because Watson had presented a discussion between Crick and his colleague Sir William Lawrence Bragg as a turf fight. The comment "there was no restraint in Francis' enthusiasm about young women" might have put Crick in a difficult position—especially at home. Maurice Wilkins lamented that Watson's memoir gave "a distorted and unfavorable image of scientists." A phrase like "Maurice had noticed that my sister was very pretty … if Maurice really liked my sister, it was inevitable that I would become closely associated with his X-ray work on DNA" was not quite gentlemanly written. Watson's comment on the DNA model provided by Linus Pauling—who had previously resolved the structure of hemoglobin—that did not stand experimental tests was not quite forgiving: "if a student had made a similar mistake, he would be thought unfit to benefit from Caltech's chemistry faculty … the bloomer was too unbelievable to keep secret for more than a few minutes." Watson's portrait of Rosalind Franklin has been labeled outright sexist. Even today, a phrase like: "though her features were strong, she was not unattractive and might have been quite stunning had she taken even a mild interest in clothes … there was never lipstick to contrast with her straight black hair, while at the age of thirty-one, her dresses showed all the imagination of English blue-stocking adolescents," might lead to raise an eyebrow.

Nonetheless, an account by a Nobel laureate could not go wasted: the book was published by *Atheneum Press* in 1968, had sold over a million copies, and had even inspired a film starring Jeff Goldblum—an American actor better known for his performances in movies such as *The Fly* and *Jurassic Park*.

Crick also wrote a personal account of the discovery of the DNA structure, but his recollection was much quieter in tone than Watson's. "I have not included detailed character sketches of my friends and colleagues," he said, "mainly because I am reluctant to write candidly about close personal relationships." Crick's memoir was more focused on the science beyond the development of the DNA model and the genetic code. It provided more insight into the discovery process of the genes' essence. Nonetheless, he commented on the discussion about his and Watson's approach to science:

> people have argued that we were overcompetitive. In some ways, I can see that we behaved insufferably … but it was not all due to competitiveness. It was because we passionately wanted to know the details of the [DNA] structure.

He also added that "rather than believe that Watson and Crick made the DNA structure, I would rather stress that the structure made Watson and Crick." In fact, these two young scientists' careers were—unsurprisingly—boosted.

Regardless of these personal recollections and the feud that might have ignited, what was left to discover was how to associate the DNA structure to the amino acids that made up the proteins and enzymes that constitute both cells and viruses. "The genetic code," Crick remarked, "was not revealed all in one go."

7.5 THE RNA TIE CLUB

It was clear from the start that three nucleotides were needed to encode the 22 amino acids used by the living organisms. "If each amino acid were coded by two bases," Crick explained, "we should only be able to code $4 \times 4 = 16$ amino acids." Gamow introduced the concept of "triplet," which is the combination of three nucleotides to encode for an amino acid. The same concept was called "codon" by Sydney Brenner. However, Gamow imagined the amino acids to fit into the major groove of the DNA helix, a model that did not stand the experimental testing. Crick commented that "Gamow … noticed that, looked in a certain way, the structure could have twenty different kinds of cavities … he boldly assumed that there was just one type of cavity for each amino acid." To support his statement, "Gamow had given his list of the magic twenty." Watson and Crick "sat and studies Gamow's letter at the Eagle," and realized that some of the provided amino acids "were unlikely and that he had left out some obvious candidates such as asparagine and glutamine."

Gamow also contributed to developing the genetic code by funding an exclusive league of twenty renowned scientists involved in nucleic acid research. The members, one for each amino acid, wore personalized woolen bowties; thus, the association was called "the RNA Tie Club." Despite its high reputation, the Club failed to solve the genetic code puzzle: while the members included the likes of Crick and Watson, non-members (Nirenberg and Matthaei) solved it. Nonetheless, the Club—like the Phage Group—stimulated cutting-edge research in the nascent field of molecular biology.

The genetic code later proposed by Francis Crick was theoretically superior and was backed up by experimental evidence. "Since," Crick asked, "there are $4 \times 4 \times 4 = 64$ different triplets of four nucleotides, why are there not 64 kinds of amino acids?" He observed that "the mathematical problem is to find the maximum number that can be coded." He assumed that certain types of triplets would not be feasible. For instance, "the triplet AAA must be nonsense, since, if it corresponded to an amino acid, α, then $\alpha\alpha$ would be AAAAAA, and this sequence can be misinterpreted by associating α with the second to fourth, or third to fifth, letters." Thus, rejecting triplets formed by the same nucleotide should be possible. Back in the day, the names of the nucleotides were not codified; thus, the names given to the four nucleotides were just A, B, C, and D. With the triplets AAA, BBB, CCC, and DDD excluded, there were 60 combinations possible. "It is easy to see that the 60 remaining triplets can be grouped into 20 sets of three, each set of three being cyclic permutations of one another," for instance, "ABC and its cyclic permutations BCA and CAB."

Ironically, the first experimental proof of the structure of the genetic code came precisely from the use of identical nucleotides. In 1961, Marshall Warren Nirenberg and Johannes Heinrich Matthaei at the National Institute of Health—but not invited to join the RNA Tie Club—developed a cell-free apparatus by isolating ribosomes from lysates of *E. coli* strain W3100, removing the bacterial DNA digestion by DNAse and feeding such an *in vitro* system with synthetic poly-uracil (U)—the RNA equivalent of the thymine present in the DNA—chains, and different radioactively labeled amino acids. They obtained a signal when they added phenylalanine (Phe) in their apparatus, demonstrating that the sequence TTT (UUU in the RNA code) was associated with this amino acid. Afterward, it was only a matter of using synthetic RNAs with different combinations of nucleic acids and seeing what amino acid gave a signal.

It is now known that the magic number of 64 combinations is reached by having three codons (TAA, TAG, and TGA) terminate the translation process (because they are not associated with any amino acids) and multiple codons encoding the same amino acid for the remaining 61 combinations—for instance, lysine is encoded not only by AAA but also by AAG and arginine by CGT, CGC, CGA, CGG, AGA, and AGG. For this reason, the genetic code is called "degenerate." It has also been recognized that all proteins start with the amino acid methionine (codon ATG), although this might be cleaved away, in some cases, in mature proteins. In addition, notwithstanding a few exceptions, such as in the mitochondrion, the code is "universal" because all organisms, from viruses to man, use the same encoding strategy (assign one codon to a specific amino acid).

Furthermore, the code is not overlapping. Given the sequence ATGGATTACACC, it will not be read in reading shift, such as in:

ATGGATTACACC → amino acid 1

A**TGGATTACACC** → amino acid 2

AT**GGATTACACC** → amino acid 3

ATG**GATTACACC** → amino acid 4

but as:

ATG GAT TAC ACC

↓ ↓ ↓ ↓

amino acid 1 2 3 4.

The four triplets of this example will correspond to the amino acids methionine, asparagine, tyrosine, and threonine (Met, Asp, Tyr, and Thr). Such a reading approach is known as comma-free because it corresponds to writing without spaces (or commas) as in batmanmadact (bat man mad act).

In 1958, Crick presented the so-called "central dogma of molecular biology," which is still in effect today. It states that genetic information flows from nucleic acids to proteins but not vice versa. Proteins are crucial in determining cell identity since they can directly construct the cell structure, indirectly affect the other molecules that comprise it, and govern the expression of all genes. Thus, bacteria have a different protein pattern than human cells, different bacterial species or strains have different protein fingerprints, and the cells of the human body's various tissues have distinct protein profiles. These profiles are defined by the nucleic acids, that is, by the genes present in the organism's genomes and the temporal expression of these genes. It is the genome that is passed from one generation to the next, not the protein profile. For instance, spermatocytes carry the chromosomes inside a small flagellated cell that differentiates in all the tissues that make up the human body after fusing with an oocyte. Bacteria or fungi can do the same: spores contain relatively few proteins not ordinarily present in a vegetative cell; after a spore germinates, the protein makeup of the original cell is restored.

The central dogma is a testament to Larmackism. Mature cells encode proteins by following the genetic profile encoded into the DNA genome: the collection of genotypes. The environment acts on these proteins to compel the genome to encode other pre-encoded proteins to adapt to mutated conditions, such as a lack of certain nutrients. However, such back modulation cannot rewrite the genome; genetic evolution does this through mutations, recombination, and acquisition of foreign genes. Therefore, Lamarck's concept that the phenotype—which is essentially made up of proteins—could determine the genotype did not stand the test of time.

Delbrück understood that the replication of the DNA molecule, which is made up of two chains linked together "in a zipper-like fashion," would result in the formation of a so-called supercoil: a twist in the molecule that would cause a physical strain in the molecule, producing enough force on the enzyme to stop the replication process. He commented, "the principal difficulty of this mechanism lies in the fact that the two chains are wound around each other." He assumed the best way to avoid supercoil formation was to cut and mend the DNA chain during the replication process. The prediction was correct: we now know that numerous enzymes (topoisomerases) play a role in supercoil resolution during DNA replication by nicking the DNA chain and subsequently rejoining the ends.

7.6 THE OUTSIDERS

The Phage Group reached a turning point where it ceased to be a happy clan and became poisoned by personal rivalry. Furthermore, Delbrück's rigidity toward considering only work based on the viruses included in the phage treaty meant that he missed some of the advances in bacterial biochemistry and structural biology that his research had set in motion. He started to refuse to look at draft papers that did not include a well-defined working hypothesis and to ignore experiments presented by researchers outside the Phage Group. A participant in the Phage Course remarked that "if an outsider tried to contribute to the phage canon, Delbrück was known to dismiss him with the comment 'but he hasn't taken the phage course'."

One of these "outsiders" was Gunter Stent, born Günter Siegmund Stensch, who interacted only indirectly with the Phage Group. Stent left Berlin after the rise of the Nazis and obtained a Ph.D. in physical chemistry in 1948 from the University of Illinois. When Stent asked Delbrück for a research position, Delbrück replied rather harshly, "we don't need someone like you." Stent did not give up and set up his phage research group, eventually not only becoming Delbrück's collaborator—for instance, they introduced the term "semiconservative replication" in a joint paper—but also a prominent figure within the Phage Group and its official "historian." Another example of phage researchers not included in the Phage Group is that of the Lederbergs: neither Joshua nor his wife participated in the Phage Course, yet they significantly advanced the field of bacterial virology. In addition, other groups around the world studied phages independently from Delbrück and associates, for instance, those led by Paul Bonét-Maury, Vladimir Sertić, and Nikolai Bulgakov in Paris, Christopher Andrewes and John Boyd in London, Hans Friederich-Freska and Wolfhard Weidel in Tübingen, and Frank Macfarland Burnet in Melbourne. There were attempts to establish a European Phage Group, but the project never gained momentum. For instance, Kellenberger organized a Phage Course in Naples in 1963 in the laboratory of Adriano Buzzati-Traverso, where 20 participants attended.

From 1953 onward, Delbrück's focus switched from phage research altogether. He wanted to hunt for new study areas because the field was growing increasingly crowded. He advised his students to "don't do fashionable research" because crowded fields give little room for new ideas. Once, he said to George Beadle, "I am still fidgeting around to settle … into research that suits me, trying all kinds of things." For instance, he characterized the structure of lipid membranes, including their electrical conductivity. In addition, he studied the properties of a chemoreceptor of the silkworm moth *Bombyx mori*.

He focused on photosynthesis, one of his first passions back in Germany. "When I took the microbiology course from van Niel," he recalled, "he drew my attention to the phototactic bacteria." Particularly, Delbrück was interested in providing experimental support for the so-called Weber-Fechner law, algebraically expressed as $\Delta I/I$, where I is the intensity of a given signal such as light. In Delbrück's words, this law, first presented in the 19th century, states that "the threshold change of this parameter which will produce excitation is proportional to the initial value of this parameter." The amount of light required to produce a biochemical response in a mold, for example, will depend on the amount of light exposure the mold has already received: a mold growing in the dark will require less light than one cultivated in the light. That is,

> these bacteria could detect intensity differences of 10%, and they could detect these intensity differences of 10% irrespective of whether the intensity was very low, or a million times higher or even a billion times higher. That fascinated me, because what could be the mechanism would have to be a very general and a very simple mechanism

"The rating that this law of Weber has received," lamented Delbrück is "a very low level in recent times, so low in fact that it is barely mentioned in many current textbooks of physiology." The fundamental premise behind this work in physiology was that the discovery of a simple solution to a complex question may have been expanded to other areas of biology, allowing him to make the dent he always sought.

This new enterprise, like the Phage Group, became a world-renowned squad. In Caltech, he organized a series of seminars (*Selected Topics in Biophysics*) that lasted another quarter of a century, more or less as long as the Phage Course at CSHL. Remarkably, the speech given during the Nobel ceremony was not about phages but rather neurophysiology, an example of epistemology. He even wrote a book about epistemology, *Mind from matter?*, published posthumously in 1985.

Nonetheless, the new Delbrück's endeavor had far less effect on the scientific world than the Phage Group's. Among the subjects characterized by Delbrück and co-workers, there was the curious feature of how *Phycomyces* could grow around obstacles in the dark without touching them, a trait first described in 1881. "*Phycomyces* … grows to the right or to the left," he remarked, and could respond to "four distinct stimuli: light, gravity, stretch, and some unknown stimulus by which it avoids solid objects." In 1972, Delbrück remarked that he was "sick at heart at the unsolved state of the problem." Even today, this feature is poorly understood and Delbrück's, despite his uneasiness, remains the best available model. He theorized that the yeast secretes a chemical that generates a gradient that allows the mold to recognize objects independently from light stimulation.

Delbrück spent two years in Cologne, Germany, at the newborn Institute for Genetics to study the photochemistry of thymine, returning to Caltech in 1963. The experience in Cologne was not merry: accustomed to American manners, he disliked European formalism and red tap. In addition, he started to have severe health annoyances, particularly chest pain. He was diagnosed with angina pectoris in 1963 and with coronaries in 1978. Nevertheless, his involvement in Europe was extended afterward with his collaboration to inaugurate the first European private University, that of Constance, in Rhineland.

Even Luria could not keep up the flag of the Phage Group for long. In 1953, he became consultant for a pharmaceutical company. The experience made him declare that

> the university is a world of scholarship and trust, where the reward for success is intellectual recognition. Industry is a world of contracts and insecurity, where pay is the reward for work … the university, for all its limitations, approximates what an ideal intellectual community should be—a microcosm where shared purpose and mutual respect counter-balance the centrifugal momentum of individual passion.

However, even the academic world was wavering. In the early 1960s, he noted that "phage research meeting attracted two or three hundred people instead of a dozen or so as ten years earlier." Luria's goal to expand the field of bacteriophage research had been fulfilled beyond expectation. However, "the field was crowded," with cutting-edge scientific discoveries diffused by "phone rather than by publication." The academic tissue was changing and Luria found "the frantic activity of my young colleagues somewhat ludicrous." Gone were the merry days of the companionship of the Phage Course. "Max Delbrück had already escaped, starting a quiet, underpopulated research program on … *Phycomyces*."

Like Delbrück, Luria needed to locate a secluded spot within the avenue of microbiology where he could conduct his research at his own pace. Luria's work on phage-driven membrane modification in *Salmonella* had already provided him with such a tranquil niche as young scientists shunned membrane biochemistry. Colicins provided an additional avenue of investigation. André Gratia discovered these proteins capable of killing bacteria in the 1920s at the PI. Initially, they were examined for their bactericidal qualities, but they proved ineffective and were abandoned. Nevertheless, Luria became intrigued by these proteins. Since it was established that these proteins were hydrophilic, they were understood to be cytoplasmic; in fact, "how did a water-soluble colicin get into or through the fatty layer of the membrane?" he asked himself.

To answer this question, Luria took a sabbatical and went to the PI, where he "was settled in a small laboratory in [Jaques] Monod's department." Back then, the PI was the recognized center for genetics. Luria had the chance to meet three of the leading PI figures: "Lwoff courtly and ironical, Jacob friendly yet guarded and self-sufficient, Monod spirited and open to intellectual given and take." Soon, Monod "became a close friend" of Luria because he "had exactly that touch of histrionic that one wants in an artist or in a friend."

When returning to the US, Luria concentrated on characterizing colicins. With the help of his assistants Celik Kaylar and Michael Weiss, Luria assessed that colicins formed pores in the bacterial membranes that neutralized "the electrical potential that exists between the outside and the inside of normal bacterial cells … blocking the [membrane] transport proteins, since this electrical potential provides the energy for transport." Luria described the effect of colicin treatment on the bacterial biochemistry in vivid terms: "there is an immediate and simultaneous arrest of DNA, RNA, and protein synthesis; a decrease of intracellular ATP levels; and an inhibition of the active transport of β-galactosides, several amino acids, and potassium, magnesium, and rubidium ions."

It is now known that colicins belong to the broader family of bacteriocins, which are small antibacterial proteins capable of forming pores (holes) through the bacterial membrane and cell wall. While colicins are expressed by *E. coli* and other members of its family (*Enterobacteriaceae*, which cluster several Gram-negative genera), bacteriocins are also expressed by Gram-positive bacteria. Like antibiotics, bacteriocins might be broad (attacking the cells of several bacterial genera) or narrow (piercing the cells of limited bacterial species) spectrum. Interestingly, colicins produced by some strains of *E. coli*—those bearing the bacteriocin-encoding plasmid—are active against the other *E. coli* strains that lack this virulence factor.

Luria remarked that understanding colicins' activity "is not just an idiosyncratic curiosity" but may "provide an informative model for the study of transmembrane protein channels." Such a prediction flew too low: with the widespread rise of antibiotic-resistant bacteria we have experienced in the last decades, bacteriocins are once again a hot topic of medical investigation as front-line antibacterials. For instance, recombinant violacein has proved active in methicillin-resistant *S. aureus*.

In addition, colicins are even used to lyse cancer cells, expanding this field of microbiology into the realm of oncotherapy.

Membrane-piercing proteins are also pivotal in the biology of phages. Although the familiar picture of phage discharge from infected cells describes the process as the "virus burst the bacteria open like a popped water bomb," the phage requires enzymes to pierce the bacterial cell wall. Bacteriolysis has been known since the time of Twort, but the isolation of these enzymes took some time. In the early 1930s, Alice Evans isolated the so-called Boston phage (now known as C1) and showed that lysates of this phage had a lytic activity on strains of streptococci that the phages themselves were unable to infect; she termed the phenomena "nascent lysis."

The phage-encoded lysins—discovered by Eliava and should not be confounded with lysine, the amino acid—are now known as endolysin to distinguish them from the other lysins present in the phage tail. While the latter are needed to access the bacterial host, the latter must get out. Endolysins and lysins are murein hydrolases that degrade the peptidoglycan polymer present in the cell wall. However, endolysins cannot be exported through the bacterial cell membrane; they require a portal to access the cell wall. Such a portal is provided by another phage protein involved in inducing bacteriolysis: holins.

Holins are phage-encoded porins that generate a pore in the bacterial membrane and allow endolysin to be exported in the outer section of the plasmalemma. Today, over 50 families of these proteins are known. The process of bacterial piercing has been described in detail for Gram-negative bacteria, and it has been shown that three classes of these proteins work in concert to cause bacteriolysis. One class of holins, known as pinholins, forms a small channel in the inner membrane of the bacteria, causing ions to leave the cytoplasm and the depolarization of the cell. Consequently, enzymatic activities such as DNA and protein synthesis inside will cease. At the same time, the actual holins will fold into their final form, aggregating into a pore that will allow the export of the endolysins in the periplasmic space below the cell wall. A third class of holins (spanins) will pierce the outer membrane afterward.

It turned out that holins are essential in determining the timing of the lytic cycle and, in turn, the burst size of the phage. Despite the usual view that the phage spawn will cause the burst of the infected cell simply by sheer pressure, phages need a system to puncture the cell and facilitate the lysis of the bacterium. Holins provides such piercing devices. Like bacteriocins, lysins have gained interest as antibacterials and provided promising experimental results.

7.7 A DIFFERENT MORAL UNIVERSE

Luria never made a secret of his political tenet: "my commitment to socialism is essentially a commitment to the idea of economic justice and equality … my commitment to political radicalism is a calculated existential choice … subject to eroding doubt … like Pascal's wager as to the existence of God." Such ideas developed first in Italy—for instance through the exemplary life of Augusto Monti and Giuseppe Levi—and in France later. By contrast, Luria noted that the Delbrücks' "attitude toward politics, especially class-oriented politics, was to consider it as a silly and even distasteful joke," although they were "eager helpers of persecuted individuals" and "among the largest contributors of Amnesty International."

Such a straight statement of personal conviction might have been commonplace in Europe but did not sit well in the American context. The tensions between the US and the USSR had created an atmosphere of overt hostility against socialist ideas in the West, so in the years following the end of the Second World War, the US society became increasingly conservative. Civil rights activities were associated with covert communist actions and were openly censored. When a Republican governor took office in the early 1950s, such censorship became a straightforward witch-hunting that became known as McCarthyism.

Joseph 'Joe' Raymond McCarthy was born on the 14th of November 1908 in Grand Chute, Wisconsin. He can be defined as a prototypical "redneck"—albeit coming from a Badger State close to Canada rather than one from the Sun Belt—for he was of peasant origin, hard worker, not exceptionally bright, and rough in manners. In addition, he was ruthless, well-determined, and very, very ambitious. McCarthy obtained a degree in law in 1935, becoming a judge just four years later and a senator in 1946. A staunch anti-communist from an early age, he started his political career with the Democratic party. However, after losing a local election in 1936, he quickly changed sides, enrolling in the Republican party. Edward Hart, a fellow Democrat who witnessed McCarthy's electoral strategy, commented, "Joe lived in a different moral universe." John Hanes, who knew McCarthy from the US State Department, described him like this: "to say he was a rough diamond was a misnomer in the sense that he was never a diamond … he was rough." McCarthy was a consummate politician who described himself as a simple "farm boy." With the onset of the war, McCarthy volunteered in the Marine Corps, being discharged with honor in 1944 as a decorated captain. He became a close friend of the FBI's first director, John Edgar Hoover.

Despite his flamboyant ascension, which made him one of the youngest senators in US history, McCarthy's career soon stalled. Nevertheless, he quickly learned how to revitalize it: like many demagogues—nowadays, preferably referred to as populists—McCarthy needed an enemy to fight. Instead of proposing laws that could improve the standard of life of his constituents, he decided to become the paladin of a group of concerned citizens who had started to spread rumors about American soldiers bullying German prisoners of war who had managed concentration camps during the war. Eventually, the battle amounted to nothing—the rumors accounted for what we today call fake news—but taught him how to find scapegoats to aggregate a political consensus and how to address the press to get the attention he desperately craved.

Soon, instead of defending Nazi prisoners, McCarthy found a much better enemy to fight, one that well-suited the middle-class mindset of American society: communists. At a public meeting in February 1950, he boasted that "I have here in my hand a list of 205 … members of the Communist Party … working and shaping policy in the State Department." The innuendo was echoed by a local newspaper with the headline "McCarthy charges Reds with holding US offices." While it was not—and still is not—illegal to be a member of a Communist Party, the competition between the US and the USSR for world hegemony meant that communists were not at all well regarded by American society. Even in Europe, where Socialists and Communist parties were more diffused, communists were often excluded from the civil service.

The point was that McCarthy did not have a list of assumed traitors in his hands; it was just another fake news meant to gain the spotlight. Within a few days, the list had shrunk to 57 people whose names he never disclosed. President Harry Truman commented that "this is the first time in my experience … that I ever heard of a Senator trying to discredit his own Government." A few days after his speech, even McCarthy had to admit that "I shot my mouth off."

Nevertheless, he could not stop: he bound himself to find communists and "commies" he had to present to the public. His friend Hoover helped him by using the mammoth FBI's machine to provide names of undesirables not initially included in the list. One of the agents involved in this borderline operation, William Sullivan, remarked, "we gave McCarthy all we had, but … we didn't have enough evidence to show there was a single Communist in the State Department." Among the first victims of McCarthy's "Red Scare" were Dorothy Kenyon (a respected attorney and former municipal judge), Owen Lattimore (a renowned professor of political sciences at Johns Hopkins University and a consultant for the White House), and Gen. George Catlett Marshall (war hero, US Army Chief of Staff for Roosevelt, Secretary of State and Defense in the Truman administration, and promoter of the post-war economic support to Europe). Their only crime was that unidentified sources leaked their names. However, McCarthy's defamation campaign forced them to resign from their esteemed positions. On the other hand, McCarthy became a millionaire through individual donations to the cause.

Like so many populists before and after him, McCarthy did not introduce any new concept: he simply borrowed the wholesale fear of communism that was deeply rooted in American society. The Un-American Activities Committee (HCUA) was established in 1938 by the Democratic senator Martin Dies—thus the somewhat ominous moniker "Dies Committee"—with the intent of investigating extremism on both ranges of the political arc, but then quickly settling on the persecution of alleged communists. Even Truman established the Permanent Subcommittee on Investigations (PSI) in 1941 and the "Loyalty Order" act in 1947, which enacted a background security check on five million citizens to ensure that communists could not be employed as civil servants. One of the top caliber victims of this hunt was the conviction for perjury of Alger Hiss, a high-ranking officer at the State Department who Richard Nixon prosecuted.

Just like Aryanism, un-American was an empty container that could mean anything and its opposite. For instance, in the late 19th century, US Senator David Hill described the income tax as "un-American," a concept restated by Ronald Reagan in his inaugural speech as US President in 1981. Yet, the income tax is today one of the primary sources of income for capitalist states worldwide, including the USA. Franklyn Delano Roosevelt, one of the most revered US Presidents, officially claimed that "no American citizen ought to have a net income … of more than $25,000 a year." Nonetheless, he had never been pointed out as a communist spy.

McCarthy joined and led PSI; thus, his activity was framed within a well-defined and widely accepted context. It has been estimated that hundreds of people were incarcerated, and over 10,000 lost their jobs in the decade that McCarthy presided over PSI. He created a climate of witch-hunting that affected thousands of people by promoting the popping up of committees engaged in investigating alleged communist infiltrations. The most remarkable case was that of Julius Rosenberg—a Manhattan Project's low-rank employee—and his wife Ethel, both sentenced to death as alleged Soviet spies, albeit the proofs against them were fragile. Another sensational victim of the Red Scare was Robert Oppenheimer, who suffered public humiliation mainly for personal and professional acrimony with Lewis Strauss, a consultant for the White House on atomic energy; Edward Teller was a witness for the prosecution. Even Albert Einstein was among those questioned by McCarthy.

Even Luria could not escape the American spirit of the time. He was investigated by the FBI in January 1950, with several colleagues questioned during the case. Although it became clear that Luria was politically very active, the procedure was closed in July without evidence demonstrating Luria's membership to the Communist Party. Nonetheless, he was once again put under FBI surveillance in September due to his friendship with Pontecorvo, who had flown to the USSR that year. Even in this case, the investigators found no evidence against Luria. In 1952, Ruth Shipley, the head of the US passport department, denied Luria an exit visa to attend a meeting of the General Society of Microbiology in England and visit his mother—who had recently become a widow—in Italy because of alleged anti-American activities.

Eventually, McCarthyism ended not because of a solid legal system that protected the citizens' civil rights but simply because McCarthy lost popularity. In 1954, McCarthy sued no less than the US Army, claiming that communists had infiltrated a base in New Jersey. However, at the hearing, which was broadcast on national TV, McCarthy did not present himself nicely—on many occasions, he even looked drunk—and was even mocked by the US Army attorney, Joseph Welch, with the famous words, "Have you no sense of decency, sir?" In 1956, Harry Slochower, a professor at Brooklyn College, lost his job after a PSI investigation. Still, he appealed to the US Supreme Court, which supported Slochower and censored McCarthy, claiming that "the practice of imputing a sinister meaning to the exercise of a person's constitutional right" must end. Soon after, McCarthy was again censored by a bipartisan movement within the parliament, losing the PSI's direction. He died in 1957, probably due to the effect of alcoholism, widely criticized and away from the much-beloved spotlights.

Ironically, the communist ideals McCarthy fought so viciously have been developed to improve the quality of life of farmer boys such as himself. Yet, the communist dictators did nothing to prove McCarthy wrong. The USSR had indeed spies within the US Administration—whom McCarthy did not uncover—but what is worse is that, in the Soviet block, one humanitarian crime came after another. In the USSR itself, Stalin ordered the deportation and execution of countless people, but in the other countries led by the communist parties, things were not much better. In Hungary, for instance, the communist party obtained only 17% of the consent in the 1945 elections. However, the communists obtained the Ministry of Interior and used law enforcement to imprison thousands of political opponents. Unsurprisingly, in the 1947 elections, the communist party had risen to 60%, and in 1949, it was the only legally allowed party to participate in the polls. Stalin died on the 5th of March 1953 and was succeeded by Nikita Khrushchev after an internal power struggle.

Although Khrushchev denounced—in secrecy—Stalin's crimes against humanity at the 20th congress of the party and declared the democratization of the Soviet society ("destalinization"), he soon ordered the Red Army to repress workers' protests in Georgia (March), Germany (May/June), and Hungary (1956).

Back in the US, despite the clear mainstream opposition toward civil rights campaigners—who were wholesale considered Soviet spies—Luria remained actively engaged in political initiatives. Already in 1946, he co-signed a letter to convince John Haldane to denounce Lysenko, and in 1948, he joined the nascent Progressive Party because it "seemed a genuine enough new political movement against the social retrenchment of the Truman years and the beginning of the Cold War." However, the party "failed to capture the imagination of moderate Americans" and was dissolved soon after. In 1957, Luria signed Pauling's open letter against the use of nuclear warheads, and shortly after, officers from the Immigration and Naturalization Service appeared at Luria's residence "asking all sort of questions, in what I interpreted as an attempt at intimidation," he recalled.

In 1959, Luria obtained a position at MIT in Boston, where he found a more open and politically active community. Even his brief tenure at Columbia University in the early 1940s must have helped him grasp American society, considering the high level of political activism shown by Columbia's students to this day. In Boston, he befriended several renowned intellectuals there, such as the political activist and economist Noam Chomsky.

There were plenty of occasions for political activism in those years. In fact, one may argue that hatred toward liberal campaigners stemmed from an underlying sense of shame as the United States did not fare much better in terms of human rights than the Soviet Union. In the quest to maintain the capitalistic machine running, the US had to endorse and nurture despots around the world. Thus, in March 1952, the Truman Administration helped the Cuban general Fulgencio Batista to conquer power and install a brutal dictatorship on the island. Batista went so far that Eisenhower, who succeeded Truman as President in November, was forced to suspend the military support to his junta. Without backing from the United States, the Batista regime crumbled in January 1959 under the push of a handful of peasants led by a Cuban lawyer (Fidel Castro) and an Argentinian practitioner (Ernesto 'Che' Guevara). Initially, Castro asked for economic support from the US. However, the Eisenhower administration was uneasy about supporting a government of socialist inclination—mainly because the powerful American corporation United Fruit Company was nationalized soon after the uprise—so Castro aligned Cuba with the USSR.

Cuba remained a thorn in the flesh for the USA—a socialist regime just a few kilometers away from the heart of capitalism—and the CIA devised a plan for the invasion of the island that Eisenhower did not implement, but that was inherited by John Fitzgerald Kennedy, who won the 1960 presidential elections against Richard Nixon with 50.01% of the votes. Kennedy approved the plan, which was launched in April 1961 with paramilitary troops landing in Cuba's Bay of Pig. However, the invading forces were immediately blocked by the Cuban army, and Kennedy withdrew military support to the invasion, writing one of the most shameful pages of American history. Luria publicly denounced the US attempted invasion of Cuba, claiming that the "Bay of Pigs invasion" was "a cowardly, immoral action rather than just an incompetent fiasco."

Another problem facing the Kennedy administration was the conflict in Vietnam, which turned out to be a more extreme version of the Korean War. Like Korea, Vietnam had been divided into an American-backed South and a Communist North. As in Cuba, the US supported a golpe led by Ngo Dinh Diem at the end of 1963, who took control of South Vietnam, but the regime was so brutal that the US plotted another golpe against their own puppet. The counter-plot failed, and the North Vietnamese took advantage of the chaos that followed to invade their neighbor. The US sent more and more troops until it was forced to reintroduce conscription to support what was perceived as an imperialist invading force. The American involvement in Vietnam was a complete failure—eventually, the North conquered the South, and the whole of Vietnam remains a communist country to this day—and triggered social unrest within the US. Both Luria and Delbrück had opposed the Vietnam War, with the former openly joining anti-war rallies "as a mobilizer of faculty members and publicizer of protest."

Luria also wrote a series of political articles on the class structure of the Italian school system and on the history of Italian anti-Semitism. He joined the Boston Area Faculty Group on Public Issues (BAFGOPI), which bought pages—nicknamed "ad-itorials," that is, "advertisements" and "editorials" at the same time—within popular newspapers such as *The New York Times*, *The Washington Post*, and *The Boston Globe* to spread civil rights ideas. "Our ads were topical," Luria recalled, "they did not go into political analysis ... were never revolutionary in content." Among the most successful BAFGOPI activities was a campaign against New York Governor Nelson Rockefeller, who wanted to establish a nationwide building plan for antinuclear shelters. After the ad-itorials were published by *The New York Times*, Rockefeller's proposal lost support and was never ratified. Other BAFGOPI ad-itorials were aimed against Edward Teller—one of the major paladins of nuclear proliferation—and the bombing of Vietnam. One of the last BAFGOPI initiatives was to oppose the re-election of the Republican President Richard Nixon in November 1972, who instead won against the Democratic opponent George McGovern by a landslide. Afterward, BAFGOPI "ended up in mothballs," as Luria put it.

Nixon—who served as vice-president under Gen. Dwight D. Eisenhower, another champion of moderatism—was one of the most conservative US presidents in modern times, with a personal hate for the wealthy and intellectuals exemplified by John F. Kennedy. After Kennedy was assassinated in November 1963, the presidency went automatically to vice-president Lyndon Baines Johnson, but Nixon won his first term in January 1969 against the Democrat Hubert Humphrey. Nixon won a second term in 1972 but did not enjoy the position for long because he was embroiled in one of the greatest scandals connected with a US president until recently.

In the dead of the night on the 17th of June 1972, the night guardian Frank Wills was carrying out his routine control at the Watergate complex—a huge building containing hotels, apartments, and offices including those of the Democratic Party—in Washington, DC. At 1:47, Wills reported a suspected burglary to the police. When the officers came, they found five people—later dubbed "the plumbers"—rummaging in the office of the Democratic Party, armed with electronic bugs, cameras, and walkie-talkies.

The burglars surrendered immediately and were promptly identified: four were Cuban exiles involved in the Bay of Pigs, and the latter was James McCord, a CIA specialist for electronic surveillance. The police soon discovered that McCord's handler was Gordon Liddy, a Nazi sympathizer and FBI agent. On the night of the burglary, Liddy was near the Watergate complex to coordinate the burglary with Howard Hunt, a spy story novelist and Security consultant for the White House. When the police arrived, Liddy and Hunt slipped out.

Initially, the case was handled simply as a burglary, but a journalist at *The Washington Post*, Robert Woodward, was spoon-fed by a whistler-blower named "Deep Throat." Woodward met Deep Throat under rather melodramatic conditions—at 2 AM in an underground parking lot—on the 9th of October, but what he learned was pure dynamite. It turned out that Deep Throat was no less than Mark Felt, the deputy director of the FBI, who resented how the acting director, Patrick Gray, had kept informing the Nixon Administration of the details of the Watergate investigation in order to hush the case and shield the President from the judiciary. In addition, Felt hoped to put Gray in a bad light and get his position as director general of the FBI. Felt told Woodward that the burglary was a covert operation to gather information to use against the Democratic Party to favor Nixon.

Further investigation showed that Liddy had presented the plan, Operation Gemstone, to John Mitchell and Jab Magruder—heads of the Committee for the Re-Election of the President, which was known by the ominous acronym CREEP—as well as John Dean, a personal aide to Nixon. The original plan included the employment of hookers to compromise high cadres of the Democratic Party, akin to the Russian "kompromat." Nixon had given the order—an implicit one, Mafia-style—to damage the Democrats, and his inner circle immediately carried out the secret mandate. It was not the first time. Liddy and Hunt had already irrupted into the office of a psychiatrist who had as a patient another whistler-blower—Daniel Ellsberg—hoping to find compromising evidence. Ironically, the illegally gathered evidence was handed over to the FBI, which later surrendered it to the commission investigating the Watergate burglary. On another occasion, Nixon wanted to obtain Lyndon Johnson's notes on the diplomatic effort with Vietnam to obtain a kompromat against the then vice-president. The notes were kept at the Brookings Institution near Washington, and Nixon shouted, "Goddamn it, get in and get those files!" to his trusted aides, Robert Haldeman and John Ehrlichman. Even in this case, Liddy was contacted and devised a plan worthy of a *Mission Impossible* plot: his team would cause a fire in the Institution and then infiltrate the building dressed as firefighters. The plan was dismissed because the Presidential Office did not want to pay for a secondhand fire truck.

Eventually, the investigation charged Liddy with four and a half years of prison, Hunt to three and a half, Haldeman and Ehrlichman to one and a half months, Magruder seven months, Mitchell nine months, and Dean obtained four months of house arrest thanks to his collaboration with the investigators. Although Nixon had to withdraw from his position as President of the United States—which he so badly craved—he served not even a day of prison, was pardoned by his successor, the Democrat Gerald Ford, and obtained a full national State funeral upon his death in 1994.

The remarkable fact was not that a politician would do everything in his power to keep the gained power but that the justice system had prevailed over the most powerful man in the Country. To paraphrase another US President, Joe Biden, "the American principle that no one is above the law was reaffirmed." In the Soviet Union and other dictatorships around the globe, it was accepted as a rule of nature that the leaders would be all-powerful, and even today, the justice system shows severe difficulties in prosecuting the crimes committed by the administration's top brass.

If Kennedy was responsible for the failed invasion of Cuba and the Vietnam War—which Nixon resolved—Nixon was liable for fostering another brutal dictatorship, that of Pinochet. In 1970, Salvador Allende legitimately won the election in Chile and started a social democratic government. The US was unhappy with the nationalization policies and the rise of the minimal wage that Allende implemented. Thus, the CIA supported a golpe Gen. Augusto Pinochet that started on the 9th of September 1973. On the 11th, Allende was killed during the defense of the Presidential palace from the invading forces. Luria recalled that after "the overthrow of the Allende government in Chile, engineered with the open approval and the covert assistance of the US government," he "had the sad honor … of introducing his widow to a Boston audience." Chile was not an isolated case of American interference in international politics: the US had actively supported golpes in Guatemala (1954) and Brazil (1964). Chile remained outstanding, though, because Pinochet remained in power until 1990—although he was arrested for human rights violations in 1998, then released in 2000 (thanks also to the pleas from UK Prime Minister Margaret Thatcher and US President George Bush), and then indicted again in 2005.

Luria's activism against the Vietnam War was not limited to the BAFGOPI's ad-itorials. In 1967, he wrote two controversial articles: in one, he compared the devastation of Vietnam to the Holocaust; in the other, he went against the physicist Ernest Pollard, a British émigré who encouraged the American scientists to assemble in a sort of novel Manhattan Project to boost the war in Vietnam, allowing the US to win the Communists as the original endeavor did against the Nazi. Luria, together with the co-signatory of the later article, Albert Szent-Györgyi, reminded his fellow scientists "to review our present professional activities in order to make sure that they do not unnecessarily contribute to the waging and prolongation of that war."

7.8 GROWING INVOLVEMENT

The climate of hostility toward human rights fighters, perceived as Soviet agent provocateurs, did not end with McCarthy. One day, toward the end of 1969, Luria found himself on the front page of *The New York Times*, but not because he had won the Nobel Prize. Together with 47 other top medical scientists, he had been banned from obtaining executive jobs at the National Institute of Health. In short, he had been black-listed, most likely, he imagined, for his "political beliefs or personal associations."

The New York Times had always been at the front line of investigative journalism—for instance, it published an article by David Burham on the New York police corruption after the revelations of the detective Frank Serpico—and, even today, remains one of the pillars of the liberal fourth estate. Among the scoops provided by

The New York Times at the time, there was a piece by William Beecher revealing how Nixon—on advice from the national security adviser Henry Alfred Kissinger—had authorized the covert bombing of North Vietnamese bases in Cambodia (Operation Menu, March 1969 to April 1970), which officially was not involved in the war with South Vietnam. On the 13th of June 1971, *The New York Times* published the article *Vietnam Archive: Study Traces 3 Decades of Growing US Involvement* by Neil Sheehan, the first of a series of installments known as the Pentagon papers.

In this series, Sheehan quoted classified documents from the US Department of Defense (DoD) leaked by Daniel Ellsberg, a consultant. The government sued Ellsberg for treason, but at the end of the month, the Supreme Court decreed that these papers were coherent with the Fifth Amendment that protected the freedom of speech. Today, Ellsberg would have had a worse time defending his right to inform the public: for instance, Julian Assange, who divulged information on illegal activities carried out by the US government, had been prosecuted as a spy and condemned to five years of prison, having already served five years.

Luria also had a critical attitude toward the newly established State of Israel. "In the present world of rampant aggressive nationalism the Holocaust itself has become not a source of human solidarity and of revulsion against racial oppression, but, perversely, a justification for a new oppression" adding that "it is depressing to see the government of Israel act like the ruler of a power-driven master race, using military superiority for policies that border on the genocidal." These words prompted his excommunication by "a trio of rabbis." Since then, the situation in Israel and its conflict with Palestine has only worsened.

Luria was also a critic of the nascent movement of environmentalism, claiming that "the environment is a bourgeois prejudice." While he favored the "removal of industrial pollution that threatens people's health and lives," Luria despised "the concern for natural surroundings, natural foods, natural ways of life are expression of middle-class narcissism" of many middle-class environmentalists. In our modern phrasing, one might say that organic food is feasible only for those who can afford it; workers at minimum wage can hardly access it. Also, soliciting for an unpolluted air while driving in an SUV is a plain contradiction. In Luria's words, "I especially detest those environmentalists who smoke two packs of cigarettes a day while complaining of exhaust fumes"—and he did not live long enough to see vape smokers.

Luria was an active member of the Genetic Society of America, the American Society of Arts and Sciences, the National Academy of Sciences, but also the Massachusetts Political Action for Peace (MASSPAX). He was elected President of the American Society for Microbiology in 1968. The following year, he won the prestigious Louisa Gross Horwitz Prize, and a few months later, he shared the Nobel Prize. He learned he won the Nobel Prize the day after participating in a significant anti-Vietnam war, "Moratorium Day," and he had a peace pin on his tuxedo during the ceremony.

He co-authored a virology textbook with James Darnell, whose aim was to present "virology as a biological science," meaning that the manual focused more on the molecular biology of viruses rather than their medicinal implications, unlike previous titles in the field. In 1972, Luria became MIT's Center for Cancer Research (CCR) director. This Institution was sparked by the private initiative of Mary

Woodard Lasker—a wealthy art dealer from New York and widow of the advertising mogul Albert Lasker, who died of lung cancer—who funded the Citizens Committee for the Conquest of Cancer. The committee intended to "conquer" cancer like NASA had conquered the Moon in 1969. Richard Nixon initially approved the ambition but then slashed the funds for public research in the Spring of 1971. Luria provided his esteemed opinion on the status of US cancer research at government hearings. He highlighted the lack of funding for fellowships to support young researchers in the US. After several weeks of political discussion, Nixon changed his previous entrenchment with a new law, the National Cancer Act, that allowed for more funding for cancer research on the 23rd of December. Under this ordinance, 15 new research centers were established within the USA, including the CCR. As director of a brand new institution, Luria had to supervise all aspects of establishing the CCR, including equipping the laboratories and recruiting the chief scientists.

While his involvement in American society increased—in scientific and political terms—he never collaborated with the Italian government since, as he remarked, this one never apologized for the racial laws that forced him and many others to leave Italy in the late 1930s. He sustained that the Italian government assumed the habit of using accomplished emigrants such as himself—what we call today "brain drain"—as propaganda to promote the Italian brand.

Luria was a fervent antiracist and a paladin of the equal separation of housework between husband and wife; thus, he contributed to the groceries, cleaning, and cooking—activities that fell only on the wives at the time. He was one of the first people to avoid using the term "Man" to refer to "Humankind" because the former was sexist. In later years, Luria stressed the importance of science in improving society. However, his perception of society was not rosy: he later commented that

> our society is permeated not by science, but by an exploitative distortion of science-based technology, as irrational as the irrational aspects of religion.

To contrast this harmful tendency, Luria appealed to Albert Camus, an influential philosopher and political activist who won the Nobel Prize for literature in 1957. Luria noted that Camus, "my favorite writer," raised the issue of responsibility, "that is, the recognition of the consequences and implications of one's activities and the willingness to face them." He remarked that "it is painfully clear that the findings of science can all too easily be employed, not to enrich the human experience but to render it more painful … the development of nuclear energy and nuclear weapons is the clearest case in history." Luria encouraged scientists, microbiologists, in particular, to be part of the solution rather than the problem:

> it is important for biologists to be aware of these possibilities [the opportunities and risks of research], so that they will call them to the attention of the public and help society reach wise decisions as to relative priorities in the uses of science and its products.

8 Homecoming

8.1 AN ETHICAL LEGACY

"The United States are now, par excellence, the holy land of bacteriophages," wrote Lwoff in 1953. Campbell stressed his words further by defining those years as the "Golden Age of Phage." These statements had aged, perhaps as a good wine, but they had aged nonetheless. The Phage Group departed with the purpose of studying the secrets of the cosmos beyond the Sun-gene, and after reaching perihelion, it did its job but eventually disappeared from view. The perihelion of the Phage Group was the discovery of the DNA structure, the carrier of the genetic information; this had been the primary goal of the Group from the very beginning. Afterward, the knowledge of genetics increased, but the role of phages—the "space probe"—dwindled in that domain until it faded altogether away.

Nowadays, genetics is studied with recombinant proteins, a complete knowledge of the sequences of the genomes, and mammalian cell lines. In other words, it is possible to research genetics acting on the genes directly; phages are no longer needed as a surrogate. Today, we know so much about genetics that we can infer diseases by generating personalized genetic profiles for each patient as well as study microbial interactions by quantifying their genomes in specific samples. Such achievements can be traced to the origin of molecular biology provided by the Phage Group but went beyond the original goals of the musketeers of bacteriophage. If the knowledge in genetics sprung from the investigation of phages—which fostered the production of recombinant insulin—phage research also provided deep insights into evolution. Moreover, the members of the Phage Group behave in a gregarious manner that is hardly a feature of modern research groups.

The wane of the Phage Group occurred gradually. When Delbrück died of myeloma in March 1981, the Phage Church lost its Pope, its spiritual guide. The rest of the "Trinity" disappeared within the next decade: Luria died of a heart attack in 1991, and Hershey from heart failure in 1997. However, the founders of the Phage Church had already given up on the field they founded earlier than that point. Delbrück had moved to mycology, Luria to bacteriology, and Hershey retired in 1972, well before the other two musketeers, the year Luria had become the director of the Cancer Center.

The fluctuation test provided one of the first concrete proofs of Darwinism, putting a large nail in the coffin of creationism. More than that, by delivering the evidence with the formalism typical of physics, Luria and Delbrück provided a novel approach that is still the setting stone of biology today. In the preceding centuries, scientific diatribes were settled by direct argumentation between the leaders of the different schools. Among the most remarkable cases there were those of Newton against Leibniz regarding the paternity of differential equations, the legal struggle of Guglielmo Marconi and Nikola Tesla for the primacy of the invention of the radio, the diatribe between d'Hérelle and Bordet on the matter of the nature of bacteriolysis,

DOI: 10.1201/9781003657712-8

and the "Black Hole War" between Stephen Hawking and Leonard Susskind regarding the mechanism of information loss from a singularity. The Phage Group introduced the gold standard that scientific issues must be resolved only by solid evidence and mathematical inference.

Since we live in a "post-truth" world today, it is safe to say that such a criterion is no longer absolute. The harsh scrutiny of attendees' scientific results during the Phage School, with Delbrück's famous line, "I don't believe a word of it"—although, in the preceding chapters, we have seen that the tone of such words should be considered less grueling than it sounds—assured that the data provided by a scientist would have been appropriately checked and foolproof. Consequently, there were fewer results in the literature, but of higher quality.

Today, instead, we are submerged by so much data, not only scientific but of all sorts, that the World Health Organization has coined the neologism "infodemic," a pandemic of information. The problem is not only that the number of scientists has increased exponentially in the last century, reaching 8.8 million in 2020—being a few thousand in the 1940s—but that the amount of scientific publications has followed the same trend, doubling every nine years. With such a swath of data, there is a "Spencerian"—we have seen in the previous pages that the term "survival of the fittest" was Herbert Spencer's coinage—fight for endurance. The motto "publish or perish" has gained a literal meaning like never before. Such a struggle forms multiple fronts.

On the one side, since scientists need publications to obtain grants, they are pressed to publish bombastic articles that can be published quickly and in high-ranked journals. Such haste is linked to the so-called "replication crisis": it has been shown that another cannot replicate a significant number of experiments published by one laboratory. Like in the Middle Ages, knowledge has become an accumulation of erudite texts referencing each other with lesser and lesser grasp on reality. One feature is real not because it has been proved in the laboratory—as was the norm at the time of the Phage Group—but because it has been published, preferably in a fancy scientific outlet. This crisis is worsened by the "citation bias," the tendency of scientists to quote mostly the literature that supports their findings.

On the other hand, journals must make money. At the time of Delbrück and Luria, journals sold subscriptions to institutions regardless of the number of articles presented. Thus, publishers had no pressure to throw a large mass of results to the public. Today, publishers sell individual articles. Hence, the more, the better. To sell an article, it has to have a hype, following the journalistic catchphrase: "if it bleeds, it leads." Hence, scientific articles with positive results are more likely to be published. A recent study by Turner and colleagues at the Oregon Health and Science University showed that the publications associated with clinical trials registered through the American Food and Drug Administration reported positive results at a rate of 94%. However, many of the results related to these trials were never actually published in peer-reviewed journals, and by analyzing the original data, the authors found that the actual positive rate was 51%. These results paint an entirely different picture of the intervention's effectiveness.

On another level, scientific publishing is a market, and like other economic sectors, it is exploited for revenue. Until a few decades ago, publishing—scientific or otherwise—was controlled by a few well-established enterprises that increased their audiences by providing high-quality products. The rise of the Internet in the 1990s led to a surge in online publishing outfits. At the time, such blooming was celebrated as a milestone for freedom, promising to liberate knowledge from the greedy hands of a few capitalists. Instead, this step triggered the proliferation of predatory journals interested only in publishing as much as possible in the shortest possible time and with minimal peer-review processing. It has been estimated that, as of 2014, 2.5 million scientific articles were published on acknowledged outfits, generating a revenue of 250 million dollars. In contrast, half a million articles were published in predatory journals, generating 74 million dollars of revenue. These figures highlight that about 10% of scientific results are thrown into the public domain with little reality check. The situation is aggravated by the fact that publication costs have shifted from the publisher to the authors in the last few decades. Thus, scientists are prompted to shop their publications, choosing the journals that offer the highest discounts.

Moreover, the way scientific articles are accessed gives additional complexity. We have seen through this book how relevant scientific information was given personally. Delbrück, Timofeev-Ressovskij, and Zimmer printed extra copies of their green pamphlet that were distributed to trusted friends, stimulating the work of other scientists such as Schrödinger. The Phage Group even had a dedicated mailing list to diffuse relevant information among the members. Apart from the additional prints, published articles were physically embedded in hard copies of a magazine. A scientist had to browse the different items of an issue before finding the wanted article, becoming exposed to results that might influence thoughts. Today, scientists use search engines that selectively pinpoint only the articles of interest. Thus, such collateral exposure is becoming extinct. In addition, search engines are shifting their algorithms to provide not the most relevant articles matching the keyword's queries—as in the case of *PubMed*—but those that have been most visited—as in the case of *Google Scholar*. Consequently, scientific knowledge is more and more often based on the most popular articles, and the more these are cited, the more they will be mentioned. As Jevin West and Carl Bergstrom at the University of Washington have commented, an accentuated "form of the Matthew Effect."

This cursory outlook shows us how the epoch in which the Phage Group operated is no longer ours, giving Lwoff's remark a literary sense that reverberates in the words of David Goodstein, California Institute of Technology, that "the period 1950–1970 was a true golden age for American science" because "young Ph.D's could choose among excellent jobs, and anyone with a decent scientific idea could be sure of getting funds to pursue it." That age is now gone. Even Luria acknowledged the loss, as far back as 1968, by remarking that "at the moment we are faced in this country with a mood of slowing down the expansion of fundamental research … in the name of illusory demands that scientists concentrate on immediate practical goals … as though research without immediate practical goals were something of a sinful luxury." Would molecular biology have risen without basic research in phage genetics? This question remains an unsolved conundrum of modern research since plans to cut further public funding in research and channel the limited grants to politically chosen practical applications are a leitmotiv of the present-day conservative agenda.

The commercialization of the research, the systematic reduction of public funding, and the exponential expansion of the scientific workforce established our current

Iron Age, where researchers had to fiercely compete—rather than collaborate—with each other. Yesterday, like today, researchers struggled to extend their contracts, and even Delbrück and Luria had to rely on friends, being Born or Fermi, to obtain their entry-level positions. Today, they probably would have had problems finding a job even with the interest of their mentors. Luria clearly expressed this change of climate: "new techniques had made my own rather limited skills obsolete," he said, adding that he felt uncomfortable with faster pace of scientific communication in the latter years of his career. "I recognize," he stated, "the need to 'sell' science to the public through technology just as one must sell novels through advertisements … yet I wish one could let science grow on its own momentum."

The ever-increasing Matthew effect promotes ruthless scientists. Were the members of the Phage Church merciless? Would Delbrück doctor his results if he were to be here today? Would Luria—who claimed that "except for a few emotionally disturbed individuals, people in science do not cheat," adding that "anyone reporting a falsified result is certain to be promptly caught if the experiment is of any significance"—jeopardize patients' lives for a publication on *Nature*, Macchiarini-style? Or would Macchiarini have behaved more like a Hershey if he had operated 50 years earlier?

Despite personal conflicts, such as the dispute between Joshua Lederberg and Esther Zimmer or the controversy sparked by Watson's biography, the members of the Phage Group upheld academic and personal integrity that serves us as a beacon in the current scientific landscape. It is difficult to imagine that Hershey, who was uncomfortable putting his name on a paper he did not work extensively on, would have doctored a table or photoshopped a picture to get a better publication. One can hardly imagine that Delbrück, known for his rigorous approach to verifying experimental results, would have chosen to publish in a predatory journal to avoid the peer-review process. Of course, one might object that it is easy for a Nobel laureate to be choosy about journals: The most renowned outfits would be accepting anything from such high-ranked scientists. It is the rest of the not-so-famous researchers who have to struggle in the arena of modern science. Nonetheless, the members of the Phage Group adhered to the quality paradigm well before winning prestigious prizes.

The proportion of biomedical articles retracted has indeed increased in the past two decades due to misconduct such as reporting unreliable (Italy being the most involved in this process) or fabricated (Great Britain excelling in the field) data. Such a trend has been nurtured by the "publish or perish" philosophy that underlines today's scientific research and the lassitude in sanctioning misconduct. Still, it is also true that individual scientists have lost sight of the moral beacon given by the members of the Phage Church, another legacy of those Golden Years that we should cherish.

8.2 PHAGE RENAISSANCE

After the Phage Group faded away, the use of phages as a genetic tool also declined within the scientific community, much like using slide rules to solve mathematical problems among engineers. The last edition of the Phage Course at CSHL was presented in 1970. Today, the CSHL's course offer includes topics such as "Chromatin, Epigenetics & Gene Expression," "Advanced Sequencing Technologies & Bioinformatics Analysis," and "Quantitative Imaging: From Acquisition to Analysis." The days of phages as probes to guess the carrier molecule of the genes and tinkering with a transmission electron microscope to get an idea of the morphology of the phages are long gone. Today, we can investigate nucleic acid's intimate structure and track virus replication cycles in real time.

Phage courses are still offered but for reasons not related to genetics. It has been a century since Félix d'Hérelle introduced phage therapy. Ironically, the employment of phages is nowadays primarily that of adjuvant for antibiotics, the same molecules that wiped out phage therapy in the West about seven decades ago. Due to the increasing spread of antimicrobial-resistant bacteria, phage therapy has returned to the medical spotlight. As a result, current phage therapy courses are centered around the isolation and application of phages as antibacterial agents.

Despite the potential for medical applications, phage therapy has not yet been proven to be an effective substitute for antibiotics. However, in Eastern countries such as Georgia, phage therapy is a cutting-edge procedure, offering hope for its potential in the West, whereas, in Western countries, it is currently used only for compassionate purposes. There is still no consensus on how to categorize phage preparations. Legally, they are not considered drugs, but they do not fit into the category of nutritional supplements either. Establishing phage therapy in Western countries will first require establishing a novel class of medicine that could accommodate the complexity of phage preparations.

Nonetheless, if the application for medical purposes is challenging, phages are now used more and more extensively in veterinary medicine. For instance, *Bafasal*™ from *Proteon Pharmaceuticals*, a Polish firm, has been approved by the European Food Safety Authority to treat poultry against *Salmonella*. Another phage preparation against *Salmonella* is *SalmoFree*™, developed at the University of Bogota. Phages are also still used to the typization of bacteria but phage applications have expanded to fields virtually unknown to d'Hérelle and Eliava. For instance, phage preparations are used to eliminate bacteria directly from food to reduce the impact of foodborne diseases. Thus, *EcoShield*™, developed by *American Intralytix*, can be sprayed on red meat or vegetables to reduce the *E. coli* bacterial load and sterilize surfaces just like any other cleaning product.

Furthermore, phages are used to expose peptides in a process known as phage display, a procedure introduced in the late 1980s by George Pearson Smith, who won the Nobel Prize for this idea in 2018. By engineering the genome of a phage such as M13 (*Inoviridae*), amino acids on the virus' surface can be exposed and recognized by a chosen protein such as an antibody or a surface receptor. The initial population of the recombinant viruses contains thousands of combinations of amino acid residues. Still, after a few rounds of purification (biopanning), isolating the viruses with the highest affinity against the protein of interest is possible. Sequencing the genome of the isolated phages provides the string of residues involved in the attachment (Figure 8.1). Such technique is applied chiefly in drug discovery but also in fields such as the determination of ligands to DNA and tumor antigens.

Moreover, phages are currently employed as genetic delivery systems for bacteria and scaffolds for forming nanostructures such as nanotubes to carry electricity. In other cases, phages have been used as advanced diagnostic tools by simultaneously

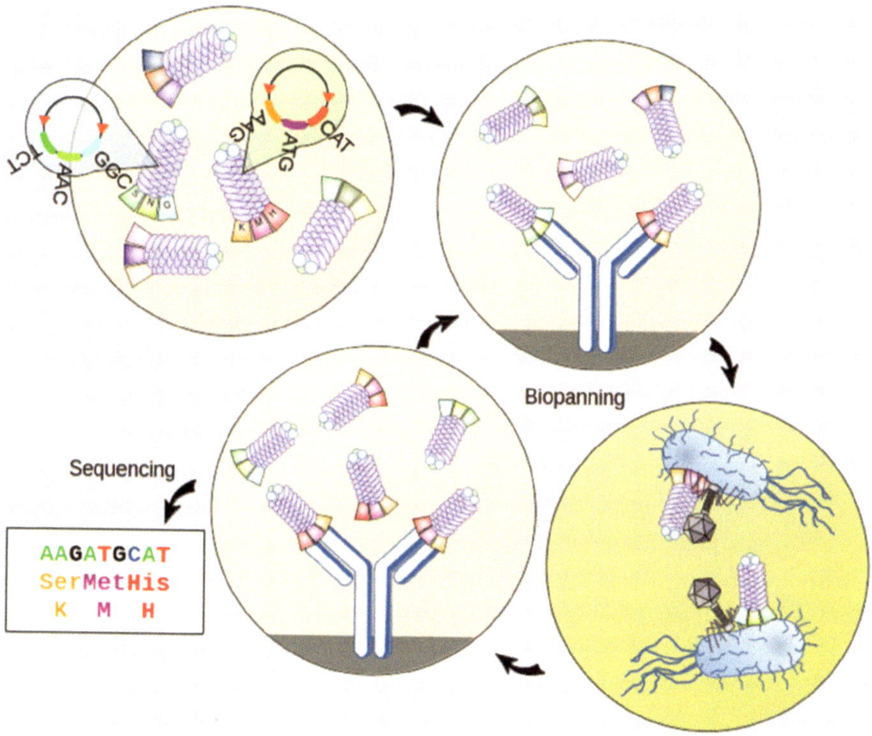

FIGURE 8.1 *Biopanning with phage display.* Biopanning is a procedure for isolating molecules that bind with high affinity to a specific target, such as an antibody. Phage display enables rapid selection of such binding molecules. Currently, kits containing a combination of phage M13 (*Inoviridae*) or phage fd (also *Inoviridae*) are the most commonly available for purchase. M13 has a helical capsid made up of 2700 capsomers of p8. One end of the capsid is closed by the capsomers p7 and p9, while at the other end, a complex of p3 (surface-exposed protein) and p6 (exposed accessory protein) is responsible for attaching to the host. The viruses have been genetically modified so that typically, segment p3 contains random amino acid residues (depicted in the figure by gray-shaded wedges) that are exposed on the surface of the capsid. These residues are involved in attaching to the target of choice. Additionally, the genome of the viruses contains sequences that are recognized by primers for sequencing (►◄). The molecule of interest, in this case, an antibody, is immobilized on a surface such as a microtitre plate. The phages are added, and those most affine to the receptor will attach to their targets, whereas the others will be washed away. After forcing the detachment from the target, the immobilized phages can be expanded in a bacterium of choice, usually *E. coli*, using a helper virus (depicted by a tailed phage in the figure). Several cycles of binding and expansion are carried out to obtain viruses whose capsid binds with the highest affinity to the protein of choice. Finally, sequencing allows determining the nucleic acid sequence of the displayed amino acid residues.

binding gold nanoparticles—which can be detected with high sensitivity—and pathogen surfaces. Overall, phages might not be today's genetic tool of choice, but they have found their way into many other aspects of human activities. Phages have yet to have their finest hour.

8.3 DEMANDING A BETTER FUTURE

The present book lingered quite a lot on historical features. Many might turn their noses up at this approach, but the present essay aimed to contextualize the work of the Phage Group in an extraordinary period that shaped our current times. Phage-based molecular research benefited from the juxtaposition of historical elements that concern not only biology but also physics and geopolitics. Such a conjunction might be lost in books focusing only on biology or history, and its elements required a proper introduction to understand better how the Phage Church developed.

First, the Phage Group can be considered a quantum mechanics spin-off. Not only were both Delbrück and Luria—but also Szilard—particle physics scholars, but the objective approach of mathematical sciences heavily influenced their approach to biology. We have seen how the authority of a specific school mostly drove the biology of the late 19th century. For instance, Bordet engaged in a furious battle of ideas with d'Hérelle because the latter dared to disagree with the former. Delbrück and Luria changed this attitude by proposing facts rather than ideas. Given the importance of quantum mechanics in this process, it was deemed worth introducing this subject properly, even because the evolution of particle physics in the first half of the 20th century was directly linked, through the Manhattan Project, to the end of the Second World War and the beginning of the Cold War.

Second, what would the Phage Group have been like without the fascistization of Europe? It is conceivable that Delbrück would have continued working with Meitner and collaborated with Timoféeff-Ressovsky and Karl Zimmer on radiation biology. Similarly, Luria might have pursued a career in radiology and teamed up more extensively with Geo Rita. Since Luria was aware of Delbrück's work while in Italy, it is possible they could have still formed a European duo, but most likely, it would have remained a long-distance affair, with a simple exchange of ideas through correspondence. It is implausible that Hershey would have moved to Europe to meet Delbrück and Luria as he did when going to Vanderbilt University. Certainly, the European duo would not have been exposed to the unusual concentration of brilliant minds clustered around CalTech, New York, and other American universities. Of course, genetics would have evolved even without a Phage Group: of all the people involved in discovering the DNA structure, for instance, only Watson was linked to the Phage Group, albeit loosely. Today, there might have been a "Neurospora Group" or a "BALB/c Group," but it can be said that the Phage Group acted as a catalyst that sped up the rate of genetic discoveries.

Delbrück and Luria were considered the high priests of the Phage Church. Would they have been so influential had they lived in Europe? The US scientists' obliviousness to Ruska's work on electron microscopy hints at how to answer this question, although one should factor in the fact that such oblivion was a consequence of fascism. Thus, it is doubtful that a Phage Course proposed in, say, Heligoland would have been as influential as the one in Cold Spring Harbor. Furthermore, European research on phages would probably have remained under the control of the Pasteur Institute and stayed within the medical microbiology field without trespassing into genetics—although, thanks mainly to the work of Lwoff and others, the Institute became a stronghold of the Phage Group in the old continent.

All in all, fascism has been one of the worst pages in modern human history, and yet it had the power to boost human knowledge by concentrating thousands of brilliant scientists in a single and amenable place. The US not only acted as a receptacle for intellectuals displaced by fascism but also provided resources, and it was keen to accept the ideas brought by these immigrants without the prejudices rooted in Europe. The payoff for this daring approach—giving away billions of dollars to foreigners may sound like outright madness—was an advancement of scientific knowledge still alive today. This point should be considered since migration is one of the hottest topics in today's politics. Neither fascism nor the Second World War emerged spontaneously out of thin air. Both were complex events resulting from juxtaposing socioeconomic features and people, and both needed a proper introduction.

The third reason to dwell on history is that sociological aspects influenced the development of the Phage Group. For instance, Joshua Lederberg studied genetics as a response to Lysenko's nefarious sway. Since the discoveries of conjugation and transduction were fortuitous happenings, what would be today's state of microbiology if Lederberg had decided to study, say, American literature? Yet, Lysenkoyism is a concept that has virtually disappeared from public knowledge and does not appear in the syllabus of biological sciences. Nonetheless, Lysenkoyism remains a potent reminder of how the degeneration of science could affect society. Such a concept was worth mentioning and needed an introduction. Also, lesser-known events in the Second World War, such as bacteriological warfare, were worth mentioning to readers interested in microbiology. Such topics might not have influenced the Phage Group directly, but they were also crucial in shaping society back then and, by reflex, our world.

Does such a historical contextualization teach us anything? Did the fellows of the Phage Group tell us something valuable for today's problems? One striking feature in common in the lives of the scientists presented in the previous chapters is that most of these sciences were wholly disconnected from the political occurrences of their time. Specifically, it is remarkable how they were surprised by the rise of fascism and the speed at which it became mainstream. Surely, such a circumstance would not happen today, would it?

To quote *The New York Times*, a journal dear to Luria, "historical lessons … fade after three generations." Today, like one century ago, we are facing, as US President Biden has remarked, a "battle between democracy and autocracy." Today, like yesterday, far-right parties have suddenly and unexpectedly gained power, and demagogues can openly declare their aspirations to establish dictatorships. Are we ready for such events? The slip back into authoritarianism is indeed swift. Even today, in Italy, it is possible to lose all civil rights—including the right to hold a passport—just as at the time of the confinement. Today, as in the 1930s, the procedure is somewhat subjective and preventive—that is, it does not require carrying out a crime—but might involve even a peaceful protest, such as writing slogans with chalk on the pavement. Today, the procedure is known as "special surveillance" and does not require deportation into another village: the suspect simply cannot leave the city of residence.

The Phage Group members were all researchers, some more talented than others, but researchers nonetheless, much like the researchers conducting scientific work today. They had their own lives and problems and would have likely been as happy as the next Joe spending time at political rallies. Nevertheless, if they had followed the political happenings of their times with more interest and had publicly spoken their minds *en masse*, there would probably have been less suffering in the world. Scientists can change the world through their discoveries; thus, they also have the moral duty of demanding a better future.

Such a lesson was given with particular strength by Luria, who faced almost four decades of FBI scrutiny—from the onset of the Second World War to the onset of the Gulf War in 1991—for his involvement in civil rights movements. Involvement in politics does not have to be violent: Luria, for instance, did not place bombs on telephone poles as the Italian publisher Giangiacomo Feltrinelli did. Nevertheless, activism—as opposed to the mere execution of experiments, with no concern about the implication of the results—can ensure that our reason will not sleep.

8.4 CONCLUSIONS

The Phage Group, or Phage Church as it was more intimately called, was a gathering of geneticists with various scientific backgrounds, including physics and biochemistry, who used phages as biological models. The Group, which had strong connections with the Rockefeller Foundation, was active in the US during the second half of the 20th century but established relations with other teams, particularly at the Pasteur Institute. The Group held regular meetings, the most important being a summer school held at Cold Spring Harbor in New York. The Group characterized the biology of phages in great detail but also contributed to the expansion of notions in the field of genetics that culminated with the discovery of the DNA structure. One of the main advancements sparked by the Phage Group was the development of recombinant DNA technology, which allows for the production of synthetic proteins. The Group has also set high-quality scientific research standards and promoted scientists' ethical behavior within society.

Bibliography

PROLOGUE

1980, Wichita

Watson, J.D., Gilman, M., Witkowski, J. and Zoller, M.W.H. (1992) "Recombinant DNA". *Freeman and Co.* (2nd edition).

Sapra, A. and Bhandari, P. (2022) "Diabetes mellitus", in *StatPearls, Treasure Island. StatPearls Publishing.*

Goeddel, D.V., Kleid, D.G., Bolivar, F. et al. (1979) "Expression in *Escherichia Coli* of chemically synthesized genes for human insulin". *Proceedings of the National Academy of Sciences of the United States of America*, vol. 76, pp. 106–110.

Ladisch, M.R. and Kohlmann, K.L. (1992) "Recombinant human insulin". *Biotechnology Progress*, vol. 8, pp. 469–478.

Guthrie, R.A. and Guthrie, D.W. (2004) "Pathophysiology of diabetes mellitus". *Critical Care Nursing Quarterly*, vol. 27, pp. 113–125.

Bommer, C., Sagalova, V., Heesemann, E. et al. (2018) "Global economic burden of diabetes in adults: projections from 2015 to 2030" *Diabetes Care*, vol. 41, pp. 963–970.

"Harvard Medical School" on *Wikipedia* website (https://en.wikipedia.org/wiki/Harvard_Medical_School#Notable_alumni)

"Harvard University" on *Wikipedia* website (https://en.wikipedia.org/wiki/Harvard_University#Notable_people)

"University of Oxford" on *Wikipedia* website (https://en.wikipedia.org/wiki/University_of_Oxford#Notable_alumni)

"University of Kansas School of Medicine" on *Wikipedia* website (https://en.wikipedia.org/wiki/University_of_Kansas_School_of_Medicine)

"History of the School of Medicine" on the *University of Kansas* website (https://www.kumc.edu/school-of-medicine/about/history.html)

"On this day in 1980 recombinant insulin was first given to an American" on *The Louisville Courier-Journal* website (https://www.doublearrowmetabolism.com/blog/on-this-day-in-1980-recombinant-insulin-was-first-given-to-an-american)

"Data and patient accounts reveal 'no mercy' for civilians in Ukraine war" on *Medicine Sans Frontieres* website (https://www.msf.org/data-and-patient-accounts-reveal-indiscriminate-attacks-against-civilians-ukraine-war)

"The race to develop human insulin" on *Harvard Business School* website (https://silo.tips/download/the-race-to-develop-human-insulin)

Restriction online tool (http://insilico.ehu.es/" works: it is the portal of the analysis)

Fraser, L. (2016) "Cloning insulin" on *Genentech* website (https://www.gene.com/stories/cloning-insulin)

"The history of a wonderful thing we call insulin" on *American Diabetes Association* communication (https://www.diabetes.org/blog/history-wonderful-thing-we-call-insulin)

"Recombinant DNA and the birth of biotech", on *Smithsonian Museum* website (https://americanhistory.si.edu/collections/object-groups/birth-of-biotech/the-business-of-biotech)

New England Biolabs website (https://international.neb.com/)

"GMOs in medicine and research" on *Britannica* website (https://www.britannica.com/science/genetically-modified-organism/GMOs-in-medicine-and-research)

"GMOs revealed: food, medicine and more" on *Center Forward* website (https://center-forward.org/gmos-revealed-food-medicine-and-more/)

"Richard Guthrie obituaries" communication (https://greatplainsdiabetes.com/wp-content/uploads/2020/12/Guthrie-Info.pdf, https://www.legacy.com/us/obituaries/kansas/name/richard-guthrie-obituary?id=8688397)

1999, Topeka

Singer, C. (1941) "A short history of science to the nineteenth century". *Oxford University Press.*

Farrington, B. (1953) "Greek science". *Penguin.*

Oparin, A. (1953) "Origin of life". *Dover Publications.*

Boyer, C.B. (1968) "A history of mathematics". *John Wiley & Sons* (1991 edition).

Cairns-Smith, G. (1985) "Seven clues to the origin of life". *Cambridge University Press.*

Abbagnano, N. and Fornero, G. (1986) "Filosofi e filosofie nella storia". *Paravia* (1992 edition).

Garrett, D. (1996) "The Cambridge companion to Spinoza". *Cambridge University Press.*

Davies, P. (1998) "The origin of life". *Penguin Press* (2003 edition).

Foster, J.B., Clark, B. and York, R. (2008) "Critique of intelligent design". *Monthly Review Press.*

Cannon, W.F. (1961) "John Herschel and the idea of science". *Journal of the History of Ideas*, vol. 22, pp. 215–239.

White, J. (1999) "Kansas Board of Education removes evolution from science curriculum" on *World Socialist Website* blog (https://www.wsws.org/en/articles/1999/08/kan-a13.html)

Belluck, P. (1999) "Board for Kansas deletes evolution from curriculum" on *The New York Times* website (https://www.nytimes.com/1999/08/12/us/board-for-kansas-deletes-evolution-from-curriculum.html)

Riedel, C. (2005) "Kansas education board downplays evolution" on *NBC News* website (https://www.nbcnews.com/id/wbna9967813)

"Kansas restores evolution standards for science classes" on *CNN News* website (https://edition.cnn.com/2001/US/02/14/kansas.evolution.02/index.html)

"Science report: A third of U.S. schools don't teach evolution" on *CNN News* website (https://edition.cnn.com/2000/NATURE/09/21/evolution.enn/index.html)

"Creation and evolution in public education" on *Wikipedia* website (https://en.wikipedia.org/wiki/Creation_and_evolution_in_public_education)

"Intelligent design" on *Wikipedia* website (https://en.wikipedia.org/wiki/Intelligent_design)

"Creationism by country" on *Wikipedia* website (https://en.wikipedia.org/wiki/Creationism_by_country#South_Korea)

Villanueva, J.C. (2009) "The Big Crunch: the end of our universe?" on *Universe Today* blog (https://www.universetoday.com/37018/big-crunch/)

Atkinson, N. (2008) "No 'Big Rip' in our future: Chandra provides insights into dark energy" *Universe Today* blog (https://www.universetoday.com/22382/no-big-rip-in-our-future-chandra-provides-insights-into-dark-energy/).

Matzke, N. (2005) "Missing link discovered!" on *Evolution Education and the Law* blog (https://web.archive.org/web/20070114121029/https://www2.ncseweb.org/wp/?p=80)

"Charles Thaxton, Fellow – CSC" on *Discovery Institute* website (https://www.discovery.org/scripts/viewDB/index.php?command=view&id=96&isFellow=true)

"Charles Darwin" on *Wikipedia* webpage (https://en.wikipedia.org/wiki/Charles_Darwin#Publication_of_the_theory_of_natural_selection)

"Giant's Causeway" on *Wikipedia* website (https://en.wikipedia.org/wiki/Giant's_Causeway#Legend)

2011, Stockholm

Macchiarini, P., Jungebluth, P., Go, T. et al. (2008) "Clinical transplantation of a tissue-engineered airway". *Lancet*, vol. 372, pp. 2023–2030. [Article retracted].

Jungebluth, P., Alici, E., Baiguera, S. et al. (2011) "Tracheobronchial transplantation with a stem-cell-seeded bioartificial nanocomposite: a proof-of-concept study". *Lancet*, vol. 378, pp. 1997–2004.

Gonfiotti, A., Jaus, M.O., Barale, D. et al. (2014) "The first tissue-engineered airway transplantation: 5-year follow-up results". *Lancet*, vol. 383, pp. 238–44. [Article retracted].

Editorial Board. (2015) "Paolo Macchiarini is not guilty of scientific misconduct". *Lancet*, vol. 386, p. 932.

Editorial Board. (2018) "The final verdict on Paolo Macchiarini: guilty of misconduct". *Lancet*, vol. 392, p. 2.

Karolinska Institute communication (https://news.ki.se/first-successful-transplantation-of-a-synthetic-tissue-engineered-windpipe) [Page removed from server; last access 2023].

Kremer, W. (2016) "Paolo Macchiarini: a surgeon's downfall" on *BBC World Service* website (https://www.bbc.com/news/magazine-37311038)

Ciralsky, A. (2016) "The celebrity surgeon who used love, money and the Pope to scam an NBC news producer" on *Vanity Fair* website (https://www.vanityfair.com/news/2016/01/celebrity-surgeon-nbc-news-producer-scam).

Cyranoski, D. (2016) "Nobel official resigns over Karolinska surgeon controversy" on *Nature* website (https://www.nature.com/articles/nature.2016.19332)

Cowell, A. and Grady, D. (2008) "Europeans announce pioneering surgery" on *The New York Times* website (https://www.nytimes.com/2008/11/20/health/research/20stemcell.html)

Schneider, L. "Claudia's trachea" on *Hipertextual* website (https://hipertextual.com/especiales/macchiarini-transplante-traquea-claudia-castillo). [Page removed from server; last access 2023].

Schneider, L. (2020) "Macchiarini victim to Swedish prosecutor: 'Claudia is alive and must be called to testify'" on *For Better Science* website (https://forbetterscience.com/2020/10/20/macchiarini-victim-to-swedish-prosecutor-claudia-is-alive-and-must-be-called-to-testify/)

"The Macchiarini case: timeline" on *Karolinska Institute* website (https://news.ki.se/the-macchiarini-case-timeline)

McCook, A. (2015) "Misconduct found in 7 papers by Macchiarini, says English write-up of investigation" on *Retraction Watch* blog (https://retractionwatch.com/2015/05/28/misconduct-found-in-7-papers-by-macchiarini-says-english-write-up-of-investigation/)

"Seven researchers responsible for scientific misconduct in Macchiarini case" on Karolinska Institute website (https://news.ki.se/seven-researchers-responsible-for-scientific-misconduct-in-macchiarini-case)

Vogel, G. (2016) "Panel: Swedish hospital should never have hired 'star surgeon'" on *Science* website (https://www.science.org/content/article/panel-swedish-hospital-should-never-have-hired-star-surgeon)

Schneider, L. (2018) "How Macchiarini was recruited to Karolinska" on *For Better Science* website (https://forbetterscience.com/2018/08/06/how-macchiarini-was-recruited-to-karolinska/)

Warren, M. (2018) "Disgraced surgeon is still publishing on stem cell therapies" on *Science* (https://www.science.org/content/article/disgraced-surgeon-still-publishing-stem-cell-therapies)

Macchiarini's deed inspired a *Netflix* documentary (*Bad surgeon*) that was released several months after this chapter was written.

A more recent CV is available online (https://www.circare.org/info/pm/pm-kijobapplic-2014-Dnr%202-1097-2014.pdf), where Macchiarini declares only one PhD from the University of Franche-Cômpte in Beçanson and results tenured professor only at the Kuban State Medical University, Russia.

New York, the Present Day

Malthus, T.R. (1826) "An essay on the principle of population". *John Murray London* (6th edition).
Meyrink, G. (1915) "The Golem". *Dedalus* (2017 edition; translated by Mike Mitchell).
Gaynes, R.P. (2011) "Germ Theory". *ASM Press*.
Belton, C. (2020) "Putin's people". *William Collins*.
Ritchie, S. (2020) "Science fictions". *Vintage*.
Korber, B., Muldoon, M., Theiler, J. et al. (2000) "Timing the ancestor of the HIV-1 pandemic strains". *Science*, vol. 288, pp. 1789–1796.
Salemi, M., Strimmer, K., Hall, W.W. et al. (2001) "Dating the common ancestor of SIVcpz and HIV-1 group M and the origin of HIV-1 subtypes using a new method to uncover clock-like molecular evolution". *FASEB Journal*, vol. 15, pp. 276–278.
Goetz, L.H. and Schork, N.J. (2018) "Personalized medicine: motivation, challenges and progress". *Fertility and Sterility*, vol. 109, pp. 952–963.
Gómez-Márquez, J. (2021) "What is life?". *Molecular Biology Reports*, vol. 48, pp. 6223–6230.
Zrimec, J., Kokina, M., Jonasson, S. et al. (2021) "Plastic-degrading potential across the global microbiome correlates with recent pollution trends". *mBio*, vol. 12, p. e02155.
Lyall, S. (2022) "Johnson's eternal swagger finally fails him". *The New York Times* No. 59,539.
Cyranoski, D. (2023) "How a cloning fraud reinvented himself". *The New York Times* No. 43,630.
"American Phage Group" on *Wikipedia* website (https://commons.wikimedia.org/wiki/American_Phage_Group#Notable_scientists_associated_with_the_Phage_Group)
Scientific misconducts can be viewed on Wikipedia's dedicated page (https://en.wikipedia.org/wiki/List_of_scientific_misconduct_incidents) and in the dedicated *Retraction Watch* blog (https://retractionwatch.com/)
Thomas, D. (2022) "Theranos scandal: Who is Elizabeth Holmes and why was she on trial?" on *BBC News* (https://www.bbc.com/news/business-58336998). By the way, Ms. Holmes was sentences 20 years for a financial fraud, without having caused any deaths.
Parr, C. (2014) "Imperial College professor Stefan Grimm 'was given grant income target'" on *The Higher Education* website (https://www.timeshighereducation.com/news/imperial-college-professor-stefan-grimm-was-given-grant-income-target/2017369.article)
Grove, J. (2015) "Stefan Grimm inquest: new policies may not have prevented suicide" on *The Higher Education* website (https://www.timeshighereducation.com/news/stefan-grimm-inquest-new-policies-may-not-have-prevented-suicide/2019563.article)
Greenhill, S. (2014) "Professor found dead after cash row: cancer scientist said to have told fellow academics that chiefs treated him 'like s***'" on *The Daily Mail* website (https://www.dailymail.co.uk/news/article-2861588/Professor-dead-cash-row-Cancer-scientist-said-told-fellow-academics-chiefs-treated-like-s.html)
Carrington, D. (2021) "Bugs across globe are evolving to eat plastic, study finds" on *The Guardian* website(https://www.theguardian.com/environment/2021/dec/14/bugs-across-globe-are-evolving-to-eat-plastic-study-finds)
Migdon, B. "New study shows microbes are evolving to eat plastic" on *The Hill* website (https://thehill.com/changing-america/sustainability/environment/585802-new-study-shows-microbes-are-evolving-to-eat/)
"Blaming thc Poor for Their Poverty" on the *Economist's View* blog (https://economistsview.typepad.com/economistsview/2006/10/blaming_the_poo.html)
"Global issues: ending poverty" on the *United Nations* website (https://www.un.org/en/global-issues/ending-poverty)
"Sustainable Development Goals" on the *World Health Organization* website (https://www.who.int/europe/about-us/our-work/sustainable-development-goals)
"The Human Genome Project" on the *National Human Genome Research Institute* website (https://www.genome.gov/human-genome-project)
"1000 Genomes, a deep catalog of human genetic variation" on *The International Genome Sample Resource* (https://www.internationalgenome.org/)

LONDON AND PARIS

UNDERFUNDED

Twort, A. (1993) "In focus, out of step: a biography of Frederick William Twort, FRS, 1977-1950". *Alan Sutton Publishing*.

Klein, P.H. (1996) "Microbiology". *Wm C. Brown Publishers* (3rd edition).

Bauman, R.W. (2011) "Microbiology with disease by taxonomy". *Pearson* (3rd edition).

Gaynes, R.P. (2011) "Germ theory". *ASM Press*.

Ritchie, S. (2020) "Science fictions". *Vintage*.

Enquist, L. and Racaniello, V. (2013) "Virology: from contagium fluidum to virome", in *Fields Virology* edited by David Knipe and Peter Howley. *Lippincott Williams & Wilkins* (6th edition).

Editorial Board. (1871) "The Brown Institution". *Nature*, vol. 5, pp. 138–140.

Koch, R. (1882) "Die Aetiologie der Tuberkulose". *Berliner Klinische Wochenschrift*, vol. 15, pp. 428–445.

Koch, R. (1890) "An address on bacteriological research". *British Medical Journal*, vol. 2, pp. 380–383.

Blaxall, F.R. (1902) "Glycerinated calf lymph". *The British Medical Journal*, vol. 2, pp. 38–40.

Twort, F.W. (1915) "An investigation on the nature of ultra-microscopic viruses". *The Lancet*, vol. 186, pp. 1241–1243.

Editorial Board. (1931) "Dr. A. T. MacConkey obituary". *Nature*, vol. 127, pp. 980–981.

Gordon, F.P. (1951) "Frederick William Twort 1877-1950 — obituary note". *Fellows of the Royal Society*, vol. 7, pp. 505–517. (https://royalsocietypublishing.org/doi/10.1098/rsbm.1951.0016)

McIntosh, J. (1941) "Obituaries: Prof. William Bulloch, F.R.S.". *Nature*, vol. 147, pp. 504–505.

Bañuls, A.L., Sanou, A., Van Anh, N.T. and Godreuil, S. (2015) "Mycobacterium tuberculosis: ecology and evolution of a human bacterium". *Journal of Medical Microbiology*, vol. 64, p. 1261–69.

Barberis, I., Bragazzi, N.L., Galluzzo, L. and Martini, M. (2017) "The history of tuberculosis: from the first historical records to the isolation of Koch's bacillus". *Journal of Preventive Medicine and Hygiene*, vol. 58, pp. e9–e12.

Saleem, A. and Azher, M. (2013) "The next pandemic – tuberculosis: the oldest disease of mankind rising one more time". *BMC*, vol. 6, p.a615.

Gradmann, C. (2014) "A spirit of scientific rigour: Koch's postulates in twentieth-century medicine". *Microbes and Infection*, vol. 16, pp. 885–892.

Theves, C., Biagini, P. and Crubezy, E. (2014) "The rediscovery of smallpox". *Clinical Microbiology Infection*, vol. 20, pp. 210–218.

Dow, C.T. and Alvarez, B.L. (2022) "Mycobacterium paratuberculosis zoonosis is a One Health emergency". *EcoHealth*, vol. 19, pp. 164–174.

Marongiu, L., Burkard, M., Lauer, U.M. et al. (2022) "Reassessment of historical clinical trials supports the effectiveness of phage therapy". *Clinical Microbiology Reviews*, vol. 35, p. e0006222.

Campbell Water, A. "Brown Institute for Animals: Vauxhall History" on *Brown Institute* website (https://vauxhallhistory.org/brown-institute-animals/)

"Christian Gottfried Ehrenberg" on *Wikipedia* website (https://en.wikipedia.org/wiki/Christian_Gottfried_Ehrenberg)

"Names for Tuberculosis – Phthisis, Scrofula, King's Touch" on *Tbfacts.org* website (https://tbfacts.org/tb-names/)

"Petri dish" on *Wikipedia* website (https://en.wikipedia.org/wiki/Petri_dish#History)

Frith, J. "History of tuberculosis. Part 1 – phthisis, consumption and the white Plague" on *JMVH* website (https://jmvh.org/article/history-of-tuberculosis-part-1-phthisis-consumption-and-the-white-plague/)

Howard, J. "Plague was one of history's deadliest diseases—then we found a cure" on *National Geographic* website (https://www.nationalgeographic.com/science/article/the-plague)

Smith, K. "The Origin of MacConkey Agar" on the *American Society for Microbiology* website (https://asm.org/Articles/2019/October/The-Origin-of-MacConkey-Agar)

"About Paratuberculosis" on the *International Association for Paratuberculosis* website (https://www.paratuberculosis.com/johnes.php)

Thomas, G.H. "Frederick William Twort: not just bacteriophage" on the *Microbiology Society* website (https://microbiologysociety.org/publication/past-issues/world-war-i/article/frederick-william-twort-not-just-bacteriophage.html)

"The Filter of Life" by Bert Hansen on *Science History Institute* website (https://www.science-history.org/distillations/the-filter-of-life)

Instrument of Recovery

Summers, W.C. (1999) "Félix d'Hérelle and the origins of molecular biology". *Yale Press University*.

Bauman, R.W. (2011) "Microbiology with disease by taxonomy". *Pearson* (3rd edition).

DePorte, M.E. and Vanderleck, J. (1917) "Studies on *Coccobacillus acridiorum* d'Hérelle and on certain intestinal organisms of locusts". *Annals Entomological Society of America*, vol. 10, pp. 47–62.

Keith Cressman "Preventing the spread of desert locust swarms" on *PreventionWeb* website (https://www.preventionweb.net/news/preventing-spread-desert-locust-swarms)

"How can we control locust swarms?" on *World Economic Forum* website (https://www.weforum.org/agenda/2015/11/how-can-we-control-locust-swarms/)

"The locust crisis and the World Bank Group" on *The World Bank* website (https://www.worldbank.org/en/topic/agriculture/brief/the-locust-crisis-and-the-world-bank-group)

"Approaches to the biological control of insect pests" on *The Connecticut Agricultural Experiment Station* website (https://portal.ct.gov/CAES/Fact-Sheets/Entomology/Approaches-to-the-Biological-Control-of-Insect-Pests)

"Biological pest control" on *Wikipedia* website (https://en.wikipedia.org/wiki/Biological_pest_control)

A Corner of a Table

Lewis, S. (1924) "Arrowsmith". *Signet Classics Printing* (1961 edition).

D'Hérelle, F. (1925) "The bacteriophage, its rôle in immunity". *Forgotten Books* (2015 edition, translated by George H. Smith).

Echeverria-Valencia, G., Flores-Villalva, S. and Espitia, C.I. (2018) "Virulence factors and pathogenicity of *Mycobacterium*", in *Mycobacterium* edited by Ribón W. *IntechOpen*.

Frey, J. (2007) "Biological safety concepts of genetically modified live bacterial vaccines". *Vaccine*, vol. 25, pp. 5598–5605.

Dublanchet, A. (2007) "The epic of phage therapy". *The Canadian Journal of Infectious Diseases and Medical Microbiology*, vol. 18, pp. 15–18.

Leggio, C., Galantini, L. and Pavel, N.V. (2008) "About the albumin structure in solution: cigar expanded form versus heart normal shape". *Royal Society of Chemistry*, vol. 10, pp. 6741–6750.

Forrellad, M.A., Klepp, L.I., Gioffré, A. et al. (2013) "Virulence factors of the *Mycobacterium tuberculosis* complex". *Virulence*, vol. 4, pp. 3–66.

Bull, J.J. (2015) "Evolutionary reversion of live viral vaccines: can genetic engineering subdue it?". *Virus Evolution*, vol. 1, p. vev005.

Jimenez-Guardeño, J.M., Regla-Nava, J., Nieto-Torres, J.L. et al. (2015) "Identification of the mechanisms causing reversion to virulence in an attenuated SARS-CoV for the design of a genetically stable vaccine". *Plos Pathogens*, vol. 11, p. e1005215.

Myelnikov, D. (2018) "An alternative cure: the adoption and survival of bacteriophage therapy in the USSR, 1922-1955". *Journal of the History of Medicine and Allied Sciences*, vol. 73, pp. 385–411.

Summers, W.C. (2021) "The discovery of bacteriophages and the historical context", in *Bacteriophages: biology, technology, therapy* edited by David R. Harper, Stephen T. Abedon and Benjamin H. Burrowes. *Springer*.

Donald, P., Kaufmann, S., Thee, S. et al. (2022) "Pathogenesis of tuberculosis: the 1930 Lübeck disaster revisited". *European Respiratory Review*, vol. 31, p. 220046.

"André Gratia (1893–1950)" on the *Belgian Society for Microbiology* website (https://belsoc-microbio.be/famous-belgian-microbiologists/andre-gratia/)

"Albert Calmette" on *Wikipedia* website (https://en.wikipedia.org/wiki/Albert_Calmette)

"Camille Guérin" on *Wikipedia* website (https://en.wikipedia.org/wiki/Camille_Gu%C3%A9rin)

"Vaccine Types" on *HHS.gov* website (https://www.hhs.gov/immunization/basics/types/index.html)

"Understanding six types of vaccine technologies" on *Pfizer* website (https://www.pfizer.com/news/articles/understanding_six_types_of_vaccine_technologies)

From the Lab-Bench to the Bedside

Depew, D.J. and Weber, B.H. (1997) "Darwinism evolving". *MIT Press* (2nd edition).

Neal, D. (2019) "Introduction to population biology". *Cambridge University Press*.

Beevor, A. (2022) "Russia: revolution and civil war, 1917–1921". *Viking/Penguin Random House*.

Gillispie, C.C. (1958) "Lamarck and Darwin in the history of science". *American Scientist*, vol. 46, pp. 388–409.

Simpson, G.G. (1961) "Lamarck, Darwin and Butler". *The American Scholar*, vol. 30, pp. 238–249.

Summers, W.C. (1999) "Félix d'Hérelle and the origins of molecular biology". *Yale Press University*.

Soyfer, V.N. (2001) "The consequences of political dictatorship for Russian science". *Nature Reviews in Genetics*, vol. 2, pp. 723–729.

La Scola, B., Audic, S., Robert, C. et al. (2003) "A giant virus in amoebae". *Science*, vol. 299, p. 2033.

Roll-Hansen, N. (2008) "Wishful science: the persistence of T. D. Lysenko's agrobiology in the politics pf science". *Osiris*, vol. 23, pp. 166–188.

Mougari, S., Sahmi-Bounsiar, D., Levasseur, A. et al. (2019) "Virophages of giant viruses: an update at eleven". *Viruses*, vol. 11, p. 733.

Aatifa Rasool, A., Mansoor, S., Bhat, K.M. et al. (2020) "Mechanisms underlying graft union formation and Rootstock Scion interaction in horticultural plants". *Frontiers in Plant Science*, vol. 11, p.590847.

Tsaballa, A., Xanthopoulou, A., Madesis, P. et al. (2021) "Vegetable grafting from a molecular point of view: the involvement of epigenetics in Rootstock-Scion interactions". *Frontiers in Plant Science*, vol. 11, p. 621999.

"Demographics of Europe" on *Wikipedia* website (https://en.wikipedia.org/wiki/Demographics_of_Europe#Vital_statistics)

"Lamarck vs. Darwin" on *New England Complex System Institute* website (https://necsi.edu/lamarck-vs-darwin)

"Explain the difference between Lamarck's and Darwin's theory of evolution? Why was Darwin' s more successful?" on *MyTutor* website (https://www.mytutor.co.uk/ answers/2529/GCSE/Biology/Explain-the-difference-between-Lamarck-s-and-Darwin-s-theory-of-evolution-Why-was-Darwin-s-more-successful/)

"Lysenkoism" on *Wikipedia* website (https://en.wikipedia.org/wiki/Lysenkoism)

"Vernalization" on *Wikipedia* website (https://en.wikipedia.org/wiki/Vernalization)

"Principles of Grafting and Budding" on *Aggie Horticulture* website (https://aggie-horticulture.tamu.edu/faculty/davies/pdf%20stuff/ph%20final%20galley/M11_DAVI4493_08_SE_C11.pdf)

Chan, A. (2024) "Why the pandemic probably started in a lab, in 5 key points" on *The New York Times* website (https://www.nytimes.com/interactive/2024/06/03/opinion/covid-lab-leak.html?smid=nytcore-android-share)

Happy Years

"Joseph Stalin" on *Wikipedia* webpage (https://en.wikipedia.org/wiki/Joseph_Stalin#1878%E2%80%931899:_Childhood_to_young_adulthood)

"1907 Tiflis bank robbery" on *Wikipedia* webpage (https://en.wikipedia.org/wiki/1907_Tiflis_bank_robbery)

CPI Inflation Calculator (https://www.in2013dollars.com/us/inflation/1913?amount=147000000000)

Iron Curtain

Fleming, A. (1929) "On the antibacterial action of cultures of a penicillum, with a special reference to their use in the isolation of *B. influenzae*". *British Journal Experimental Pathology*, vol. 10, pp. 226–236.

Eaton, M.D. and Bayne-Jones, S. (1934) "Bacteriophage therapy: review of the principles and results of the use of bacteriophages in the treatment of infections". *Journal of the Medical Association*, vol. 103, pp. 1847–1853.

Ruska, H., von Borries, B. and Ruska, E. (1940) "Die Bedeutung der Übermikroskopie für die Virusforschung". *Archiv für die gesamte Virusforschung*, vol. 1, pp. 155–169.

Krüger, A.P. and Scribner, E.J. (1941) "The bacteriophage: its nature and its therapeutic use". *Journal of the Medical Association*, vol. 116, pp. 2160–2167.

Payne, R.J. and Jansen, V.A. (2001) "Understanding bacteriophage therapy as a density-dependent kinetic process". *Journal of Theoretical Biology*, vol. 208, pp. 37–48.

"Gerhard Domagk" on *Science History Institute* website (www.sciencehistory.org/historical-profile/gerhard-domagk)

BERLIN, GÖTTINGEN, AND COPENHAGEN

Stargazing

Fischer, E.P. and Lipson, C. (1988) "Thinking about science". *W. W. Norton & Company.*

Hoskin, M. (1997) "Cambridge illustrated history of astronomy". *Cambridge University Press.*

Shampo, M.A. and Kyle, R.A. (1999) "Max Delbrück and molecular genetics". *Mayo Clinic Proceedings*, vol. 74, p. 1124.

Edgar, R.S. (1982) "Max Delbrück". *Annual Review of Genetics*, vol. 16, pp. 501–505.

Luft, F.C. (2007) "Who was Max Delbrück?". *Journal of Molecular Medicine*, vol. 85, pp. 207–211.

Luft, F.C. (2017) "Berlin wild-and the Max Delbrück Center for Molecular Medicine (MDC) Berlin-Buch". *Journal of Molecular Medicine*, vol. 95, pp. 567–570.
Fischer, E.P. (2007) "Max Delbrück". Genetics, vol. 177, pp. 673–676.
Kitchen, M. (1996) "Cambridge illustrated history of Germany". *Cambridge University Press.*
Evans, D. and Jenkins, J. (1999) "Years of Weimar and the third Reich". *Hodder Murray.*
Ray, M. "German-Danish War" on *Britannica* website (https://www.britannica.com/event/German-Danish-War)
"Economic history of World War I" on *Wikipedia* website (https://en.wikipedia.org/wiki/Economic_history_of_World_War_I)

Quantum Leap

Snow, C.P. (1959) "The two cultures". *Cambridge University Press* (2010 edition).
Baggot, J. (2011) "The quantum story: a history in 40 moments". *Oxford University Press.*
"Isn't the standard proof of a classical electron's orbit invalid?" on *Physics Stack Exchange* website (https://physics.stackexchange.com/questions/527541/isnt-the-standard-proof-of-a-classical-electrons-orbit-invalid)
"Does the orbit of the ISS decay?" on *Physics Stack Exchange* website (https://space.stackexchange.com/questions/1753/does-the-orbit-of-the-iss-decay)
"Calculate the orbital period of the electron in the first excited state of hydrogen atom" on *Doubtnut* website (https://www.doubtnut.com/question-answer-physics/calculate-the-orbital-period-of-the-electron-in-the-first-excited-state-of-hydrogen-atom-571108406)
Olsen, J.D. and McDonald, K.T. (2005, 2017 edition) "Classical lifetime of a Bohr atom". (http://kirkmcd.princeton.edu/examples/orbitdecay.pdf)
Schönhammer, K. "Max Born, Göttingen and quantum mechanics" (https://indico.cern.ch/event/1146259/contributions/4811237/attachments/2459992/4258479/Kschoenhammer_Text.pdf)
"Max Born Biographical" (2022) on *Nobel Price* website (https://www.nobelprize.org/prizes/physics/1954/born/biographical/)
Bisno, A. (2020) "The Einstein-Szilard Refrigerator" on *Lemelson Center for the Study of Invention and Innovation* website (https://invention.si.edu/einstein-szilard-refrigerator)
Segrè, G. (2011) "Ordinary geniuses". *Penguins Books.*
Baggott, J. (2009) "Atomic". *Icon Books* Ltd (2015 edition).

Back Home

Hürter, T. (2021) "Too big for a single mind". *The Experiment LLC* (translated by David Shaw).
Müller, H.J. (1928) "The production of mutations by X-rays". *Proceedings of the National American Society*, vol. 14, pp. 714–726.
Müller, H.J. (1930) "The frequency of translocation produced by X-ray in Drosophila". *Genetics*, vol. 15, pp. 283–311.
Jarlskog, G. and Jönsson, L. (1973) "Measurement of Delbrück scattering and observation of photon splitting at high energies". *Physical Review* D, vol. 8, pp. 3813–3823.
Delbrück, M. (1978) "The arrow of time – beginning and end". *Engineering and Science* Sept.-Oct, pp. 5–9.
Sansare, K., Khanna, V. and Karjodkar, F. (2011) "Early victims of X-rays: a tribute and current perception". *Dentomaxillofacial Radiology*, vol. 40, pp. 123–125.
Nüsslin, F. (2020) "Wilhelm Conrad R¨ontgen: The scientist and his discovery". *Physica Medica*, vol. 79, pp. 65-68.
"Lise Meitner" on *Atomic archive* website (https://www.atomicarchive.com/resources/biographies/meitner.html)

"Lise Meitner" on *Britannica* website (https://www.britannica.com/biography/Lise-Meitner)

Kniefacz, K. (2022) "Women at the University of Vienna" on *University of Vienna* website (https://geschichte.univie.ac.at/en/topics/women-university-vienna)

Ball, P. (2019) "Lize Meitner: the nuclear pioneer who escaped the Nazis" on *Science focus* website (https://www.sciencefocus.com/science/lise-meitner-the-nuclear-pioneer-who-escaped-the-nazis/)

"Uranium-235 Fission" on *Hyperphysics* website (http://hyperphysics.phy-astr.gsu.edu/hbase/NucEne/U235chn.html)

"World's Most Dangerous Toy? Radioactive Atomic Energy Lab Kit with Uranium (1950)" on *Bulletin of the Atomic Scientists* website (https://thebulletin.org/virtual-tour/worlds-most-dangerous-toy-radioactive-atomic-energy-lab-kit-with-uranium-1950/)

"What is a Geiger Counter?" on *United States Nuclear Regulatory Commission* website (https://www.nrc.gov/reading-rm/basic-ref/students/science-101/what-is-a-geiger-counter.html)

"How Do Dosimeters Work?" on *Sciencing* webside (https://sciencing.com/do-dosimeters-work-5167635.html)

"Genetic Effects of Radiation" on *National Library of Medicine* website (https://www.ncbi.nlm.nih.gov/books/NBK218706/)

"Marfan syndrome" on *Mayo Clinic* website (https://www.mayoclinic.org/diseases-conditions/marfan-syndrome/symptoms-causes/syc-20350782)

"Thomas Hunt Morgan: the fruit fly scientist" on *Nature Education* website (https://www.nature.com/scitable/topicpage/thomas-hunt-morgan-the-fruit-fly-scientist-6579789/)

"Thomas Hunt Morgan and sex linkage" on *Nature Education* website (https://www.nature.com/scitable/topicpage/thomas-hunt-morgan-and-sex-linkage-452/)

"Mutant fruit flies" on *Exploratorium* website (https://annex.exploratorium.edu/exhibits/mutant_flies/mutant_flies.html)

"Henri Becquerel Biographical" on *The Nobel Prize* website (https://www.nobelprize.org/prizes/physics/1903/becquerel/biographical/)

"WilhelmConradRöntgen"onthe*UniversityofWürzburg*website(https://www.uni-wuerzburg.de/en/uniarchiv/personalities/eminent-scholars/wilhelm-conrad-roentgen/)

"Hermann J. Muller" on *The Cold Spring Harbor Laboratory* website (https://www.cshl.edu/personal-collections/hermann-j-muller/)

"Hermann J. Muller" on *The Nobel Prize* website (https://www.nobelprize.org/prizes/medicine/1946/muller/facts/)

Gleason, K.M. (2017) "Hermann Joseph Muller's Study of X-rays as a Mutagen, (1926–1927)". *Embryo Project Encyclopedia* (https://embryo.asu.edu/handle/10776/11441)

Elmer, N. (2016) "A look at the life of geneticist Hermann J Muller – part I" on *The University of Texas* website (https://integrativebio.utexas.edu/about/history/hermann-j-muller-part-i)

Elmer, N. (2016) "A look at the life of geneticist Hermann J Muller – part II" on *The University of Texas* website (https://integrativebio.utexas.edu/about/history/hermann-j-muller-part-ii)

Carlson, E.A. (2009) "Hermann Joseph Muller, a biographical memoir" on *The National Academy of Sciences* website (https://www.nasonline.org/publications/biographical-memoirs/memoir-pdfs/muller-hermann.pdf)

Neves, S.F. (2017) "What bizarre flies have taught us" on *The European Molecular Biology Laboratory* website (https://www.embl.org/news/science/what-we-learned-from-fruit-flies/)

"Laboratory B in Sungul" on *Scholarly Community Encyclopedia* website (https://encyclopedia.pub/entry/33463)

Oleynikov, P.V. (2000) "German scientists in the Soviet atomic project". *The nonproliferation Review* (https://www.nonproliferation.org/wp-content/uploads/npr/72pavel.pdf)

"Pascual Jordan" on *Britannica* website (https://www.britannica.com/biography/Pascual-Jordan)

The Green Pamphlet

Schrödinger, E. (1944) "What is life?". *Cambridge University Press* (1967 edition).
Blackburn, G.M. and Gait, M.J. (1996) "Nucleic acids in chemistry and biology". *Oxford University Press.*
Watson, J.D. and Crick, F.H.C. (1953) "Molecular structure of nucleic ACIS". *Nature*, vol. 171, pp. 737–738.
Hergersberg, M. (1991) "Biological aspects of cytosine methylation in eukaryotic cells". *Experientia*, vol. 47, pp. 1171–1185.
Friedberg, E.C., Walker, G.C., Siede, W. et al. (2006) "DNA repair and mutagenesis". *American Society for Microbiology Press.*
Pray, L.A. (2008) "Discovery of DNA structure and function: Watson and Crick". *Nature Education*, vol. 1, p. 100.
Lister, R., Ecker, J.R. (2009) "Finding the fifth base: genome-wide sequencing of cytosine methylation". *Genome Research*, vol. 19, pp. 959–966.
Piovesan, A., Pelleri, M.C., Antonaros, F. et al. (2019) "On the length, weight and GC content of the human genome". *BMC Research Notes*, vol. 12, p. 106.
"Rosalind Franklin" on *Britannica* website (https://www.britannica.com/biography/Rosalind-Franklin)
"The Discovery of the Double Helix, 1951–1953" on the *National Library of Medicine* website (https://profiles.nlm.nih.gov/spotlight/sc/feature/doublehelix)
"The Discovery of DNA's Structure" on *PBS* website (https://www.pbs.org/wgbh/evolution/library/06/3/l_063_01.html)
"Roentgen (unit)" on *Wikipedia* website (https://en.wikipedia.org/wiki/Roentgen_(unit))
"Calculating the sum formula" on *Biotage* website (https://atdbio.com/nucleic-acids-book/Molecular-weight-and-mass)

ROME AND LOS ANGELES

A Vessel to the Promised Land

Brecht, B. (1957) "The resistible rise of Arturo Ui". *Bloomsbury* (2007 edition, translated by Ralph Manheim).
Mack Smith, D. (1979) "Italy, a modern history". *Editori La Terza* (translated by Alberto Aquarone).
Trotsky, L. (1979) "Problemi della rivoluzione in Europa". *Oscar Saggi Mondadori* (Omnibus edition edited by Livio Maitan)
Luria, S.E. (1984) "A slot machine, a broken test tube". *Harper & Row Publishers.*
Paxton, R.O. (2005) "The anatomy of fascism". *Penguin Books.*
Alighieri, D. (2012) "The vision; or hell, purgatory and paradise". *Ulan Press* (re-issue of Alighieri's "Divine Comedy" published in 1472 from an original of 1321. Cantica I "Inferno", canto XXVI. Translated by Rev. Henry Francis Cary).
Eco, U. (2017) "Il fascismo eterno". *La nave di Teseo.*
Guarnieri, P. (2020) "L'emigrazione intellettuale dell'Italia fascista". *Firenze University Press.*
Ruth Ben-Ghiat, R. (2021) "Strongmen – Mussolini to the present". *Norton.*
Siddiqi, O. (1991) "In memoriam: Salvador Edward Luria". *Journal of Genetics*, vol. 70, no. 1, pp. 65–68.
Guglielmetti, L. (2012) "Adesione, dissenso, pragmatismo. Il mondo della cultura negli anni Trenta di fronte al fascismo e al nazismo". *Rivista di Studi Politici Internazionali*, vol. 79, pp. 306–312.
Kokkinidis, T. (2022) "*Nea Hellas*: the ship that brought thousands of Greeks to the US" on *Greek Reporter* website (https://greekreporter.com/2022/07/24/nea-hellas-ship-brought-greeks-us/)

"Remembering the TSS *Nea Hellas*" on *Weebly* website (https://memoriesoftheneahellas.weebly.com/)
"The *Nea Hellas* in World War Two" on *Weebly* website (https://memoriesoftheneahellas.weebly.com/the-nea-hellas-in-world-war-ii.html)
Grignolio, A. (2013) "Luria, Salvador Edward" on *Enciclopedia Treccani* (https://www.treccani.it/enciclopedia/salvador-edward-luria_%28Il-Contributo-italiano-alla-storia-del-Pensiero:-Scienze%29/)
"Una via di mezzo tra esilio e prigionia: il confino, l'arma di repressione silenziosa del regime fascista" on *Parentesi Storiche* website (https://parentesistoriche.altervista.org/confino-fascismo/)
Ianuale, N. "Il confino politico ai tempi del fascismo" on *Vanilla Magazine* website (https://www.vanillamagazine.it/il-confino-politico-ai-tempi-del-fascismo/)
Tommasi, M. "Storia e significato del confino. L'esperienza dei confinati durante il fascismo" on Studenti Storia website (https://www.studenti.it/storia-e-significato-confino-esperienza-confinati-durante-il-fascismo.html)
Ravveduto, M. "I 12 professori che rifiutarono il giuramento fascista" on *Fanpage* website (https://www.fanpage.it/cultura/i-12-professori-che-85-anni-fa-rifiutarono-il-giuramento-fascista/)
"Solo 12 professori universitari, su oltre 1200, rifiutarono di giurare fedeltà al fascismo" on MeDiaRe website (https://www.me-dia-re.it/solo-12-professori-universitari-su-oltre-1200-rifiutarono-di-giurare-fedelta-al-fascismo/)
Papotti, P. "I magnifici dodici" on *Patria Indipendente* website (https://www.patriaindipendente.it/idee/formazione/promemoria-7-i-magnifici-dodici/)
The brave dozen was formed by the following University professors: Ernesto Buonaiuti (Rome); Mario Carrara (Turin); Gaetano De Sanctis (Rome); Giorgio Errera (Pavia); Giorgio Levi Della Vida (Rome); Fabio Luzzatto (Macerata); Piero Martinetti (Milan); Bartolo Nigrisoli (Bologna); Francesco Ruffini (Turin); Edoardo Ruffini Avondo (Perugia); Lionello Venturi (Turin); e Vito Volterra (Rome). (https://www.swas.polito.it/services/poli_flash/foto/I%20dodici%20professori%20che%20non%20hanno%20giurato.pdf)
"Arnold of Brescia" on *Britannica* website (https://www.britannica.com/biography/Arnold-of-Brescia)
Del Vecchio, N. (2021) "Arnaldo fascista ribaldo" on *Tag43* website (https://www.tag43.it/arnaldo-mussolini-fratello-duce-parco-latina/)
"Mussolini, Arnaldo" on *Treccani* website (https://www.treccani.it/enciclopedia/arnaldo-mussolini_%28Dizionario-Biografico%29/)
"Una breve storia di Fiume" on *Città di Fiume* website (https://www.rijeka.hr/it/amministrazione-cittadina/storia-di-fiume/)
"Impresa di Fiume" on *Treccani* website (https://www.treccani.it/enciclopedia/impresa-di-fiume_%28Dizionario-di-Storia%29/)
"Gabriele D'Annunzio e l'impresa di Fiume" on Mondadori Education website (https://www.mondadorieducation.it/media/contenuti/universita/roccucci_storia_contemporanea/assets/documents/cap_12_1.pdf)
Simonelli, F.C. "Fiume e d'Annunzio. I cento anni di un caso storiografico ancora aperto" on *Laboratorio di Storia* website (https://storicamente.org/simonelli-fiume-e-d-annunzio)

All the Way to Rome

Rita, G. (1968) "The interferons: an international symposium". *Academic Press.*
Onofri, N.S. (1980) "La strage di palazzo d'Accursio". *Feltrinelli Editore.*
Bolker, B.M. (2008) "Ecological models and data in R". *Princeton University Press.*
Luria, S. (1939) "Sur l'unité lytique du bactériophage". *Comptes rendus des séances de la Societé de biologie et de ses filiales*, vol. 130, pp. 904–907.

Rita, G. (1940) "Sul comportamento del batteriofago nelle acque del Tevere". *Bollettino dell'Istituto Sieroterapico Milanese*, vol. 29, pp. 27–33.

Rita, G. and Cammarella, C. (1940) "Ricerche batteriologiche e osservazioni epidemiologiche sulla dissenteria batterica infantile a Roma". *Annali d'Igiene*, vol. 43, p. 33.

Rita, G. and Silvestri, L.G. (1950) "Studio morfologico della modificazione del cosiddetto nucleo in cellule batteriche infettate con virus (batteriofagi)". *Bollettino della Società italiana di biologia sperimentale*, vol. 26, pp. 954–957.

Rita, G. and Fantuzzi, L.F. (1954) "Ricerche sul processo di lisogenizzazione di E. coli. I.". *Nuovi annali d'igiene e microbiologia*, vol. 5, pp. 222–225.

Saladini Pilastri, F. (2017) "21 Novembre 1920: l'eccidio di Palazzo D'Accursio come nascita del fascismo" on *Bibliomanie* website (https://www.bibliomanie.it/?p=297)

"Le barricate dell'agosto 1922 a Parma" on Senti le rane che cantano website (https://sites.google.com/site/sentileranechecantano/schede/antifascismo-e-lotta-di-liberazione/le-barricate-antifasciste-dell-agosto-1922-a-parma)

"Agosto 1922 – L'offensiva squadrista e le barricate di Parma" on *Sinistra Classe Rivoluzione* website (https://www.rivoluzione.red/agosto-1922-loffensiva-squadrista-e-le-barricate-di-parma/)

"Alle radici della guerra civile a Pisa e nella provincia" on *Toscana Novecento* website (https://www.toscananovecento.it/custom_type/alle-radici-della-guerra-civile-a-pisa-e-nella-provincia/)

Dasti, R. "I primi passi del fascismo a Crema e nel Cremasco (1919–1921)" on *Comune di Cremona* webiste (https://www.comunecrema.it/sites/default/files/insula_fulcheria_2010_16_28_contro_i_tori_infuriati_divenuti_conigli_i_primi_passi_del_fascismo_a_crema_e_nel_cremasco_1919-1921.pdf). Last accessed 2023.

"A 100 anni dall'assalto degli squadristi guidati da Italo Balbo contro la federazione delle cooperative di Ravenna" on *Ethica Societas* website (https://www.ethicasocietas.it/27Luglio1922-assaltocooperative/)

"Il fascismo al potere" on *Storia Disegno* website (https://storia.disegno.it/app/books/CPAC00_DGMESTOR52/html/170) . Last accessed 2023.

"La violenza fascista (1921–1926)" on *CGIL* website (https://www.cgil.it/la-cgil/la-nostra-storia/la-violenza-fascista-1921-1926-ug9rjiwe

"Un sindacalista, un militare in carriera, un giornalista e un capo dei reduci: chi erano i quadrumviri fascisti della marcia su Roma" on *il Dolomiti* website (https://www.ildolomiti.it/societa/2022/un-sindacalista-un-militare-in-carriera-un-giornalista-e-un-capo-dei-reduci-chi-erano-i-quadrumviri-fascisti-della-marcia-su-roma)

Ottolenghi, G. (2022) "Quadrumviri, le guide fidate di Mussolini nella marcia su Roma" on *L'incontro* website (https://www.lincontro.news/quadrumviri-le-guide-fidate-di-mussolini-nella-marcia-su-roma/)

"Mussolini: soldi, sesso, segreti" on *Youtube* website (https://www.youtube.com/watch?v=G3cd9SSyFEU)

Calderoni, M.R. "Come gli Agnelli hanno rapinato l'Italia lungo un intero secolo" on *Cobas* website (https://www.sialcobas.it/2016/03/come-gli-Agnelli-hanno-rapinato-litalia-lungo-un-intero-secolo/)

"La Fiat tra giolittismo e fascismo" on *Conflitti & Strategie website* (https://www.conflittiestrategie.it/la-fiat-tra-giolittismo-e-fascismo)

Forgione, A. (2016) "Storia (poco narrata) degli Agnelli, i veri sovrani del Novecento italiano" on *Il Blog di Angelo Forgione* website (https://angeloforgione.com/2016/05/25/juventus_agnelli_dlv/)

"Enrico Fermi, Biographical" on *Nobel Prize* website (https://www.nobelprize.org/prizes/physics/1938/fermi/biographical/)

"Enrico Fermi, Italian-American physicist" on *Britannica* website (https://www.britannica.com/biography/Enrico-Fermi)

"Enrico Fermi (1901–1954)" on *the Atomic Archive* website (https://www.atomicarchive.com/resources/biographies/fermi.html)
"Enrico Fermi: la scienza e il potere" on *Dima* website (https://www.senato.it/documenti/repository/presidente/discorsi/EnricoFermi.pdf)
"Bruno Pontecorvo" on *Britannica* website (https://www.britannica.com/biography/Bruno-Pontecorvo)
"L'eredità di Bruno Pontecorvo" on *Media INAF* website (https://www.media.inaf.it/2013/09/11/leredita-di-bruno-pontecorvo/)
Polcaro, V.F. "Scienziati e fascismo. Storia di un controverso rapporto" on *Patria Indipendente* website (https://www.patriaindipendente.it/terza-pagina/leonardo/scienziati-e-fascismo-storia-di-un-controverso-rapporto/)
"Slow neutrons favour fission" on *Radioactivity* website (https://radioactivity.eu.com/phenomenon/slow_neutrons)
Del Testa, A.M. "Un illustre scienziato fanese d'adozione: Geo Rita" (https://www.sistemabibliotecariofano.it/fileadmin/grpmnt/5596/NSF_4_1989_Mei_Del_Testa_Un_il.pdf)
Grignolio, A. (2016) "Rita, Geo". *Enciclopedia Treccani* (https://www.treccani.it/enciclopedia/geo-rita_%28Dizionario-Biografico%29/)
"Probability and statistics" on *Britannica* website (https://www.britannica.com/science/probability)
"Cumulative distributions with dice" on *NMSU* website (https://web.nmsu.edu/~pbaggett/Statistics/Lessons/TTD/P4CCWD/P4CCWD.html)
"Casino Royale: Roll the Dice" on *Informit* website (https://www.informit.com/articles/article.aspx?p=2469049&seqNum=2)
Surbhi, S. (2017) "Difference between binomial and Poisson distribution" on *Key Differences* website (https://keydifferences.com/difference-between-binomial-and-poisson-distribution.html)
"How to demonstrate amount of virus required to lyse all cells using Poisson distribution?" on *CrossValidated* website (https://stats.stackexchange.com/questions/611608/how-to-demonstrate-amount-of-virus-required-to-lyse-all-cells-using-poisson-dist)

Nuclear Fission

Hürter, T. (2021) "Too big for a single mind". *The Experiment LLC* (translated by David Shaw).
Ruth Ben-Ghiat, R. (2021) "Strongmen – Mussolini to the present". *Norton.*
Segrè, G. (2011) "Ordinary geniuses" *Penguins Books.*
Meitner, L. and Frisch, O.R. (1939) "Disintegration of uranium by neutrons: a new type of nuclear reaction. *Nature*, vol. 143, pp. 239–240.
Deutsch, K.W. (1945) "Anti-semitic ideas in the Middle Ages: international civilizations in expansion and conflict". *Journal of the History of Ideas*, vol. 6, pp. 239–251.
Bernardini, G. (1977) "The origins and development of racial Anti-Semitism in Fascist Italy". *The Journal of Modern History*, vol. 49, pp. 431–453.
Guzzi, D. (2012) "The myth of the 'good Italian', the antisemitism and the colonial crimes". *Constelaciones, Revista de Teoría Crítica.* (https://constelaciones-rtc.net/article/view/790)
Yu, G., Duh, Y.S., Yang, X. et al. (2022) "Holistic case study on the explosion of ammonium nitrate in Tianjin Port". *Sustainability*, vol. 14, 3429.
Nesseler, C., Gomez-Gonzalez, C., Parshakov, P., and Dietl, H. (2023) "Examining discrimination against Jews in Italy with three natural field experiments". *Journal of Behavioral and Experimental Economics*, vol. 106, p. 102045.
"Nuclear fission" on *Academic Brooklyn* website (ttps://www.britannica.com/science/nuclear-fission)
"The periodic table" (https://hobart.k12.in.us/ksms/PeriodicTable/abc.htm)

"How many stars are there in the Universe?" on *European Space Agency* website (https://www.esa.int/Science_Exploration/Space_Science/Herschel/How_many_stars_are_there_in_the_Universe)

"2015 Tianjin explosions" on *Wikipedia* website (https://en.wikipedia.org/wiki/2015_Tianjin_explosions)

"How powerful was the Beirut blast?" on *Reuters* website (https://www.reuters.com/graphics/LEBANON-SECURITY/BLAST/yzdpxnmqbpx/)

"Worked out example" on *SimplyScience* website (https://www.simply.science/popups/30004.html)

"Neutron" on *Britannica* website (https://www.britannica.com/science/neutron)

"The equivalence of mass and energy" on *Stanford Encyclopedia of Philosophy* website (https://plato.stanford.edu/entries/equivME/)

"TNT equivalent" on *Wikipedia* website (https://en.wikipedia.org/wiki/TNT_equivalent)

"Otto Hahn" on *Wikipedia* website (https://de.wikipedia.org/wiki/Otto_Hahn)

"Discovery of nuclear fission" on *Wikipedia* website (https://en.wikipedia.org/wiki/Discovery_of_nuclear_fission)

"Otto Hahn and the Mainau Declaration of 1955" on *Max Planck Society* website (https://www.mpg.de/18357917/otto-hahn-mainau-declaration)

"Ettore Majorana: il mistero del genio scomparso" on *ZetaTiElle* website (https://www.zetati-elle.com/ettore-majorana-mistero-scomparsa-fisico-nucleare/)

"Medieval antisemitism" on *The Holocaust Explained* website (https://www.theholocaustex-plained.org/anti-semitism/medieval-antisemitism/)

The Emperor's New Clothes

Evans, D. and Jenkins, J. (1999) "Years of Weimar and the third Reich". *Hodder Murray.*

Paxton, R.O. (2004) "The anatomy of fascism". *Penguin Books.*

Baggott, J. (2009) "Atomic". *Icon Books* Ltd (2015 edition).

Hürter, T. (2021) "Too big for a single mind". *The Experiment LLC* (translated by David Shaw).

Harding, C. (1979) "Interview with Max Delbrück". *California Institute of Technology Oral History Project.*

Evans, D. and Jenkins, J. (1999) "Years of Weimar and the third Reich". *Hodder Murray.*

Goodrick-Clarke, N. (1985) "The occult roots of Nazism". *Tauris Parke Publishing* (2004 edition).

Phelps, R.H. (1963) "'Before Hitler Came': Thule Society and Germanen Orden". *The Journal of Modern History*, vol. 35, pp. 245–261.

Chazan, G. (2022) "German police thwart extremist plot to seize power". *The Financial Times*, No. 41192.

Feuer, A., Hakim, D. and Protess, B. (2023) "Where Trump stands in 4 criminal inquiries". *The New York Times*, No. 43527.

Klein, E. (2025) "Now is the time of monsters". *The New York Times*, 12/01/2025 (https://www.nytimes.com/2025/01/12/opinion/ai-climate-change-low-birth-rates.html?smid=nytcore-android-share)

"Emergency Quota Act" on *Wikipedia* website (https://en.wikipedia.org/wiki/Emergency_Quota_Act)

"Refugee Crisis: European Commission takes decisive action" (2015) *European Commission Press Room* (https://ec.europa.eu/commission/presscorner/detail/en/IP_15_5596)

Šabić, Š. (2017) "The relocation of refugees in the European Union". *Friederich Ebert Stiftung* website (https://library.fes.de/pdf-files/bueros/kroatien/13787.pdf)

Fiske, J. (1881) "Who are the Aryans?". *The Atlantic* website (https://www.theatlantic.com/magazine/archive/1881/02/who-are-the-aryans/521367/)

"Zend-Avesta" on *Encyclopedia Britannica 1911* website (https://theodora.com/encyclopedia/z/zendavesta.html)
"Aryan" on *Holocaust Encyclopedia* website (https://encyclopedia.ushmm.org/content/en/article/aryan-1)
Reucher, G. (2021) "Richard Wagner and the Jews" on *Die Welt* website (https://www.dw.com/en/jews-and-richard-wagner/a-58646578)
Eylon, L. "Anti-Semitism: the controversy over Richard Wagner" on *Jewish Virtual Library* website (https://www.jewishvirtuallibrary.org/richard-wagner-controversy)
Land, G. (2021) "The Brownshirts: the role of the Sturmabteilung (SA) in Nazi Germany" on *History Hit* website (https://www.historyhit.com/hitlers-bullyboys-the-role-of-the-sa-in-nazi-germany/)
Campion, M.J. (2014) "How the world loved the swastika: until Hitler stole it" on *BBC News* website (https://www.bbc.com/news/magazine-29644591)
Sunder, K. (2021) "The ancient symbol that was hijacked by evil" on *BBC Culture* website (https://www.bbc.com/culture/article/20210816-the-ancient-symbol-that-was-hijacked-by-evil)
"Beer Hall Putsch" on *Britannica* website (https://www.britannica.com/event/Beer-Hall-Putsch).
"French indemnity" on *Wikipedia* website (https://en.wikipedia.org/wiki/French_indemnity)
"World War I reparations" on *Wikipedia* website (https://en.wikipedia.org/wiki/World_War_I_reparations)
Borger, J. (2021) "Insurrection Day: when white supremacist terror came to the US Capitol" on *The Guardian* website (https://www.theguardian.com/us-news/2021/jan/09/us-capitol-insurrection-white-supremacist-terror)
Savage, C. (2021) "Was the Jan. 6 Attack on the Capitol an Act of 'Terrorism'?" on *The New York Times* website (https://www.nytimes.com/2022/01/07/us/politics/jan-6-terrorism-explainer.html?searchResultPosition=6)
"What were the Nuremberg Race Laws?" on *Holocaust Encyclopedia* website (https://encyclopedia.ushmm.org/content/en/article/the-nuremberg-race-laws)
"The Enabling Act of 23 March 1933" on *Deutscher Bundestag* website (https://www.bundestag.de/resource/blob/189778/d0f948962723d454c536d24d43965f87/enabling_act-data.pdf)

A Visit to the Basement

Ellis, E.L. and Delbrück, M. (1939) "The growth of bacteriophages". *Journal of General Physiology*, vol. 22, pp. 365–284.
Bachman, B. (1972) "Pedigrees of some mutant strains of *Escherichia coli* K-12". *Bacteriological Reviews*, vol. 36, pp. 525–557.
Summers, W.C. (1993) "How bacteriophage came to be used by the phage group". *Journal of History of Biology*, vol. 26, pp. 255–267.
Robin, A., Weiss, R.A. and Vogt, P.K. (2011) "100 years of Rous sarcoma virus". *Journal of Experimental Medicine*, vol. 208, pp. 2351–2355.
Ruiz, N. and Silhavy, T.J. (2022) "How *Escherichia coli* became the flagship bacterium of molecular biology". *Journal of Bacteriology*, vol. 204, pp. 1–9.
"Emory Ellis" on *Wikiwand* website (https://www.wikiwand.com/en/Emory_Ellis)
Andrewes, C. (1979) "Richard Edwin Shope". *Biographical Memoirs of the National Academy of Sciences* (https://nap.nationalacademies.org/read/573/chapter/14)
"How *E. coli* rose to prominence" on *Small Things Considered* website (https://schaechter.asmblog.org/schaechter/2021/11/how-ecoli-rose-to-prominence.html)
"How did *E. coli* get named K-12?" on *Small Things Considered* website (https://schaechter.asmblog.org/schaechter/2016/01/how-did-e-coli-get-named-k-12.html)
"*Escherichia coli* in molecular biology" on *Wikipedia* website (https://en.wikipedia.org/wiki/Escherichia_coli_in_molecular_biology)

Flying Solo

Fischer, E.P. and Lipson, C. (1988) "Thinking about science". *W. W. Norton & Company.*

Beevor, A. (2012) "The Second World War". *Weidenfeld & Nicholson* (2014 edition).

Krüger, A.P. and Northrop, J.H. (1930) "The kinetics of the bacterium-bacteriophage reaction". *The Journal of General Physiology*, vol. 14, pp. 223–254.

Northrop, J.H. (1937) "Concentration and purification of bacteriophage". *The Journal of General Physiology*, vol. 21, pp. 335–366.

Northrop, J.H. (1939) "Increase in bacteriophage and gelatinase concentration in cultures of *Bacillus megatherium*". *The Journal of General Physiology*, vol. 23, pp. 335–366.

Delbrück, M. (1940) "The growth of bacteriophage and lysis of the host". *The Journal of General Physiology*, vol. 23, pp. 643–660.

Yasuoka, S., Ohnishi, T., Kawano, S. et al. (1997) "Purification, characterization and localization of a novel trypsin-like protease found in the human airway". *American Journal of Respiratory Cell and Molecular Biology*, vol. 16, pp. 300–308.

Krüger, M. and Linke, W.A. (2011) "The giant protein titin: a regulatory node that integrates myocyte signaling pathways". *The Journal of Biological Chemistry*, vol. 286, pp. 9905–9912.

Caro, L.G. (1965) "The molecular weight of lambda DNA". *Virology*, vol. 25, pp. 226–236.

Yap, M.L. and Rossmann, M.G. (2014) "Structure and function of bacteriophage T4". *Future Microbiology*, vol. 9, pp. 1319–1327.

Payne, R.J. and Jansen, V.A. (2001) "Understanding bacteriophage therapy as a density-dependent kinetic process". *Journal of Theoretical Biology*, vol. 208, pp. 37–48.

Delbrück, M. (1940) "Adsorption of bacteriophage under various physiological conditions of the host". *The Journal of General Physiology*, vol. 23, pp. 631–642.

Delbrück, M. (1940) "Statistical fluctuations in autocatalytic reactions". *Journal of Chemical Physics*, vol. 8, pp. 120–124.

Pauling, L. and Delbrück, M. (1940) "The nature of the intermolecular forces operative in biological process". *Science*, vol. 92, pp. 77–79.

Hadas, H., Einav, M., Fishov, I. and Zaritsky, A. (1997) "Bacteriophage T4 development depends on the physiology of its host *Escherichia coli*". *Microbiology*, vol. 143. pp. 179–185.

Nabergoj, D., Modic, P. and Podgornik, A. (2017) "Effect of bacterial growth rate on bacteriophage population growth rate". *MicrobiologyOpen*, vol. 7. p. e558.

"Bryan, D., El-Shibiny, A. Hobbs, Z. et al. (2016) "Bacteriophage T4 infection of stationary phase *E. coli*: life after log from a phage perspective". *Frontiers in Microbiology*, vol. 7, p. 1391.

"List of countries by population in 1939" on *Wikipedia* website (https://en.wikipedia.org/wiki/List_of_countries_by_population_in_1939)

"Vichy France (Divided States of America)" on *Alternative History* website (https://althistory.fandom.com/wiki/Vichy_France_(Divided_States_of_America)

Quantum Biology

Marx, K. (1841) "Difference between the Democritean and Epicurean philosophy of nature", in *The first writing of Karl Marx* edited by Paul Schafer, *IG Publishing*, 2006.

Harding, C. (1979) "Interview with Max Delbrück". *California Institute of Technology Oral History Project.*

Phelps, R.H. (1963) "Before Hitler Came: Thule Society and Germanen Orden". *The Journal of Modern History*, vol. 35, pp. 245–261.

Beyler, R.H. (1996) "Targeting the organism". *Isis*, vol. 87, pp. 248–273.

Basran, J., Patel, S., Sutcliffe, M.J. and Scrutton, N.S. (2001) "Importance of barrier shape in enzyme-catalyzed reactions. Vibrationally assisted hydrogen tunneling in tryptophan tryptophylquinone-dependent amine dehydrogenases". *Journal of Biological Chemistry*, vol. 276, pp.6234–6242.

Schroer, B. (2003) "Pascual Jordan, his contribution to quantum mechanics and his legacy in contemporary local quantum physics". (https://doi.org/10.48550/arXiv.hep-th/0303241)

Masgrau, L., Roujeinikova, A., Johannissen, L.O. et al. (2006) "Atomic description of an enzyme reaction dominated by proton tunneling". *Science*, vol. 312, pp. 237–341.

Friedberg, E.C., Walker, G.C., Siede, W. et al. (2006) "DNA repair and mutagenesis". *ASM Press* (2nd edition).

Engel, G.S., Calhoun, T.R., Read, E.L. et al. (2007) "Evidence for wavelike energy transfer through quantum coherence in photosynthetic systems". *Nature*, vol. 446, pp. 752–786.

Cai, J. Guerreschi, G.G. and Briegel, H.J. (2010) "Quantum control and entanglement in a chemical compass". *Physical Review Letters*, vol. 104, p. 220502.

McFadden, J. and Al-Khalili, J. (2018) "The origins of quantum biology". *Proceedings of the Royal Society A*, vol. 474, p. 20180674.

Hutchinson, C.A. Chuang, R.Y., Noskov, V.N. et al. (2016) "Design and synthesis of a minimal bacterial genome". *Science*, vol. 351, p. 1414.

Pelletier, J.F., Sun, L., Wise, K.S. et al. (2021) "Genetic requirements for cell division in a genomically minimal cell". *Cell*, vol. 184, pp. 2430–2440.

"First self-replicating synthetic bacterial cell" on *JCVI* website (https://www.jcvi.org/research/first-self-replicating-synthetic-bacterial-cell)

"Göttingen Manifesto" on *Wikipedia* website (https://en.wikipedia.org/wiki/G%C3%B6ttingen_Manifesto)

NASHVILLE AND BLOOMINGTON

Homo Scientificus

Luria, S.E. (1984) "A slot machine, a broken test tube". *Harper & Row Publishers*.

Fischer, E.P. and Lipson, C. (1988) "Thinking about science". *W. W. Norton & Company*.

Segrè, G. (2011) "Ordinary geniuses". Penguins Books.

Beevor, A. (2012) "The Second World War". *Weidenfeld & Nicholson* (2014 edition).

Ireland, T. (2023) "The good virus". *Hodder Press*.

Ruska, H. (1940) "Visualization of bacteriopage lysis in the hypermicoscope". *Naturwissenschaften*, vol. 28, pp. 45–46 (translated by Hans Ackermann, reproduced in *Bacteriophage*, vol. 1, pp. 183–187).

Luria, S.E. and Anderson, T.F. (1942) "The identification and characterization of bacteriophages with the electron microscope". *Proceedings of the National Academy of Sciences of the United States of America*, vol. 28, pp. 127–131.

Luria, S.E., Delbrück, M. and Anderson, T.F. (1943) "Electron microscope studies of bacterial viruses". *Journal of Bacteriology*, vol. 46, pp. 57–77.

Anderson, T.F. (1944) "The role of tryptophane in the adsorption of two bacterial viruses on their host, *E. coli*". *Journal of Cellular and Comparative Physiology*, vol. 25, pp. 17–26.

Hawkes, P.W. (1990) "Obituaries: Ernst Ruska". *Physics Today*, vol. 43, pp. 84–85.

Perry, R.P. (2005) "Thomas Foxen Anderson". *Biographical Memoirs*, vol. 87, pp. 1–22.

Myelnikov, D. (2018) "An alternative cure: the adoption and survival of bacteriophage therapy in the USSR, 1922–1955". *Journal of the History of Medicine and Allied Sciences*, vol. 73, pp. 385–411.

Asmodelle, E. (2015) "The collaboration of Mileva Marić and Albert Einstein". *Asian Journal of Physics*, vol. 24. (https://www.researchgate.net/publication/274263348_The_Collaboration_of_Mileva_Maric_and_Albert_Einstein)

Esterson, A. and Cassidy, C.D. (2019) "Does Einstein's first wife deserve some credit for his work? That's the wrong question to ask" on *Time* website (https://time.com/5551098/mileva-einstein-history/)

"Our Legacy" on *RCA* website (https://www.rca.com/brand/rca-legacy)
"Vladimir Zworykin" on *Britannica* website (https://www.britannica.com/biography/Vladimir-Zworykin)
"The Rape of Nanking" on *History* website (https://www.history.com/this-day-in-history/the-rape-of-nanking)
"Japanese-American incarceration during World War II" on *National Archives* website (https://www.archives.gov/education/lessons/japanese-relocation)
"Japanese internment camps" on *History* website (https://www.history.com/topics/world-war-ii/japanese-american-relocation)
"Salvador E. Luria, M.D." on *American Association of Immunologists* website (https://www.aai.org/About/History/Notable-Members/Nobel-Laureates/SalvadorELuria#gsc.tab=0)
Egororv, B. (2021) "How a cholera epidemic was defeated in Stalingrad during World War II" on *Russia Beyond* website (https://www.rbth.com/history/334467-how-cholera-epidemic-was-defeated)
"120 years since the birth of Zinaida Ermolyeva" on *X7 Research* website (https://x7cpr.com/en/120-years-since-the-birth-of-zinaida-ermolyeva/)
"'Not a single cut leg!' The feat of Zinaida Yermolyeva" on *Top War* website (https://en.topwar.ru/158482-ni-odnoj-otrezannoj-nogi-podvig-zinaidy-ermolevoj.html)
"Ermoleva, Zinaida (1898–1974)" on *Encyclopedia* website (https://www.encyclopedia.com/women/encyclopedias-almanacs-transcripts-and-maps/ermoleva-zinaida-1898-1974)
"Zinaida Yermolyeva" on *Wikipedia* website (https://en.wikipedia.org/wiki/Zinaida_Yermolyeva)
"A history of bacteriophage production and therapeutic use in Russia" on *Science First Hand* website (https://scfh.ru/en/papers/phages-attack/)

The Mutants

Cairns, J., Stent, G.S. and Watson, J.D. (1966) "Phage and the origins of molecular biology". *Cold Spring Harbor Laboratory Press* (2007 edition).
Luria, S.E. (1984) "A slot machine, a broken test tube". *Harper & Row Publishers.*
Luria, S.E. and Delbrück, M. (1943) "Mutations of bacteria from virus sensitivity to virus resistance". *Genetics*, vol. 28, pp. 491–511.
Fischer, E.P. and Lipson, C. (1988) "Thinking about science". *W. W. Norton & Company.*
Lea, D.E. and Coulson, C.A. (1949) "The distribution of the numbers of mutants in bacterial populations". *Journal of Genetics*, vol. 49, pp. 264–285.
Jones, M.E., Thomas, S.M. and Rogers, A. (1994) "Luria-Delbrück fluctuation experiments: design and analysis". *Genetics*, vol. 136, pp. 1209–1216.
Zheng, Q. (2000) "Statistical and algorithmic methods for fluctuation analysis with SALVADOR as an implementation". *Mathematical Bioscience*, vol. 176, pp. 195–213.
Demerec, M. (1945) "Production of *Staphylococcus* strain resistant to various concentration of penicillin". *Proceedings of the National Academy of Sciences of the United States of America*, vol. 31, pp. 16–24.
Witkin, E.M. (1946) "Inherited differences in sensitivity to radiation in *Escherichia coli*". *Proceedings of the National Academy of Sciences of the United States of America*, vol. 32, pp. 59–68.
Homes, C., Ghafari, M., Abbas, A. et al. (2017) "Luria-Delbrück, revisited: the classic experiment does not rule out Lamarckian evolution". *Physical Biology*, vol. 14, p. 055004.
Bonhivers, M., Plançon, L. Ghazi, A. et al. (1998) "FhuA, an *Escherichia coli* outer membrane protein with a dual function of transporter and channel which mediates the transport of phage DNA". *Biochimie*, vol. 80, pp. 363–369.
Killmann, H., Videnov, G., Jung, G. et al. (1995) "Identification of receptor binding sites by competitive peptide mapping: phages T1, T5 and ϕ80 and colicin M bind to the gating loop of FhuA". *Journal of Bacteriology*, vol. 177, pp. 694–698.

Killmann, H., Benz, R. and Braun, V. (1996) "Properties of the FhuA channel in the *Escherichia coli* outer membrane after deletion of FhuA portions within and outside the predicted gating loop". *Journal of Bacteriology*, vol. 178, pp. 6913–6920.

Endriß, F., Braun, M., Killmann, H. and Braun, V. (2003) "Mutant analysis of the *Escherichia coli* FhuA protein reveals sites of FhuA activity". *Journal of Bacteriology*, vol. 185, pp. 4683–4692.

Di Noia, J.D. and Neuberger, M.S. (2002) "Molecular mechanisms of antibody somatic hypermutation". *Annual Reviews in Biochemistry*, vol. 76, pp. 1–22.

Chi, X., Li, Y. and Qin, X. (2020) "V(D)J recombination, somatic hypermutation and class switch recombination of immunoglobulins: mechanism and regulation". *Immunology*, vol. 160, pp. 233–247.

Ishino, Y. Shinagawa, H., Makino, K. et al. (1987) "Nucleotide sequence of the *lap* gene, responsible for alkaline phosphatase isozyme conversion in *Escherichia coli* and identification of the gene product". *Journal of Bacteriology*, vol. 169, pp. 5429–5433.

Horvath, P. and Barrangou, R. (2010) "CRISPR/Cas, the immune system of bacteria and archea". *Science*, vol. 327, pp. 167–170.

Fineran, P.C. (2019) "Resistance is not futile: bacterial 'innate' and CRISPR-Cas 'adaptive' immune system". *Microbiology*, vol. 165, pp. 834–841.

Ishino, Y., Krupovic, M. and Forterre, P. (2018) "History of CRSPR-Cas from encounter with a mysterious repeated sequence to genome editing technology". *Journal of Bacteriology*, vol. 200, p. e00580-17.

Newcombe, H.B. (1949) "Origin of bacterial variants". *Nature*, vol. 164, pp. 150–151.

Mutual Exclusion

Cairns, J., Stent, G.S. and Watson, J.D. (1966) "Phage and the origins of molecular biology". *Cold Spring Harbor Laboratory Press* (2007 edition).

Hershey, A.D., Kalmanson, G. and Bronfenbrenner, J. (1943) "Quantitative methods in the study of the phage-antiphage reaction". *The Journal of Immunology*, vol. 46, pp. 267–279.

Hershey, A.D., Kalmanson, G. and Bronfenbrenner, J. (1943) "Quantitative relationships in the phage-antiphage reaction: unity and homogeneity of the reactants". *The Journal of Immunology*, vol. 46, pp. 281–299.

Delbrück, M. and Luria, S.E. (1945) "Interference between bacterial viruses I: interference between two bacterial viruses acting upon the same host and the mechanism of virus growth". *Journal of Bacteriology*, vol. 50, pp. 111–141.

Delbrück, M. (1945) "Interference between bacterial viruses III: the mutual exclusion effect and the depressor effect". *Journal of Bacteriology*, vol. 50, pp. 151–70.

Delbrück, M. (1945) "The burst size distribution in the growth of bacterial viruses (bacteriophages)". *Journal of Bacteriology*, vol. 50, pp. 131–135.

Delbrück, M. (1945) "Effects of specific antisera on the growth of bacterial viruses (bacteriophages)". *Journal of Bacteriology*, vol. 50, pp. 137–150.

Luria, S.E. (1945) "Mutations of bacterial viruses affecting their host range". *Genetics*, vol. 30, pp. 84–99.

Delbrück, M. (1948) "Biochemical mutants of bacterial viruses". *Journal of Bacteriology*, vol. 56, pp. 1–16.

Weigle, J.J. and Delbrück, M. (1951) "Mutual exclusion between an infecting phage and a carried phage". *Journal of Bacteriology*, vol. 62, pp. 301–318.

McKinney, H.H. (1929) "Mosaic diseases in the Canary Islands, west Africa and Gibraltar". *Journal of Agricultural Research*, vol. 39, pp. 557–578.

Ziegler, J.E., Lavin, G.I. and Horsfall, F.L. (1944) "Interference between the influenza viruses: II". *Journal of Experimental Medicine*, vol. 79, pp. 379–400.

Snyder, L. (1995) "Phage-exclusion enzymes: a bonanza of biochemical and cell biology reagents?". *Molecular Microbiology*, vol. 15, pp. 415–420.

Sun, X., Göhler, A., Heller, K.J. and Neve, H. (2006) "The *ltp* gene of temperate *Streptococcus thermophilus* phage TP-J34 confers superinfection exclusion to *Streptococcus thermophilus* and *Lactococcus lactis*". *Virology*, vol. 350, pp. 146–157.

Abedon, S.T. (2015) "Bacteriophage secondary infection". *Virologica Sinica*, vol. 30, pp. 3–10.

Dulbecco, R. (1952) "Mutual exclusion between related phages". *Journal of Bacteriology*, vol. 63, pp. 209–217.

Uetake, H., Luria, S.E. and Burrous, J.W. (1958) "Conversion of somatic antigens in *Salmonella* by phage infection leading to lysis or lysogeny". *Virology*, vol. 5, pp. 68–91.

Uchida, T., Robbins, P.W. and Luria, S.E. (1963) "Analysis of the serologic determinats groups of the *Salmonella* E-group O-antigens". *Biochemistry*, vol. 2, pp. 663–668.

Fielding, P.E. and Lunt, M.R. (1970) "The relation between breakdown of superinfecting virus doexyribonucleic acid and temporal exclusion by T4 and T5 bacteriophages". *Journal of General Virology*, vol. 6. pp. 333–342.

Casjens, S. (2003) "Prophages and bacterial genomics: what have we learned so far?". *Molecular Microbiology*, vol. 49, pp. 277–300.

Blasdel, B.G. and Abedon, S.T. (2013) "Superinfection immunity", in *Brenner's encyclopedia of genetics* edited by Maloy, S. and Hughes, K. *Academic Press* (2nd edition).

Bondy-Demony, J., Qian, J., Westra, E.R. et al. (2016) "Prophages mediate defense against phage infection through diverse mechanisms". *The ISME Journal*, vol. 10, pp. 2854–2866.

Hershey Heaven

Cairns, J., Stent, G.S. and Watson, J.D. (1966) "Phage and the origins of molecular biology". *Cold Spring Harbor Laboratory Press* (2007 edition).

Stahl, F.W. (2000) "We can sleep later: Alfred D. Hershey and the origins of molecular biology". *Cold Spring Harbor Laboratory Press.*

Beevor, A. (2012) "The Second World War". *Weidenfeld & Nicholson* (2014 edition).

Pua, D., Dybbro, D. and Rogers, A. (2018) "Unit 731: the forgotten Auschwitz". *Pacific Atrocities Education* (2nd edition).

Gold, H. (2019) "Japan's infamous unit 731: firsthand accounts of Japan's wartime human experimentation program". *Tuttle Publishing.*

Stahl, F.W. (1998) "Hershey". *Genetics*, vol. 149, pp. 1–6.

Watson, J. (1998) "The lives they lived: Alfred D. Hershey; Hershey Heaven". *The New York Times*, No. 51,027 (4/1/1998).

"Alfred D. Hershey, biographical" on the *Nobel Prize Foundation* website (https://www.nobelprize.org/prizes/medicine/1969/hershey/biographical/)

"Alfred Hershey" on *Wikipedia* website (https://en.wikipedia.org/wiki/Alfred_Hershey)

"Alfred D. Hershey, PhD" on *The American Association of Immunologists* website (https://www.aai.org/About/History/Notable-Members/Nobel-Laureates/AlfredDHershey)

"Jacques Jacob Bronfenbrenner (1883–1953)" on *Washington University School of Medicine* website (https://beckerexhibits.wustl.edu/mig/bios/bronfenbrenner.html)

"Le Quattro giornate di Napoli" on *Municipality of Naples* website (https://www.comune.napoli.it/flex/cm/pages/ServeBLOB.php/L/IT/IDPagina/5730)

"The Effects of Atomic Bombs on Hiroshima and Nagasaki" on *Atomic Archive* website (https://www.atomicarchive.com/resources/documents/bombing-survey/section_III.html)

"Bombings of Hiroshima and Nagasaki – 1945" on *Atomic Heritage Foundation* website (https://ahf.nuclearmuseum.org/ahf/history/bombings-hiroshima-and-nagasaki-1945/)

"Underwater Aircraft Carriers: Japan's Secret Weapon" on *Youtube* website (https://www.youtube.com/watch?v=gxyk84t4Q8w&t=603s)

Cairns, J., Stent, G.S. and Watson, J.D. (1966) "Phage and the origins of molecular biology". *Cold Spring Harbor Laboratory Press* (2007 edition).
Latarjet, R. (1948) "Intracellular growth of bacteriophage studied by Roentgen irradiation". *The Journal of General Physiology*, vol. 31, pp. 529–544.
Luria, S.E. and Latarjet, R. (1964) "Ultraviolet irradiation of bacteriophage during intracellular growth". *Proceedings of the National Academy of Sciences of the United States of America,* vol. 52, pp. 1521–1529.
Fisher, D.B. and Pardee, A.B. (1968) "The Luria-Latarjet effect studied by T4-lysozyme production". *Virology*, vol. 34, pp. 91–96.
Hyman, P. (1993) "The genetics of the Luria-Latarjet effect in bacteriophage T4: evidence for the involvement of multiple DNA repair pathways". *Genetics Research*, vol. 62, pp. 1–9.
Koprowski, H., Levy, S.B. and Benjamin, T. (1999) "Raymond Latarjet 1911–1998". *Cancer Research*, vol. 59, p. 2256.

The Pleasures of Cold Spring Harbor

Cairns, J., Stent, G.S. and Watson, J.D. (1966) "Phage and the origins of molecular biology". *Cold Spring Harbor Laboratory Press* (2007 edition).
Summers, W.C. (2022) "The American Phage Group, founders of molecular biology". *Yale University Press.*
Susman, M. (1995) "The Cold Spring Harbor Phage Course (1945–1970): a 50th anniversary remembrance". *Genetics*, vol. 139, pp. 1101–1106.
Kellenberger, E. (1995) "History of phage research as viewed by a European". *FEMS Microbiology Reviews*, vol. 17, pp. 7–24.
"Coming of phage: celebrating the fiftieth anniversary of the first phage course" on *Cold Spring Harbor Laboratory* pamphlet (https://www.virology.ws/wp-content/uploads/2012/08/phage_brochure_lowres.pdf)
"Phage Information Service, nos. 1–7, 9–14 1947–1959 [3 folders]" on *The Caltech Archive* website (https://collections.archives.caltech.edu/repositories/2/archival_objects/19860)

NEW YORK

The Musketeers of the Bacteriophages

Cairns, J., Stent, G.S. and Watson, J.D. (1966) "Phage and the origins of molecular biology". *Cold Spring Harbor Laboratory Press* (2007 edition).
Gaddis, J.L. (2005) "The Cold War". *Penguin Books* (2007 edition).
Summers, W.C. (2005) "History of phage research and phage therapy", in *Phage: their role in bacterial pathogenesis and biotechnology* edited by Waldor, M.K., Friedman, D.I. and Adhya, S.L. *ASM Press.*
Baggott, J. (2009) "Atomic". *Icon Books Ltd* (2015 edition).
Beevor, A. (2012) "The Second World War". *Weidenfeld & Nicholson* (2014 edition).
Summers, W.C. (2022) "The American Phage Group, founders of molecular biology". *Yale University Press.*
Nishihara, M., Friedman, N. and Vasken Aposhian, H. (1969) "Biological activity of 5-hydroxymethyluracil and its deoxynucleoside in noninfected and phage-infected *Bacillus subtilis*". *Journal of Virology*, vol. 3, pp. 164–170.
Mullins, N.C. (1972) "The development of a scientific specialty: the Phage Group and the origins of molecular biology". *Minerva*, vol. 10, pp. 51–82.

Kupchan, C.A. (2023) "We can't deny Kissinger's effectiveness". *The New York Time*, No. 60,009 (20/12/2023).
Hennigan, W.J. (2024) "The human toll of atomic testing" on *The New York Times* website (https://www.nytimes.com/interactive/2024/06/20/opinion/nuclear-weapons-testing.html?smid=nytcore-android-share)
"Alfred D. Hershey Nobel Lecture" on *Nobel Prize* website (https://www.nobelprize.org/prizes/medicine/1969/hershey/lecture/)
"Alfred D. Hershey Banquet speech" on *Nobel Prize* website (https://www.nobelprize.org/prizes/medicine/1969/hershey/speech/)
"Salvador E. Luria Nobel Lecture" on *Nobel Prize* website (https://www.nobelprize.org/prizes/medicine/1969/luria/lecture/)
"Max Delbrück Nobel Lecture" on *Nobel Prize* website (https://www.nobelprize.org/prizes/medicine/1969/delbruck/lecture/)
"Phage group" on *Wikipedia* website (https://en.wikipedia.org/wiki/Phage_group)
"Berlin Blockade" on *History* website (https://www.history.com/topics/cold-war/berlin-blockade)
"Fact sheet: thermonuclear weapons" on *Center for Arms Control and Non-Proliferation* website (https://armscontrolcenter.org/fact-sheet-thermonuclear-weapons/)
"Hydrogen bomb" on the *US Department of Energy* website (https://www.osti.gov/opennet/manhattan-project-history/Science/BombDesign/hydrogen-bomb.html)
"Remembering Castle Bravo 69 years later" on *Nuclear Age Peace Foundation* website (https://www.wagingpeace.org/remembering-castle-bravo-69-years-later/)
"Castle Bravo: the largest U.S. nuclear explosion" on *Brookings* website (https://www.brookings.edu/articles/castle-bravo-the-largest-u-s-nuclear-explosion/)
"Which countries have nuclear weapons?" on the *International Campaign to Abolish Nuclear Weapons* website (https://www.icanw.org/nuclear_arsenals)
"Ground zero" on *Carlos Lab* website on (https://www.carloslabs.com/projects/200712b/groundzerohtml)
"B28 nuclear bomb" on *Wikipedia* website (https://en.wikipedia.org/wiki/B28_nuclear_bomb)

The Revenants

Luria, S.E. (1947) "Reactivation of irradiated bacteriophages by transfer of self-reproducing units". *Proceedings of the National Academy of Sciences of the United States of America*, vol. 33, pp. 253–264.
Rubenstein, I., Thomas, C.A. and Hershey, A.D. (1961) "The molecular weights of T2 bacteriophage DNA and its first and second breakage products". *Proceedings of the National Academy of Sciences of the United States of America*, vol. 47, pp. 1113–1122.
Davison, P.F., Freidelder, D., Hede, R. and Levinthal, C. (1961) "The structural unity of the DNA of T2 bacteriophage". *Proceedings of the National Academy of Sciences of the United States of America*, vol. 47, pp. 1123–1129.
Dörmann, A.H. (1952) "The intracellular growth of bacteriophages I: liberation of intracellular bacteriophage T4 by premature lysis with another phage or with cyanide". *Journal of General Physiology*, vol. 35, pp. 645–656.
Sinsheimer, R.L. (1959) "Purification and properties of bacteriophage φX174". *Journal of Molecular Biology*, vol. 1, pp. 37–42.
Sinsheimer, R.L. (1959) "A single-stranded deoxyribonucleic acid from bacteriophage φX174". *Journal of Molecular Biology*, vol. 1, pp. 43–53.
"Who discovered that phage phiX174 has a single-stranded genome?" on *Stack Exchange Biology* website (https://biology.stackexchange.com/questions/114114/who-discovered-that-phage-phix174-has-a-single-stranded-genome/114122#114122)

Witkin, E.M. (1946) "Inherited differences in sensitivity to radiation in *Escherichia coli*". *Proceedings of the National Academy of Sciences of the United States of America*, vol. 32, pp. 59–68.

Hershey, A.D. and Rotman, R. (1949) "Genetic recombination between host-range and plaque-type mutants of bacteriophage in single bacterial cells". *Genetics*, vol. 34, pp. 44–71.

Lederberg, J. and Tatum, E.L. (1946) "Gene recombination in *Escherichia coli*". *Nature*, vol. 158, p. 558.

Tatum, E.L and Lederberg, J. (1947) "Gene recombination in the bacterium *Escherichia coli*". *Journal of Bacteriology*, vol. 53, pp. 673–684.

"The development of bacterial genetics" on *National Library of Medicine* website (https://profiles.nlm.nih.gov/spotlight/bb/feature/bacgen1)

Linkage Disequilibrium

Mandel, J.D. and Hershey, A.D. (1960) "A fractionating column for analysis of nucleic acids". *Analytical Biochemistry*, vol. 1, pp. 66–77.

Cairns, J. (1961) "An estimate of the length of the DNA molecule of T2 bacteriophage by autoradiography". *Journal of Molecular Biology*, vol. 3, pp. 756–761.

Burgi, E. and Hershey, A.D. (1963) "Sedimentation rate as a measure of molecular weight of DNA". *Biophysical Journal*, vol. 3, pp. 309–321.

Visconti, N. and Delbrück, M. (1958) "The mechanism of genetic recombination in phage". *Genetics*, vol. 38, pp. 5–33.

Stahl, F.W., Edgar, R.S. and Steinberg, J. (1964) "The linkage map of bacteriophage T4". *Genetics*, vol. 50, pp. 539–552.

Hershey, A.D. and Rotman, R. (1948) "Linkage among genes controlling inhibition of lysis in a bacterial virus". *Proceedings of the National Academy of Sciences of the United States of America*, vol. 34, pp. 89–96.

Stahl, F.W. and Steinberg, C.M. (1964) "The theory of formal phage genetics for circular maps". *Genetics*, vol. 50, pp. 531–538.

Streisinger, G. and Bruce, V. (1960) "Linkage of genetic markers in phages T2 and T4". *Genetics*, vol. 45, pp. 1289–1296.

Streisinger, G., Edgar, R.S. and Denhardt, G.H. (1964) "Chromosome structure in phage T4, I: circularity of the linkage map". *Proceedings of the National Academy of Sciences of the United States of America*, vol. 51, pp. 775–779.

"Punnett square calculator" on *Science Prime* website (https://scienceprimer.com/punnett-square-calculator)

"Centimorgan" on *National Human Genome Research Institute* website (https://www.genome.gov/genetics-glossary/Centimorgan)

Breaking Glass

Evans, A.C. (1934) "The prevalence of *Streptococcus* bacteriophages". *Science*, vol. 80, pp. 40–41.

Luria, S.E. and Human, M.L. (1952) "A nonhereditary, host-induced variation of bacterial viruses". *Journal of Bacteriology*, vol. 64, pp. 557–569.

Bertani, G. and Weigle, J.J. (1953) "Host controlled variation in bacterial viruses". *Journal of Bacteriology*, vol. 65, pp. 113–121.

Arber, W. and Dussoix, D. (1962) "Host specificity of DNA produced by *Escherichia coli* I: host controlled modification of bacteriophage λ". *Journal of Molecular Biology*, vol. 5, pp. 18–36.

Dussoix, D. and Arber, W. (1962) "Host specificity of DNA produced by *Escherichia coli* II: control over acceptance of DNA from infecting phage λ". *Journal of Molecular Biology*, vol. 5, pp. 37–49.

Lederberg, S. (1964) "Degradation of non-replicating bacteriophage DNA in non-accepting cells". *Journal of Molecular Biology*, vol. 8, pp. 623–528.

Meselson, M. and Yuan, R. (1968) "DNA restriction enzyme from *E. coli*". *Nature*, vol. 217, pp. 1110–1114.

Arber, W. and Linn, S. (1969) "DNA modification and restriction". *Annual Reviews in Biochemistry*, vol. 38, pp. 467–500.

Smith, H.O. and Wilcox, K.W. (1970) "A restriction enzyme from *Hemophilus influenzae* I: purification and general properties". *Journal of Molecular Biology*, vol. 51, pp. 379–391.

Kelly, T.J. Jr. and Smith, H.O. (1970) "A restriction enzyme from *Hemophilus influenzae* II: base sequence of the recognition site". *Journal of Molecular Biology*, vol. 51, pp. 393–409.

Danna, K. and Nathans, D. (1971) "Specific cleavage of Simian Virus 40 DNA by restriction endonuclease of *Hemophilus influenzae*". *Proceedings of the National Academy of Sciences of the United States of America*, vol. 68, pp. 2913–2917.

Danna, K., Sack, G.H. Jr. and Nathans, D. (1973) "Studies of Simian Virus 40 DNA VII: a cleavage map of the SV40 genome". *Journal of Molecular Biology*, vol. 78, pp. 363–376.

Murray, N.E. (2000) "Type I restriction systems: sophisticated molecular machines (a legacy of Bertani and Weigle)". *Microbiology and Molecular Biology Reviews*, vol. 64, pp. 412–434.

Lönen, W.A.M. (2003) "Tracking EcoKI and DNA fifty years on: a golden story full of surprises". I, vol. 31, pp. 7059–7069.

Bertani, G. (2004) "Lysogeny at mid-twentieth century: P1, P2 and other experimental systems". *Journal of Bacteriology*, vol. 186, pp. 595–600.

Roberts, R.J. (2005) "How restriction enzymes became the workhorses of molecular biology". *Proceedings of the National Academy of Sciences of the United States of America*, vol. 102, pp. 5905–5908.

Loenen, W.A.M., Dryden, D.T.F., Raleigh, E.A. et al. (2013) "Highlights of the DNA cutters: a short history of the restriction enzymes". *Nucleic Acids Research*, vol. 42, pp. 3–19.

Lindqvist, B.H., Haggar-Ljungqvist, E. and Calendar, R. (2015) "Giuseppe Bertani (1923–2015)". *Bacteriophage*, vol. 5, p. e1054060.

Calendar, R., Haggar-Ljungqvist, E., Lindqvist, B.H. and Finkel, S.E. (2019) "Discovery of a barrier to infection and host-controlled variation: 1952 1953", in *Restriction enzymes. a history*, edited by Wil A.M. Loenen. *Cold Spring Harbor Laboratory Press*.

"Mapping of restriction sites on plasmid DNA" on *EDVOTEK* website (https://www.edvotek.com/105.100202.pdf)

"Maxam–Gilbert sequencing: what it is and 3 modern applications" on *BiteSizeBio* website (https://bitesizebio.com/36696/maxam-gilbert-sequencing/)

The Blender Experiment

Paul, J.R. (1927) "A comparative study of smooth and rough pneumococcus colonies". *Journal of Experimental Medicine*, vol. 46, pp. 793–805.

Griffith, F. (1928) "The significance of pneumococcal types". *The Journal of Hygiene*, vol. 27, pp. 113–159.

Avery, O.T., McLeod, C.M. and McCarty, M. (1944) "Studies on the chemical nature of the substance inducing transformation of pneumococcal types". *Journal of Experimental Medicine*, vol. 79, pp. 137–158.

Hotchkiss, R.D. (1948) "The quantitative separation of purines, pyrimidines and nucleosides by paper chromatography". *Journal of Biological Chemistry*, vol. 175, pp. 315–332.

Hershey, A.D. and Chase, M. (1952) "Independent functions of viral protein and nucleic acid in growth of bacteriophage". *The Journal of General Physiology*, vol. 36, pp. 39–56.

Hernandez, V. (2019) "The Hershey-Chase experiments (1952), by Alfred Hershey and Martha Chase" on *Arizona State University* website (https://embryo.asu.edu/pages/hershey-chase-experiments-1952-alfred-hershey-and-martha-chase)

Hotchkiss, R.D. and Marmur, J. (1954) "Double marker transformations as evidence of linked factors in desoxyribonucleate transforming agents". *Proceedings of the National Academy of Sciences of the United States of America*, vol. 40, pp. 55–60.

Hotchkiss, R.D. (1965) "Oswald T. Avery: 1877–1955". *Genetics*, vol. 51, pp. 1–10.

Ernesto Garcia, E., Garcia, P. and Lopez, R. (1993) "Cloning and sequencing of a gene involved in the synthesis of the capsular polysaccharide of *Streptococcus pneumoniae* type 3". *Molecular and General Genetics*, vol. 239, pp. 188–195.

Arrecubieta, C., Lopez, R. and Garcia, E. (1996) "Type 3-specific synthase of *Streptococcus pneumoniae* (Cap3B) directs type 3 polysaccharide biosynthesis in *Escherichia coli* and in pneumococcal strains of different serotypes". *Journal of Experimental Medicine*, vol. 184, pp. 449–455.

Witkin, E.M. (2005) "Remembering Rollin Hotchkiss (1911–2004)". *Genetics*, vol. 170, pp. 1443–1447.

Cartee, R., Forsee, T.W. and Yother, J. (2005) "Initiation and synthesis of the *Streptococcus pneumoniae* type 3 capsule on a phosphatidylglycerol membrane anchor". *Journal of Bacteriology*, vol. 187, pp. 4470–4479.

Johnston, C., Martin, B., Fichant, G. et al. (2014) "Bacterial transformation: distribution, shared mechanisms and divergent control". *Nature Reviews in Microbiology*, vol. 12, pp. 181–196.

Paton, J.C. and Trappetti, C. (2018) "*Streptococcus pneumoniae* capsular polysaccharide". *Microbiology Spectrum*, vol. 7, p. GPP3-0019-2018.

"Sulfur-35" on *Princeton University* website (https://ehs.princeton.edu/book/export/html/256)

"Phosphorus-33" on *Princeton University* website (https://ehs.princeton.edu/laboratory-research/radiation-safety/radioactive-materials/radioisotope-fact-sheets/phosphorus-33)

"Photon Unit Calculator" on *KM Labs* website (https://www.kmlabs.com/en/wavelength-to-photon-energy-calculator)

Sudden Disappearance

Hershey, A.D. (1946) "Mutation of bacteriophage with respect to type of plaque". *Genetics*, vol. 1, pp. 620–640.

Lwoff, A. (1953) "Lysogeny". *Bacteriological Reviews*, vol. 17, pp. 269–337.

Incadorna, N. and Selvidge, L. (1973) "Mechanism of adsorption and eclipse of bacteriophage $\phi\chi$174". *Journal of Virology*, vol. 11, pp. 775–782.

Barksdale, L. and Arden, S.B. (1974) "Persisting bacteriophage infections, lysogeny and phage conversions". *Annual Reviews in Microbiology*, vol. 28, pp. 265–300.

Marco, R., Jazwinski, S.M. and Kornberg, A. (1974) "Binding, eclipse and penetration of the filamentous bacteriophage M13 in intact and disrupted cells". *Virology*, vol. 62, pp. 209–223.

Ripp, S. and Miller, R.V. (1997) "The role of pseudolysogeny in bacteriophage-host interactions in natural freshwater environment". *Microbiology*, vol. 143, pp. 2065–2070.

Rabinovitch, A., Fishov, I., Hadas, H. et al. (2002) "Bacteriophage T4 development in *Escherichia coli* is growth rate dependent". *Journal of Theoretical Biology*, vol. 216, pp. 1–4.

Hay, A.M., Howie, H.L. Gorham, J.D. et al. (2021) "Mouse background genetics in biomedical research: the devil's in the details". *Transfusion*, vol. 61, pp. 3017–3025.

McNeill, L. (2018) "The history of breeding mice for science begins with a woman in a barn" on *Smithsonian Magazine* website (https://www.smithsonianmag.com/science-nature/history-breeding-mice-science-leads-back-woman-barn-180968441/)

λ

Calef, E. and Ricciardello, G. (1960) "Recombination experiments on prophage host relationships". *Virology*, vol. 12, pp. 81–103.

Romig, W. and Brodetsky, A.M. (1961) "Isolation and preliminary characterization of bacteriophages for *Bacillus subtilis*". *Journal of Bacteriology*, vol. 82, pp. 135–141.

Hershey, A.D., Burgi, E. and Ingraham, L. (1963) "Cohesion of DNA molecules isolated from phage lambda". *Proceedings of the National Academy of Sciences of the United States of America*, vol. 49, pp. 748–755.

MacHattie, L.A. and Thomas, C.A., Jr. (1964) "DNA from bacteriophage lambda: molecular length and conformation". *Science*, vol. 44, pp. 1142–1144.

Kaiser, A.D. and Inman, R.B. (1965) "Cohesion and the biological activity of bacteriophage lambda DNA". *Journal of Molecular Biology*, vol. 13, pp. 78–91.

Strack, H.B. and Kaiser, A.D. (1965) "On the structure of the ends of lambda DNA". *Journal of Molecular Biology*, vol. 12, pp. 32–49.

Franklin, N.C., Dove, W.F. and Yanofsky, C. (1965) "The linear insertion of a prophage into the chromosome of *E. coli* shown by deletion mapping". *Biochemical and Biophysical Research Communications*, vol. 18, pp. 910–923.

Wu, R. and Kaiser, A.D. (1968) "Structure and base sequence in the cohesive ends of bacteriophage lambda DNA". *Journal of Molecular Biology*, vol. 35, pp. 523–537.

Marini, J.C. and Landy, A. (1977) "The isolation of restriction fragments containing the *att* site of bacteriophage λ". *Virology*, vol. 76, pp. 196–209.

Landy, A. and Ross, W. (1977) "Viral integration and excision: structure of the lambda *att* sites". *Science*, vol. 197, pp. 1147–1159.

Mano, Y., Sakai, H. and Komano, T. (1979) "Growth and DNA synthesis of bacteriophage ϕχ174 in a *dnaP* mutant of *Escherichia coli*". *Journal of Virology*, vol. 30, pp. 650–656.

Campbell, A.M. (1993) "Thirty years ago in genetics: prophage insertion into bacterial chromosomes". *Genetics*, vol. 133, pp. 433–438.

Campbell, A. (2007) "Phage integration and chromosome structure: a personal history". *Annual Reviews in Genetics*, vol. 41, pp. 1–11.

Mäntynen, S., Laanto, E., Oksanen, H.M. et al. (2021) "Black box of phage-bacterium interactions: exploring alternative phage infection strategies". *Open Biology*, vol. 11, p. 210188.

Cortes, M.G., Lin, Y., and Zeng, L. et al. (2021) "From bench to keyboard and back again: a brief history of lambda phage modeling". *Annual Reviews in Biophysics*, vol. 50, pp. 117–134.

Feiss, M., Cleary, P., Court, D. and Adhya, S. (2018) "Allan McCulloch Campbell" on the *National Academy of Sciences* website (https://www.nasonline.org/wp-content/uploads/2024/06/campbell-allan.pdf)

"Who demonstrated that phage lamba is integrated into the host chromosome?" on *Stack Exchange Biology* blog (https://biology.stackexchange.com/questions/114329/who-demonstrated-that-phage-lamba-is-integrated-into-the-host-chromosome?noredirect=1#comment211109_114329)

"How was Campbell's Model of phage lambda integration into *E. coli* definitively proven?" on *Stack Exchange Biology* blog (https://biology.stackexchange.com/questions/114827/how-was-campbells-model-of-phage-lambda-integration-into-e-coli-definitively-p)

WEST COAST

The Bacterial Sexual Revolution

De Kruif, P. (1926) "Microbe hunters". *Harcourt Inc.* (1996 edition).

Lederberg, J. and Tatum, E.L. (1946) "Gene recombination in *Escherichia coli*". *Nature*, vol. 158, p. 558.

Lederberg, J. (1947) "Gene recombination and linked segregations in *Escherichia coli*". *Genetics*, vol. 32, pp. 505–525.

Morse, M.L., Lederberg, E.M. and Lederberg, J. (1956) "Transductional heterogenotes in *Escherichia coli*". *Genetics*, vol. 41, pp. 758–778.

Low, K.B. (1972) "*Escherichia coli* K-12 F-prime factors, old and new". *Bacteriology Reviews*, vol. 36, pp. 587–607.

Lederberg, J. (1979) "Edward Lawrie Tatum". *Annual Reviews in Genetics*, vol. 13, pp. 1–5.

Lederberg, J. (1986) "Forty years of genetic recombination in bacteria". *Nature*, vol. 324, pp. 627–628.

Zuckermann, H. and Lederberg, J. (1986) "Forty years of genetic recombination in bacteria: postmature scientific discovery?". *Nature*, vol. 324, pp. 629–631.

Lederberg, J. (1987) "Genetic recombination in bacteria: a discovery account". *Annual Reviews in Genetics*, vol. 21, pp. 23–46.

Lederberg, J. (1996) "Genetic recombination in *Escherichia coli*: disputation at Cold Spring Harbor, 1946–1996". *Genetics*, vol. 144, pp. 439–443.

Lederberg, J. (1998) "Plasmid (1952–1997)". Plasmid, vol. 39, pp. 1–9.

Jayaraman, R. (2011) "William Hayes and his Pallanza bomb shell". *Resonance*, vol. 16, pp. 911–921.

Helinski, D.R. (2022) "A brief history of plasmids". *EcoSalPlus*, vol. 10, p. 10.1128.

Sapp, J. (2021) "Genes, germs and medicine: the life of Joshua Lederberg". *World Scientific Publishing*.

Birge, E.A. (1984) "Bacterial and bacteriophage genetics". *Springer* (2006 edition).

Witkin, E.M. (2002) "Chances and choices: Cold Spring Harbor 1944–1955". *Annual Reviews in Microbiology*, vol. 56, pp. 1–15.

"Transduction, plasmids and the foundation of biotechnology" on the *National Library of Medicine* website (https://profiles.nlm.nih.gov/spotlight/bb/feature/bacgen2)

Davis, B.D. (1950) "Nonfiltrability of the agents of genetic recombination in *Escherichia coli*". *Journal of Bacteriology*, vol. 60, pp. 507–508.

Perry, R.P. (2005) "Thomas Foxen Anderson". *Biographical Memoirs*, vol. 87, pp. 1–22.

Roth, T.F. and Helinski, D.R. (1967) "Evidence for circular DNA forms of a bacterial plasmid". *Proceedings of the National Academy of Sciences of the United States of America*, vol. 58, pp. 650–657.

Palchaudhuri, S., Bell, E. and Salton, M.R.J. (1975) "Electron microscopy of plasmid deoxyribonucleic acid from *Neisseria gonorrhoeae*". *Infection and Immunity*, vol. 11, pp. 1141–1114.

Cavalli, L.L. and Heslot, H. (1949) "Recombination in bacteria: outcrossing *Escherichia coli* K 12". Nature, vol. 164, pp. 1057–1058.

Lederberg, E.M. and Lederberg, J. (1953) "Genetic studies of lysogenicity in *Escherichia coli*". *Genetics*, vol. 38, pp. 51–64.

Galagan, J.E., Calvo, S.E., Borkovich, K.A. et al. (2003) "The genome sequence of the filamentous fungus *Neurospora crassa*". *Nature*, vol. 422, pp. 859–868.

Raju, N.B. (2009) "*Neurospora* as a model fungus for studies in cytogenetics and sexual biology at Stanford". *Journal of Bioscience*, vol. 34, pp. 139–159.

Lederberg, J. and Lederberg, E.M. (1951) "Replica plating and indirect selection of bacterial mutants". *Journal of Bacteriology*, vol. 63, pp. 399–406.

Ryan, F.J. (1953) "Evolution observed". *Scientific American*, vol. 189, pp. 78–83.

Weigle, J.J. and Delbrück, M. (1951) "Mutual exclusion between an infecting phage and a carried phage". *Journal of Bacteriology*, vol. 62, pp. 301–318.

"The Nobel Prize in Physiology or Medicine 1958" on *The Nobel Prize* website (https://www.nobelprize.org/prizes/medicine/1958/summary/)

Maddox, B. (2003) "The double helix and the 'wronged heroine'". *Nature*, vol. 421, pp. 407–408.

Serendipity, Again

Zinder, N.D. and Lederberg, J. (1952) "Genetic exchange in *Salmonella*". *Journal of Bacteriology*, vol. 64, pp. 679–699.

Lederberg, E.M. and Lederberg, J. (1953) "Genetic studies of lysogenicity in *Escherichia coli*". *Genetics*, vol. 38, pp. 51–64.

Morse, M.L., Lederberg, E.M. and Lederberg, J. (1956) "Transduction in *Escherichia coli* K-12". *Genetics*, vol. 41, pp. 142–156.

Morse, M.L., Lederberg, E.M. and Lederberg, J. (1956) "Transductional heterogenotes in *Escherichia coli*". *Genetics*, vol. 41, pp. 758–779.

Zinder, N.D. (1958) "'Transduction' in bacteria". *Scientific American*, vol. 199, pp. 38–43.

Zinder, N.D. (1992) "Forty years ago: the discovery of bacterial transduction". *Genetics*, vol. 132, pp. 291–294.

Gottesman, M.E. and Weisberg, R.A. (2004) "Little lambda, who made thee?". *Microbiology and Molecular Biology Reviews*, vol. 68, pp. 796.

Parkinson, J.S. (2016) "Classic spotlight: the discovery of bacterial transduction". *Journal of Bacteriology*, vol. 198, pp. 2899–2900.

Wang, C., Tu, J., Liu, J. and Molineux, I.J. (2019) "Structural dynamics of bacteriophage P22 infection initiation revealed by cryo-electron tomography". *Nature Microbiology*, vol. 4, pp. 1049–1056.

Kresge, N., Simoni, R.D. and Hill, R.L. (2011) "The molecular genetics of bacteriophages: the work of Norton Zinder". *The Journal of Biological Chemistry*, vol. 286, pp. e4–e5.

Susskind, M.M., Wright, A. and Botstein, D. (1971) "Superinfection exclusion by P22 prophage in lysogens of *Salmonella typhimurium* II: genetic evidence for two exclusion systems". *Virology*, vol. 45, pp. 638–652.

Susskind, M.M., Botstein, D. and Wright, A. (1974) "Superinfection exclusion by P22 prophage in lysogens of *Salmonella typhimurium* III: failure of superinfecting phage DNA to enter sieA$^+$ lysogens". *Virology*, vol. 62, pp. 350–366.

Susskind, M.M. and Botstein, D. (1978) "Molecular genetics of bacteriophage P22". *Microbiological Reviews*, vol. 42, pp. 385–413.

Kwoh, D.Y. and Kemper, J. (1978) "Bacteriophage P22-mediated specialized transduction in *Salmonella typhimurium*: high frequency of aberrant prophage excision". *Journal of Virology*, vol. 27, pp. 519–534.

Ranade, K. and Poteete, A.R. (1993) "Superinfection exclusion (*sieB*) genes of bacteriophage P22 and λ". *Journal of Bacteriology*, vol. 175, pp. 4712–4718.

Fillol-Salom, A., Bacigalupe, R., Humphrey, S. et al. (2021) "Lateral transduction is inherent to the life cycle of the archetypical *Salmonella* phage P22". *Nature Communications*, vol. 12, p. 6510.

Chen, J., Quiles-Puchalt, N., Chiang, Y.N. et al. (2018) "Genome hypermobility by lateral transduction". *Science*, vol. 362, pp. 207–212.

Chiang, Y.N., Penadés, J.R. and Chen, J. (2019) "Genetic transduction by phages and chromosomal islands: the new and noncanonical". *PLoS Pathogen*, vol. 15, p. e1007878.

"How did the Zinder–Lederberg experiment on transduction work?" on *Stack Exchange Biology* website (https://biology.stackexchange.com/questions/114006/how-did-the-zinder-lederberg-experiment-on-transduction-work/114054#114054)

"Transduction: generalized and specialized transduction" on *Online Biology Notes* website (https://www.onlinebiologynotes.com/transduction-generalized-and-specialized-transduction/)

"What are the differences between generalized and specialized transduction?" on *AAT Bioquest* website (https://www.aatbio.com/resources/faq-frequently-asked-questions/what-are-the-differences-between-generalized-and-specialized-transduction)

Recriminations

Crichton, M. (1969) "The Andromeda strain". *Arrow Books* Ltd (1993 edition).
Meselson, M. and Weigle, J.J. (1961) "Chromosome breakage accompanying genetic recombination in bacteriophage". *Proceedings of the National Academy of Sciences of the United States of America*, vol. 47, pp. 857–868.
Kellenberger, G. and Zichichi, M.L. (1961) "A mutation affecting the DNA content of bacteriophage lambda and its lysogenizing properties". *Journal of Bacteriology*, vol. 3, pp. 399–408.
Kellenberger, G., Zichichi, M.L. and Weigle, J.J. (1961) "Exchange of DNA in the recombination of bacteriophage λ". *Proceedings of the National Academy of Sciences of the United States of America*, vol. 47, pp. 869–878.
Kellenberger, G., Zichichi, M.L. and Epstein, H.T. (1962) "Heterozygosis and recombination of bacteriophage λ". *Virology*, vol. 17, pp. 44–55.
Zichichi, M.L. and Kellenberger, G. (1963) "Two distinct functions in the lysogenization process: the repression of phage multiplication and the incorporation of the prophage in the bacterial genome". *Virology*, vol. 19, pp. 450–460.
Cavalli-Sforza, L.L. (2008) "Joshua Lederberg, 1925–2008". *Cell*, vol. 132, pp. 724–725.
Citi, S. and Berg, D.E. (2016) "Grete Kellenberger-Gujer: molecular biology research pioneer". *Bacteriophage*, vol. 6, p. e1173168.
"Biographical overview" on the *National Library of Medicine* website (https://profiles.nlm.nih.gov/spotlight/bb/feature/biographical-overview)
"The history of artificial intelligence" on *Stanford Libraries* website (https://exhibits.stanford.edu/ai/catalog?f%5Btopic_facet%5D%5B%5D=DENDRAL)
Schindler, T.E. (2021) "A hidden legacy: the life and work of Esther Zimmer Lderberg". *Oxford University Press*.
Pisano, I. (2022) "Esther Lederberg: pioneer in microbial genetics" on *Frontiers* website (https://www.frontiersin.org/news/2022/09/27/children-in-science-esther-lederberg-pioneer-in-microbial-genetics/)
Steinmetz, K. (2022) "Esther Lederberg and her husband were both trailblazing scientists: why have more people heard of him?" on *Time* website (https://time.com/longform/esther-lederberg/)
Cobb, W. (2015) "Sexism in science: did Watson and Crick really steal Rosalind Franklin's data?" on *The Guardian* website (https://www.theguardian.com/science/2015/jun/23/sexism-in-science-did-watson-and-crick-really-steal-rosalind-franklins-data)
Cobb, W. and Comfort, N. (2023) "What Rosalind Franklin truly contributed to the discovery of DNA's structure" on *Nature* website (https://www.nature.com/articles/d41586-023-01313-5)
Lamm, L. (2023) "The Matilda effect: how women are becoming invisible in science" on *Lost Women of Science* website (https://www.lostwomenofscience.org/news-events/the-matilda-effect-how-women-are-becoming-invisible-in-science)
"Nobel Prize facts" on *Nobel Prize* website (https://www.nobelprize.org/prizes/facts/nobel-prize-facts/)

The Essence of Genes

Watson, J. (1968) "The double helix – a personal account of the discovery of the structure of DNA'. *Weidenfeld & Nicholson* (2010 edition).
Crick, F. (1988) "What mad pursuit". *Basic Books Inc. Publishers*.
Watson, J.D. and Crick, F.H.C. (1953) "Molecular structure of nucleic acids". *Nature*, vol. 171, pp. 737–738.

Kornberg, A. and Baker, T.A. (1992) "DNA replication". *W.H. Freeman and Co.*

De Chadarevian, S. (2003) "Portrait of a discovery: Watson, Crick and the double helix". *Iris*, vol. 94, pp. 90–105.

Lawrence, P.A. (2016) "Francis Crick: a singular approach to scientific discovery". *Cell*, vol. 167, pp. 1436–1439.

"The Nobel Prize in physiology or medicine 1962" on Nobel Prize website (https://www.nobelprize.org/prizes/medicine/1962/summary/)

Jones, R. (2021) "Sexism in science: was Rosalind Franklin robbed of a Nobel Prize?" on *Loyola Marymount University* website (https://lmuthisweek.lmu.edu/2021/03/22/sexism-in-science-was-rosalind-franklin-robbed-of-a-nobel-prize/)

"The Watson-Crick Nobel: was someone left out?" on *Medical Discovery News* website (https://www.utmb.edu/mdnews/podcast/episode/the-watson-crick-nobel-was-someone-left-out)

Belluz, J. (2019) "DNA scientist James Watson has a remarkably long history of sexist, racist public comments" on *Vox* website (https://www.vox.com/2019/1/15/18182530/james-watson-racist)

McKie, R. (2012) "DNA pioneer James Watson reveals helix story was almost never told". *The Guardian* website (https://www.theguardian.com/science/2012/dec/09/dna-james-watson-helix-nobel-50#:~:text=%22You%20have%20grossly%20invaded%20my,and%20unfavourable%20image%20of%20scientists%22)

"The Double Helix" on *Wikipedia* website (https://en.m.wikipedia.org/wiki/The_Double_Helix#:~:text=It%20has%20also%20been%20criticized,Double%20Helix%20in%20the%20U.S)

The RNA Tie Club

Crick, F.H.C, Griffith, J.S. and Orgel, L.E. (1957) "Codes without commas". *Proceedings of the National Academy of Sciences of the United States of America*, vol. 43, pp. 416–421.

Crick, F.H.C. (1968) "The origin of the genetic code". *Journal of Molecular Biology*, vol. 38, pp. 367–379.

Crick, F.H.C. (1958) "On protein synthesis". *Symposia of the Society for Experimental Biology*, vol. 12, pp. 138–163.

Crick, F. (1970) "Central dogma of molecular biology". *Nature*, vol. 227, pp. 561–563.

Delbrück, M. (1954) "On the replication of desoxyribonucleic acid (DNA)". *Proceedings of the National Academy of Sciences of the United States of America*, vol. 40, pp. 783–788.

Portin, P. (2014) "The birth and development of the DNA theory of inheritance: sixty years since the discovery of the structure of DNA". *Journal of Genetics*, vol. 93, pp. 293–302.

Saier, M.H. Jr. (2019) "Understanding the genetic code". *Journal of Bacteriology*, vol. 201, p. E00091-19.

Liu, Y. (2020) "A code within the genetic code: codon usage regulates co-translational protein folding". *Cell Communication and Signaling*, vol. 18, p. 145.

Hayes, B. (1998) "The invention of the genetic code" on the *American Scientist* website (https://www.americanscientist.org/article/the-invention-of-the-genetic-code)

"Cracking the genetic code" on *Genomics Education Program* website (https://www.genomicseducation.hee.nhs.uk/blog/cracking-the-genetic-code/)

Nirenberg, M.W. and Matthaei, J.H. (1961) "The dependence of cell-free protein synthesis in *E. coli* upon naturally occurring or synthetic polyribonucleotides". *Proceedings of the National Academy of Sciences of the United States of America*, vol. 47, pp. 1588–1602.

Leder, P. and Nirenberg, M.W. (1964) "RNA codewords and protein synthesis III: on the nucleotide sequence of cycteine and leucine RNA codeword". *Proceedings of the National Academy of Sciences of the United States of America*, vol. 52, pp. 1521–1529.

Leder, P. (2010) "Marshall Warren Nirenberg". *Science*, vol. 327, p. 972.
Caskey, T.C. (2010) "Marshall Nirenberg (1927–2010)". *Nature*, vol. 464, p. 44.

The Outsiders

Ireland, T. (2023) "The goof virus". *Hodder & Stoughton* Ltd. To be fair, Ireland also correctly reported that the "virus issues its final devastating genetic instruction to the hijacked cell: to metabolise 'suicide molecules'."

Cohen, R.J., Jan, Y.N., Matricon, J. and Delbrück, M. (1975) "Avoidance response, house response and wind responses of the sporangiophore of *Phycomyces*". *The Journal of General Physiology*, vol. 66, pp. 67–95.

Susman, M. (1995) "The Cold Spring Harbor Phage Course (1945–1970): a 50th anniversary remembrance". *Genetics*, vol. 139, pp. 1101–1106.

Kellenberger, E. (1995) "History of phage research as viewed by a European". *FEMS Microbiology Reviews*, vol. 17, pp. 7–24.

"In memoriam, Gunther S. Stent" on *The University of California* website (https://senate.universityofcalifornia.edu/_files/inmemoriam/html/gunthersstent.html)

Delbrück, M. (1954) "On the replication of desoxyribonucleic acid (DNA)". *Proceedings of the National Academy of Sciences of the United States of America*, vol. 40, pp. 783–788.

Steinberg, C. and Stahl, F. (1964) "The theory of formal phage genetics". *Cold Spring Harbor Symposium on Quantitative Biology*, vol. 23, pp. 42–46.

Johns, H.E., Rapaport, S.A. and Delbrück, M. (1962) "Photochemistry of thymine dimers". *Journal of Molecular Biology*, vol. 4, pp. 104–114.

Plate, C.A., Suit, J.L., Jetten, A.M. and Luria, S.E. (1974) "Effects of colicin K on a mutant of *Escherichia coli* deficient in Ca^{+2}, Mg^{+2}-activated adenosine triphosphatase". *The Journal of Biological Chemistry*, vol. 249, pp. 6138–6143.

Luria, S.E. (1975) "Colicins and the energetics of cell membranes". *Scientific American*, vol. 233, pp. 30–37.

Gillor, O., Kirkup, B.C. and Riley, M.A. (2004) "Colicins and microcins: the next generation antimicrobials". *Advances in Applied Microbiology*, vol. S4, pp. 129–146.

Schmelcher, M. and Loessner, M.J. (2021) "Bacteriophage endolysins: extending their application to tissue and the bloodstream". *Current Opinions in Biotechnology*, vol. 68, pp. 51–59.

Taherikalani, M. and Ghafourian, S. (2021) "Anticancer properties of colicin E7 against colon". *Gastroenterology Reviews*, vol. 16, pp. 364–368.

Markovic, K.G., Grujovic, M.Z., Koracevic, M.G. et al. (2022) "Colicins and microcins produced by *Enterobacteriaceae*: characterization, mode of action and putative applications". *International Journal of Environmental Research and Public Health*, vol. 19. p. 11825.

Upatissa, S. and Mitchell, R.J. (2023) "The 'cins' of our fathers: rejuvenated interest in colicins to combat drug resistance". *Journal of Microbiology*, vol. 61, pp. 145–158.

Evans, A.C. (1934) "The prevalence of *Streptococcus* bacteriophage". *Science*, vol. 80, pp. 40–41.

Trudil, D. (2015) "Phage lytic enzymes: a history". *Virologica Sinica*, vol. 30, pp. 26–32.

Fischetti, V.A. (2018) "Development of phage lysins as novel therapeutics: a historical perspective". *Viruses*, vol. 10, p. 310.

Gründling, A., Manson, M.D. and Young, R. (2001) "Holins kill without warning. *Proceedings of the National Academy of Sciences of the United States of America*, vol. 98, pp. 9348–9352.

Young, R. (2002) "Bacteriophage Holins: deadly diversity". *Journal of Molecular Microbiology and Biotechnology*, vol. 4, pp. 21–36.

Saier, M.H. Jr. and Reddy, B.L. (2015) "Holins in bacteria, eukaryotes and archaea: multifunctional xenologues with potential biotechnological and biomedical applications". *Journal of Bacteriology*, vol. 197, pp. 7–17.

A Different Moral Universe

Luria, S.E. (1984) "A slot machine, a broken test tube". *Harper & Row Publishers.*

Bird, K. and Sherwin, M.J. (2005) "American Prometheus". *Garzanti* (2023 Italian version, translated by Emanuele and Alfonso Vinassa de Regny).

Editors, C.R. (2015) "Joseph McCarthy: the controversial life and career of 20th century America's most notorious senator". *CreateSpace Independent Publishing.*

Nichols, D.A. (2017) "Ike and McCarthy". *Simon & Schunter Paperbacks.*

Saez, E. and Zucman, G. (2019) "The triumph of injustice". *W. W. Norton & Company Publishing.*

Tye, L. (2020) "Demagogue: the life and long shadow of senator Joe McCarthy". *Houghton Mifflin Harcourt.*

Dobbs, M. (2021) "King Richard: Nixon and Watergate: an American tragedy". *Scribe Publications.*

Selya, R. (2022) "Salvador Luria: an immigrant biologist in Cold War America". *Massachusetts Institute of Technology Press* (Italian version, translated by Allegra Panini, *Raffaello Cortina Ed.*, 2023).

Luria, S.E. and Szent-Györgyi, A. (1967) "Vietnam: a national catastrophe". *Science*, vol. 158, p. 47.

Luria, S.E. (1968) "The microbiologist and his times". *Bacteriological Reviews*, vol. 32, pp. 401–403.

Fanods, N. and Otterman, S. (2024) "Inside a decision at Columbia that escalated campus unrest". *The New York Times*, No. 60,134 (24/04/2024).

"Police clear building at Columbia and arrest dozens of protesters" on *The New York Time* website (https://www.nytimes.com/live/2024/04/30/nyregion/columbia-protests-college#columbia-student-protest-encampment)

"What we know about the protests and arrests at Columbia University" on *The New York Time* website (https://www.nytimes.com/2024/04/22/us/columbia-university-protests.html?searchResultPosition=3)

Marshall, K. (2015) "Rednecks: a brief history" on *JSTOR Daily* website (https://daily.jstor.org/redneck-a-brief-history/)

Kaufman, D. (2020) "The real legacy of a demagogue" on *The New Republic* website (https://newrepublic.com/article/159568/real-legacy-demagogue-joseph-mccarthy-donald-trump-book-review)

Klein, C. (2023) "How Lewis Strauss shaped the atomic age—and orchestrated Robert Oppenheimer's downfall" on *History* website (https://www.history.com/news/lewis-strauss-nuclear-program-oppenheimer)

Phippen, J.W. (2023) "In the matter of J. Robert Oppenheimer" on *Los Alamos National Laboratory* website (https://www.lanl.gov/media/publications/national-security-science/0723-in-the-matter-of-j-robert-oppenheimer)

"McCarthyism" on *Wikipedia* website (https://en.wikipedia.org/wiki/McCarthyism#cite_note-82)

"The Rosenberg trial" on the *Atomic Heritage Foundation* website (https://ahf.nuclearmuseum.org/ahf/history/rosenberg-trial/)

"The Berlin blockade" on *The North Atlantic Treaty Organization* website (https://www.nato.int/cps/en/natohq/declassified_136188.htm)

"The blockade of Berlin" on the *Harry S. Truman Library* website (https://www.trumanlibrary.gov/education/presidential-inquiries/blockade-berlin)

"The Berlin airlift, 1948–1949" on the *Office of Historian* website (https://history.state.gov/milestones/1945-1952/berlin-airlift)
"Augusto Pinochet" on *Human Right Watch* website (https://www.hrw.org/legacy/campaigns/chile98/index.htm)
Doubek, J. (2023) "The U.S. set the stage for a coup in Chile. It had unintended consequences at home" on *NPR* website (https://www.npr.org/2023/09/10/1193755188/chile-coup-50-years-pinochet-kissinger-human-rights-allende)

Growing Involvement

Luria, S.E. and Darnell, J.E. Jr. (1953) "General Virology". *Zanichelli Ed.* (2nd edition, translated by Carla Malva and Giuliana Razzino).
Salvadori, M.L. (1990) "Storia dell'età moderna e contemporanea". *Löscher Ed.* (vol. 3).
Schulman, B.J. (2001) "The Seventies: the great shift in American culture, society and politics". *Da Capo Press* (2002 edition).
Ellsberg, D. (2002) "Secrets: a memoir of Vietnam and the Pentagon papers". *Viking Publishing.*
Goodale, J.C. (2013) "Fighting for the press: the inside story of the Pentagon papers and other battles". *OR Books.*
Beecher, W. (1969) "Raids in Cambodia by U.S. unprotested". *The New York Times*, No. 40,648 (9/5/1969).
Lyons, R. (1969) "Second H.E.W. blacklist includes Nobel laureate". *The New York Times*, No. 40,812 (20/10/1969).
Burnham, D. (1971) "Serpico tells of delay on police inquiry". *The New York Times*, No. 41,598 (15/12/1971).
Gitlin, T. (2016) "Kissinger was a courtier to atrocity" on *The New York Times* website (https://www.nytimes.com/roomfordebate/2016/02/13/henry-kissinger-sage-or-pariah/kissinger-was-a-courtier-to-atrocity)
Cuddy, B. (2017) "Was it legal for the U.S. to bomb Cambodia?" on *The New York Times* website (https://www.nytimes.com/2017/12/12/opinion/america-cambodia-bomb.html)
Tan, R. and Cabato, R. (2023) "Henry Kissinger's central role in the U.S. carpet bombing of Cambodia" on *The Washington Post* website (https://www.washingtonpost.com/world/2023/11/30/henry-kissinger-cambodia-bombing-war/)
Risen, C. (2024) "William Beecher, who revealed secret Cambodia bombing, dies at 90" on *The New York Times* website (https://www.nytimes.com/2024/02/18/business/media/william-beecher-dead.html)
Savage, C. (2024) "Assange's plea deal sets a chilling precedent, but it could have been worse" on *The New York Times* website (https://www.nytimes.com/2024/06/25/us/politics/assange-plea-deal-press-freedom.html?smid=nytcore-android-share)
Kirchick, J. (2024) "The extradition of Julian Assange threatens press freedoms" on *The New York Times* (https://www.nytimes.com/2024/02/19/opinion/julian-assange-extradition-espionage.html?searchResultPosition=6)
Sony, O. and Wright, G. (2023) "Henry Kissinger's Cambodia legacy of bombs and chaos" on the *BBC* website (https://www.bbc.com/news/world-asia-67582813)
"Julian Assange" on *Britannica* website (https://www.britannica.com/biography/Julian-Assange)

HOMECOMING

An Ethical Legacy

Luria, S.E. (1984) "A slot machine, a broken test tube". *Harper & Row Publishers.*
DeLacey, G., Record, C. and Wade, J. (1985) "How accurate are quotations and references in medical journals?". *British Medical Journal*, vol. 291, pp. 884–886.

Goodstein, G. (1997) "The Big Crunch". EOS, vol. 78, pp. 329-334.

Archibald, T., Fraser, C. and Grattan-Guinness, I. (2004) "The history of differential equations, 1670–1950". *Oberwolfach Report 51/2004*.

Turner, E.H., Matthews, A.M., Linardatos, E. et al. (2008) "Selective publication of antidepressant trials and its influence on apparent efficacy". *The New England Journal of Medicine*, vol. 358, pp. 252–260.

Errington, T.M. Iorns, E., Gunn, W. et al. (2014) "An open investigation of the reproducibility of cancer biology research". *Elife*, vol. 3. p. e04333.

Jergas, H. and Baethge, C. (2015) "Quotation accuracy in medical journal articles: a systematic review and meta-analysis". *PeerJ*, vol. 3, p. e1364.

Gusenbauer, M. (2019) "Google Scholar to overshadow them all? Comparing the sizes of 12 academic search engines and bibliographic databases". *Scientometrics*, vol. 118, pp. 177–214.

Zarocostas, J. (2020) "How to fight an infodemic". *The Lancet*, vol. 395, p. 676.

Kim, L., Adolph, C., West, J.D. and Stovel, K. (2020) "The influence of changing marginals on measures of inequality in scholarly citations: evidence of bias and a resampling correction". *Sociological Science*, vol. 7, pp. 314–341.

West, J.D. and Bergstrom, C.T. (2021) "Misinformation in and about science". *Proceedings of the National Academy of Sciences of the United States of America*, vol. 118, p. e1912444117.

Carafoli, E. and Bucci, E. (2022) "Scientific fraud part II: from past to present, facts and analyses". *European Review*, vol. 30, pp. 850–871.

Freijedo-Farinas, F., Ruano-Ravina, A., Pérez-Rios, M. et al. (2024) "Biomedical retractions due to misconduct in Europe: characterization and trends in the last 20 years". *Scientometrics*, vol. 129, pp. 2867–2882.

Lozada, C. (2024) "What I learned when I read 887 pages of Project 2025" on *The New York Times* website (https://www.nytimes.com/2024/02/29/opinion/project-2025-trump-administration.html?smid=nytcore-android-share)

"Patent law rewrites history" on *Legal Language Blog* website (https://www.legallanguage.com/legal-articles/patent-law-radio/)

"Black hole information paradox" on *Wikipedia* website (https://en.wikipedia.org/wiki/Black_hole_information_paradox)

"Word of the Year 2016" on *Oxford Languages* website (https://languages.oup.com/word-of-the-year/2016/)

"Number of scientists worldwide reaches 8.8M, as global research spending grows faster than the economy" on *Science Business* website (https://sciencebusiness.net/news/number-scientists-worldwide-reaches-88m-global-research-spending-grows-faster-economy)

"90% of all the scientists that ever lived are alive today" on *Future of Life Institute* website (https://futureoflife.org/guest-post/90-of-all-the-scientists-that-ever-lived-are-alive-today/)

"Hesiod's five ages of Man" on *Though Co.* website (https://www.thoughtco.com/the-five-ages-of-man-111776)

Phage Renaissance

Wentworth, B.B. (1963) "Bacteriophage typing of the staphylococci". *Bacteriological Reviews*, vol. 27, pp. 253–272.

Korn-Wendisch, F. and Schneider, J (1992) "Phage typing: a useful tool in actinomycete systematics". *Gene*, vol. 115, pp. 243–247.

Abuladze, T., Li, M., Menetrez, M.Y. et al. (2008) "Bacteriophages reduce experimental contamination of hard surfaces, tomato, spinach, broccoli, and ground beef by *Escherichia coli* O157:H7". *Applied and Environmental Microbiology*, vol. 74, pp. 6230–6238.

Wernicki, A., Nowaczek, A. and Urban-Chmiel, R. (2017) "Bacteriophage therapy to combat bacterial infections in poultry". *Virology Journal*, vol. 14, p. 179.

Sun, Y., Passaretti, P., Hernandez, I., et al. (2020) "Nanomechanics of graphene oxide-bacteriophage based self-assembled porous composites". *Scientific Reports*, vol. 10, p.15618.

Marongiu, L., Burkard, M., Lauer, U.M. et al. (2022) "Reassessment of historical clinical trials supports the effectiveness of phage therapy". *Clinical Microbiology Reviews*, vol. 35, p. e0006222.

Jaroszewicz, W., Morcinek-Orłowska, J., Pierzynowska, K. et al. (2022) "Phage display and other peptide display technologies". *FEMS Microbiology Reviews*, vol. 46, p. fuab052.

Liu, H.Y., Wang, Z.G., Liu, S.L. et al. (2023) "Single-virus tracking with quantum dots in live cells". *Nature Protocols*, vol. 18, pp. 458–489.

Pierzynowska, K., Morcinek-Orłowska, J., Gaffke, L. et al. (2024) "Applications of the phage display technology in molecular biology, biotechnology and medicine". *Critical Reviews in Microbiology*, vol. 50, pp. 450–490.

Song, B.P.C., Ch'ng, A.C.W. and Lim, T.S. (2024) "Review of phage display: a jack-of-all-trades and master of most biomolecule display". *International Journal of Biological Macromolecules*, vol. 256, p. 128455.

Ahmed, H., Lopez, H., Boselli, F. et al. (2024) "Biomimetic plasmonic nanophages by head/tail self-assembling: gold nanoparticle/virus interactions". *American Chemical Society Nano*, vol. 18, pp. 21302–21315.

Nguyen, T.M., Kim, S.J., Ryu, D.G. et al. (2024) "Helical hybrid nanostructure based on chiral M13 bacteriophage via evaporation-induced three-dimensional process". *Nanomaterials*, vol. 14, p. 1208.

"Courses & Workshops" on *CSHL* website (https://meetings.cshl.edu/courseshome.aspx)

"International Summer School on Fundamental Phage Research" on *The Eliava Institute* website (https://eliava-institute.org/en/news/97)

"How photoshop became a verb" on *The Verge* website (https://www.theverge.com/2020/2/19/21143794/photoshop-30th-anniversary-adobe-verb-origin-story)

Demanding a Better Future

Wang, H., Yang, Y., Xu, Y. et al. (2024) "Phage-based delivery systems: engineering, applications, and challenges in nanomedicines". *Journal of Nanobiotechnology*, vol. 22, p. 365.

Barker, P. (2024) "Trump's own rhetoric fuels claims of fascism" on *The New York Times* No. 44,049 (30/10/2024).

Goldmacher, S. and Lerer, L. (2024) "His return ushers in a new era of uncertainty" on *The New York Times* No. 44,056 (7/11/2024).

Brooks, D. (2024) "The authoritarians have the momentum" on *The New York Times* website (https://www.nytimes.com/2024/05/16/opinion/trump-liberals-authoritarians.html?smid=nytcore-android-share)

Liptak, A. (2024) "Supreme court says Trump has some immunity in election case" on *The New York Times* website (https://www.nytimes.com/2024/07/01/us/politics/supreme-court-trump-immunity.html?smid=nytcore-android-share)

Cohen, R. (2024) "Just how dangerous is Europe's rising far right?" on *The New York Times* website (https://www.nytimes.com/2024/05/05/world/europe/europe-far-right.html?smid=nytcore-android-share)

Baker, P. (2024) "Biden denounces 'reckless' G.O.P. efforts to discredit Trump conviction" on *The New York Times* website (https://www.nytimes.com/2024/05/31/us/politics/biden-trump-guilty-verdict.html?smid=nytcore-android-share)

Bellotti, D. (2024) "Ribelle coi gessetti 'pericolosa, merita la sorveglianza'". *L'Espresso*, No. 16, year 70, 19/04/2024.

"Giangiacomo Feltrinelli" on *Wikipedia* website (https://en.wikipedia.org/wiki/Giangiacomo_Feltrinelli)

Index

A

B

C

D

E

Y

Z